MXene Nanocomposites

MXenes offer single-step processing, excellent electrical conductivity, easy heat dissipation behavior, and capacitor-like properties and are used in photodetectors, lithium-ion batteries, solar cells, photocatalysis, electrochemiluminescence sensors, and supercapacitors. Because of their superior electrical and thermal conductivities, these composites are an ideal choice in electromagnetic interference (EMI) shielding. *MXene Nanocomposites: Design, Fabrication, and Shielding Applications* presents a comprehensive overview of these emerging materials, including their underlying chemistry, fabrication strategies, and cutting-edge applications in EMI shielding.

- Covers modern fabrication technologies, processing, properties, nanostructure formation, and mechanisms of reinforcement.
- Discusses biocompatibility, suitability, and toxic effects.
- Details innovations, applications, opportunities, and future directions in EMI shielding applications.

This book is aimed at researchers and advanced students in materials science and engineering and is unique in its detailed coverage of MXene-based polymer composites for EMI shielding.

MXene Nanocomposites
Design, Fabrication, and Shielding Applications

Edited by
Poushali Das, Andreas Rosenkranz,
and Sayan Ganguly

CRC Press
Taylor & Francis Group
Boca Raton London New York

CRC Press is an imprint of the
Taylor & Francis Group, an **informa** business

First edition published 2023
by CRC Press
6000 Broken Sound Parkway NW, Suite 300, Boca Raton, FL 33487-2742

and by CRC Press
4 Park Square, Milton Park, Abingdon, Oxon, OX14 4RN

CRC Press is an imprint of Taylor & Francis Group, LLC

ISBN: 978-1-032-25092-2 (hbk)
ISBN: 978-1-032-25093-9 (pbk)
ISBN: 978-1-003-28151-1 (ebk)

DOI: 10.1201/9781003281511

Typeset in Times
by KnowledgeWorks Global Ltd.

Contents

Editor Biographies

Dr. Poushali Das is a senior postdoctoral research fellow at McMaster University in Canada. She completed a PhD from the Indian Institute of Technology in Kharagpur. She has more than 45 research publications in reputed international journals.

Dr. Andreas Rosenkranz is Professor of Materials-Oriented Tribology and New 2D Materials in the Department of Chemical Engineering, Biotechnology and Materials at the University of Chile. He has published more than 100 peer-reviewed journal publications, is a fellow of the Alexander von Humboldt Foundation, and acts as a scientific editor for different well-reputed scientific journals, including *Applied Nanoscience* and *Frontiers in Chemistry*.

Dr. Sayan Ganguly is a senior postdoctoral researcher at University of Waterloo, Canada. He obtained his PhD from the Indian Institute of Technology in Kharagpur. He has published more than 70 papers and chapters in international journals and books.

Contributors

Prashant S. Alegaonkar
Department of Physics
Central University of Punjab
Ghudda, India

Bona Elizebath Baby
Post Graduate and Research
    Department of Chemistry,
    Government College for Women
University of Kerala
Trivandrum, India

Sayani Biswas
Department of Physics
Central University of Punjab
Ghudda, India

Madhurya Chandel
Faculty of Materials Science and
    Engineering
Warsaw University of Technology
Warsaw, Poland

Poushali Das
School of Biomedical Engineering
McMaster University
Ontario, Canada

Narayan Chandra Das
Rubber Technology Centre
Indian Institute of Technology
Kharagpur, India

Sayan Ganguly
Bar-Ilan Institute for Nanotechnology
    and Advanced Materials,
    Department of Chemistry
Bar-Ilan University
Ramat-Gan, Israel

Suman Kumar Ghosh
Rubber Technology Centre
Indian Institute of Technology
Kharagpur, India

Demudu Babu Gorle
Materials Research Centre
Indian Institute of Science
Bangalore, India

Sonam Gupta
Associate Scientific Writer
Indegene
Bangalore, India

Agnieszka Maria Jastrzębska
Faculty of Materials Science and
    Engineering
Warsaw University of Technology
Warsaw, Poland

Nasima Khatun
Semiconducting Oxide Materials,
    Nanostructures, and Tailored
    Heterojunction (SOMNaTH)
    Lab
Department of Physics and 2D
    Materials and Innovation Group
Indian Institute of Technology
    Madras
Chennai, Tamil Nadu, India

Anand Krishnamoorthy
Department of Basic
    Sciences
Amal Jyothi College
    of Engineering
Kanjirappally, Kerala, India
and
Apcotex Industries Limited
MIDC Industrial Area
Taloja, Kerala, India

Indu Kumari
Department of Biotechnology
Chandigarh Group of Colleges
Landran, Mohali, India

K. Kumari
Department of Chemical
  Engineering
SLIET Longowal
Punjab, India

P. P. Kundu
Department of Polymer Science
  and Technology
University of Calcutta
Kolkata, India
and
Department of Chemical
  Engineering
Indian Institute of Technology
Roorkee, India

Krishnendu Nath
Rubber Technology Centre
Indian Institute of Technology
Kharagpur, India

Srikanth Ponnada
Sustainable Materials and Catalysis
  Research Laboratory (SMCRL),
  Department of Chemistry
Indian Institute of Technology
  Jodhpur
Karwad, Jodhpur, India

Amin Reza Rajabzadeh
School of Biomedical Engineering
and
W Booth School of Engineering
  Practice and Technology
McMaster University
Ontario, Canada

Battula Venkateswara Rao
Department of Engineering
  Chemistry
Andhra University College of
  Engineering (A)
Andhra University
Visakhapatnam, India

Somnath C. Roy
Semiconducting Oxide Materials,
  Nanostructures, and Tailored
  Heterojunction (SOMNaTH) Lab
Department of Physics and 2D
  Materials and Innovation Group
Indian Institute of Technology Madras
Chennai, Tamil Nadu, India

Rakesh K. Sharma
Sustainable Materials and Catalysis
  Research Laboratory (SMCRL),
  Department of Chemistry
Indian Institute of Technology Jodhpur
Karwad, Jodhpur, India

Vineeta Shukla
Department of Physics
Indian Institute of Technology
  Kharagpur
Kharagpur, West Bengal, India

Amandeep Singh
Department of Polymer Science and
  Technology
University of Calcutta
Kolkata, India

Seshasai Srinivasan
School of Biomedical Engineering
and
W Booth School of Engineering
  Practice and Technology
McMaster University
Ontario, Canada

Anita Wojciechowska
Faculty of Materials Science and
  Engineering
Warsaw University of Technology
Warsaw, Poland

Sarita Yadav
Department of Chemistry
National Institute of Technology
Warangal, Telangana, India

1 Introduction and Background of MXenes

Srikanth Ponnada
Sustainable Materials and Catalysis Research
Laboratory (SMCRL), Department of Chemistry
Indian Institute of Technology Jodhpur
Karwad, Jodhpur, India

Sarita Yadav
Department of Chemistry
National Institute of Technology
Warangal, Telangana, India

Demudu Babu Gorle
Materials Research Centre
Indian Institute of Science
Bangalore, India

Indu Kumari
Department of Biotechnology
Chandigarh Group of Colleges
Landran, Mohali, India

Battula Venkateswara Rao
Department of Engineering Chemistry
Andhra University College of Engineering (A)
Andhra University
Visakhapatnam, India

Rakesh K. Sharma
Sustainable Materials and Catalysis Research
Laboratory (SMCRL), Department of Chemistry
Indian Institute of Technology Jodhpur
Karwad, Jodhpur, India

DOI: 10.1201/9781003281511-1

CONTENTS

1.1   INTRODUCTION

The world of 2D materials is enlarged day by day and their applicability has become vast, and thus they have become an unavoidable part of most of our daily gadgets. They began a new era in the electronics industry and have driven tremendous research for more than a decade. These 2D materials are versatile with their extraordinary properties, especially electronic, electrochemical and mechanical [1–5]. The inventions have enhanced the performance of many conventional devices and resulted in an electronically driven world. Graphene is a signature material, among other 2D materials, and has its own place in the modern electronics industry. Since the emergence of graphene, a large group of 2D materials, like phosphorene, transition-metal dichalcogenides, graphitic carbon nitride and boron nitrides, has been discovered and established in diverse applications, including energy conversion/storage, solar cells and photodetectors [5–10]. These 2D materials are diverse in physical and chemical properties, which totally differ from their bulk state [11]. Transition-metal carbides and nitrides, called MXenes, are the new members of the 2D family, and the first reported MXene is Ti_3C_2 synthesized from the selective etching of Al from Ti_3AlC_2 [12]. MXenes have a general formula of $M_{n+1}X_nT_x$ (n = 1–3), where M stands for the early transition metal; X is carbon/nitrogen; T_x represents surface terminations, which are O, OH, Cl and F; and x stands for the number of surface terminations [13, 14]. The ratio of M and X elements in the MXene structure is an important factor in determining the property. Currently, more than 30 MXene compositions are being synthesized, and a lot more are theoretically predicted [13]. Figure 1.1 shows the variety of M elements with T and the theoretically predicted M elements for the formation of MXenes in the periodic table. The atomic representation of different MXenes is shown at the bottom of Figure 1.1.

In MXenes, n layers of carbon/nitrogen are covered with n + 1 layers of transition metal M [15]. The different structures of MXenes are shown in Figure 1.2A. They are typically derived from their MAX phase through the selective etching of the A layer using etchant hydrofluoric acid (HF) [12]. This MAX phase belongs to layered ternary nitrides or carbides, with a general formula of $M_{n+1}AX_n$ (n = 1–3), where A is a group that includes 13–16 elements. The three main structures of MXenes are M_2X, M_3X_2 and M_4X_3. Once the MXene is exfoliated from the bulk MAX phase, its optical, electronic and electrochemical properties are quite different, depending

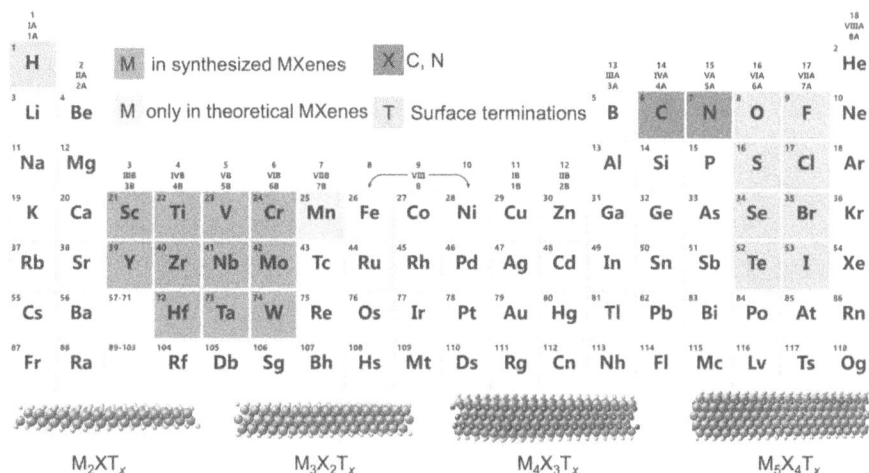

FIGURE 1.1 Periodic table showing elements used to build MXene. The schematics of four typical structures of MXene are shown at the bottom.

(Image adapted with permission from Ref. [13] © 2019, American Chemical Society.)

on the surface transition metal M and surface functionality [16–19]. These A-layer atoms are joined with the early transition metal M layer through the metallic bond. The M–X bond in the structure is of a mixed ionic/covalent/metallic nature and is stronger than the M–A bond. In the unit cell, the early transition metal (M-ion) has a coordination number of six (M6X) and this will make six chemical bonds with the nearby C/N atom and the surface termination. They are blessed with structural stiffness and electrical and thermal conductivity, and more than 70 MAX-phase compounds have been experimentally synthesized till now [19, 31].

With the tremendous research development in MXenes, they now represent the biggest 2D family known so far [20]. More than 30 stoichiometric MXenes have been reported, and more than a hundred are expected to exist (shown in Figure 1.2). They have a reported electronic conductivity of up to 20,000 S/cm, which is the highest among the 2D family, and the Fermi level of MXenes can be modified by external stimuli [14, 20]. MXenes are blessed with other vibrant properties like large interlayer spacing, outstanding biocompatibility, hydrophilicity due to the existence of surface-terminated groups, tunable surface chemistry and excellent intercalation ability and applicability, and they are not limited to optoelectronic devices, energy conversion/storage devices or photo and electro-catalysis [21–25].

Generally, MXenes are three or more atomically thick layers, and the word "ene" at the end of MXene denotes the 2D structure of the material [26]. The properties of the MXene structure greatly depend on its particle size, lamellar spacing and layer number. By altering these, the desirable properties can be achieved. For example, by decreasing the number of layers of MXenes, the conductivity can be improved, and it has been observed that this will improve the photovoltaic features [26]. Additionally, the selection of M and X elements affects the electronic properties of the materials;

(A)

M_2X M_3X_2 M_4X_3

Hydrogen   Oxygen

Carbon   Titanium

(B)

	M_2X			M_3X_2				M_4X_3			

Mono-transition metal MXenes

Solid solution MXenes

Ordered double-transition metal (M) MXenes

Ordered divacancy MXenes   N/A   N/A

M_2X				M_3X_2				M_4X_3			
Sc_2C	Ti_2C	Ti_2N	Zr_2C	Ti_3C_2	Ti_3N_2	$Ti_3(CN)$	Zr_3C_2	Ti_4N_3	V_4C_3	Nb_4C_3	Ta_4C_3
Zr_2N	Hf_2C	Hf_2N	V_2C	$(Ti,V)_3C_2$	$(Ti,Nb)_3C_2$	$(Ti,Ta)C_2$	$(Ti,Mn)C_2$	$(Ti,Nb)_4C_3$	$(Nb,Zr)_4C_3$	$(Ti,Nb)_4C_3$	$(Ti,Ta)_4C_3$
V_2N	Nb_2C	Ta_2C	Cr_2C	Hf_3C_2	$(Hf_2V)C_2$	$(Hf_2Mn)C_2$	$(V_2Ti)C_2$	$(V_2Ti_2)C_3$	$(V_2Nb_2)C_3$	$(V_2Ta_2)C_3$	$(Nb_2Ta_2)C_3$
Cr_2N	Mo_2C	$Mo_{1.3}C$	$Cr_{1.3}C$	$(Cr_2Ti)C_2$	$(Cr_2V)C_2$	$(Cr_2Nb)C_2$	$(Cr_2Ta)C_2$	$(Cr_2Ti_2)C_3$	$(Cr_2V_2)C_3$	$(Cr_2Nb_2)C_3$	$(Cr_2Ta_2)C_3$
$(Ti,V)_2C$	$(Ti,Nb)_2C$	W_2C	$W_{1.3}C$	$(Mo_2Sc)C_2$	$(Mo_2Ti)C_2$	$(Mo_2Zr)C_2$	$(Mo_2Hf)C_2$	$(Mo_2Ti_2)C_3$	$(Mo_2Zr_2)C_3$	$(Mo_2Hf_2)C_3$	$(Mo_2V_2)C_3$
Mo_2N	$Nb_{1.3}C$	$Mo_{1.3}Y_{0.2}C$		$(Mo_2V)C_2$	$(Mo_2Nb)C_2$	$(Mo_2Ta)C_2$	$(W_2Ti)C_2$	$(Mo_2Nb_2)C_3$	$(Mo_2Ta_2)C_3$	$(W_2Ti_2)C_3$	$(W_2Zr_2)C_3$
M: Sc, Y, Ti, Zr, Hf, V, Nb, Ta, Cr, Mo, W, Mn X: C, N				$(W_2Zr)C_2$	$(W_2Hf)C_2$			$(W_2Hf_2)C_3$			

Theoretical   Experimental   Solid solution double-M   Ordered double-M   Ordered divacancy

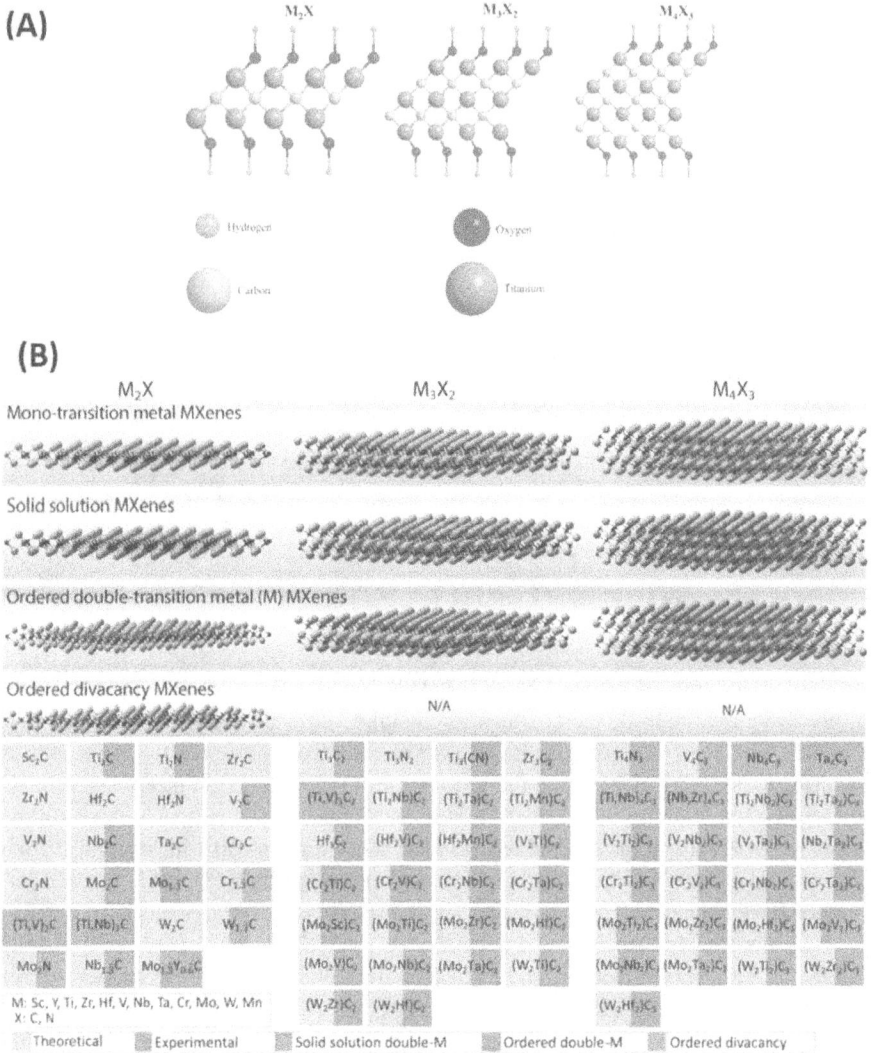

FIGURE 1.2 (A) Crystal structure of various layered MXenes with surface terminations as a hydroxyl group (OH). (Adapted with permission from **Ref**. [59]. © 2019 Elsevier B.V. All rights reserved.) (B) MXenes reported so far. MXenes can have at least three different formulas: M_2X, M_3X_2 and M_4X_3, where M is an early transition metal and X is carbon and/ or nitrogen. They can be made in three different forms: mono-M elements (for example, Ti_2C and Nb_4C_3); a solid solution of at least two different M elements (for example, $(Ti, V)_3C_2$ and $(Cr, V)_3C_2$); or ordered double-M elements, in which one transition metal occupies the perimeter layers and another fills the central M layers (for example, Mo_2TiC_2 and $Mo_2Ti_2C_3$, in which the outer M layers are Mo and the central M layers are Ti). Solid solutions on the X site produce carbonitrides. (Adapted with permission from **Ref**. [16]. © 2017, Springer Nature.)

for example, Mo_2TiC_2 exhibits a semiconducting nature, while Ti_2C_3 shows a metallic nature.

They exist in different forms according to the arrangement of one or more transition metals in the structure, called the solid-solution phase and the ordered phase. Density Functional Theory (DFT) studies proved that ordered MXenes are energetically more stable than their solid-solution counterparts for some combinations of M elements in the structure [27]. Several theoretical and experimental studies have proved the chemical and mechanical stability of MXenes due to their ceramic nature and their existence in different forms [28–30]. MXenes may also be used in surface-state engineering because of their adjustable surface functions, which can be altered by alkylation or heat treatments [31]. MXenes have a lot more advantages compared to other 2D-layered structures. The first one is its bulk production in water and its non-toxicity nature, and they are normally derived from earth-abundant elements. In addition, having high surface area and high electrical conductivity helps in MXenes' applicability in batteries, supercapacitors and biosensors [32–35].

After the discovery of MXenes, they were extensively studied for use in energy storage applications, like Li-ion, Na-ion, Mg-ion, Al-ion and Li-S batteries; textile and flexible energy storage devices; different energy harvesting devices; and electrochemical capacitors, including micro-capacitors [36–39]. MXenes have been used in energy storage applications due to their high surface area, fast electrolyte ion transportation, spontaneous cation intercalation properties and reversible redox reaction at the surface. Recently, MXenes have been explored in biosensors and in cancer therapy treatment because of their outstanding support for protein immobilization [40, 41]. Their electrocatalytic activity, including hydrogen evolution, oxygen evolution and CO_2 reduction, is reported worldwide [42–45]. Researchers have found the applicability of MXenes in different types of electronics due to their high electrical, thermal and various magnetic properties related to energy storage, conversion and sensor applications. Still, the majority of the research has focused on the first discovered MXene, $Ti_3C_2T_x$. A large group of unmapped MXenes and their combinations has still not been studied; thus, huge possibilities for a variety of applications still exist.

The $M_{n+1}AX_n$ (n = 1,2,3) phase is the building block for MXene and is layered in hexagonal structures with P63/mmc space group, classified as 211, 312 and 413 [14, 46]. Examples of various MAX phases are Ti_2AlN, Zr_3AlC_2 and V_4AlC_3. Here, the transition-metal layer is closely packed, and the C/N atom fills the octahedral site in the MAX phase. An A-layer atom is incorporated with $M_{n+1}X_n$ structures (M_2X, M_3X_2 and M_4X_3 shown in Figure 1.2B) [47, 48]. Most of them belong to the 211 phase, followed by 312 and 413 phases. These phases are differing from the number of M layers attached to the A layers. This can be synthesized from different solid solutions of M, A, X elements. In addition to the traditional MAX phase ($M_{n+1}AX_n$), according to the stoichiometry of M′ and M″, A and X elements, two different types of MAX phase exist: an out-of-plane ordered double-transition metal phase called o-MAX and an in-plane ordered double-transition metal phase called i-MAX. These crystal structures are shown in Figure 1.3A–C [54]. $Mo_2Ti_2AlC_3$ and Mo_2ScAlC_3 belong to the o-MAX group, and they are experimentally synthesized and exfoliated to form the corresponding MXene. Each M layer in the i-MAX (($M'_{2/3}M''_{1/3}$)2AX)

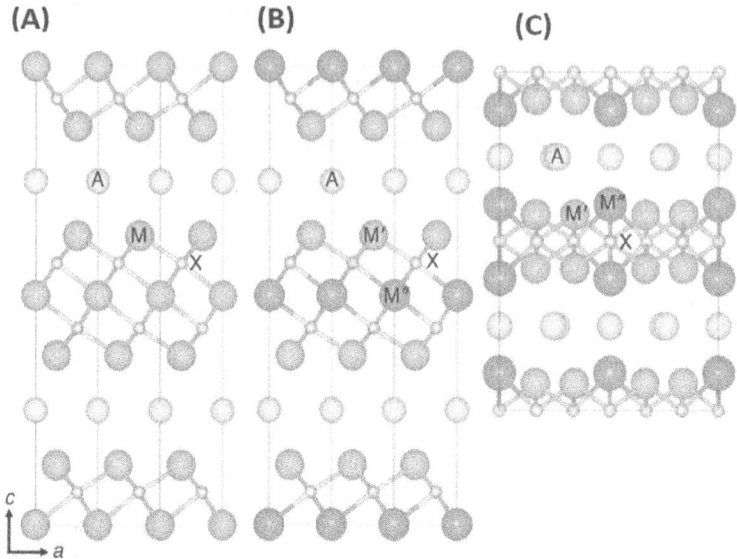

FIGURE 1.3 Crystal structure of (A) M_3AX_2, (B) out-of-plane ordered double-transition metal $M'_2M''AX_2$ and (C) in-plane double-transition metal $(M'_{2/3}M''_{1/3})2AX$. Solid line indicates the border of the unit cell.

(Image adapted with permission from Ref. [54]. © 2019 Elsevier Ltd. All rights reserved.)

phase contains M′ and M″ elements; hence, the i-MAX phase differs from the traditional MAX phase and o-MAX phase with a space group symmetry of C2/c. The examples of the i-MAX phase are $(V_{2/3}Zr_{1/3})2AlC$ and $(Cr_{2/3}Zr_{1/3})2AlC$. These MAX phases are chemically exfoliated to obtain corresponding MXenes [49–54]. These crystal structures are shown in Figure 1.3.

More than 30 stoichiometric MXenes have been synthesized experimentally (Figure 1.4) [55], and many more stoichiometric compositions have been predicted computationally. The exceptional properties of MXenes, such as high electrical conductivity, high volumetric electrochemical capacitance, optical transparency and electrochromism in thin films, thermal stability and mechanical strength, have led to extensive research [56–58]. Accordingly, MXenes are suggested for application in a variety of fields, including energy storage and electronic components, along with sensors, CO_2 reduction, gas adsorption and biomedical applications.

MXenes can be mainly categorized into two separate groups based on their transition-metal composition, that is, mono-transition-metal MXenes and double-transition-metal MXenes.

1.2 MONO-TRANSITION-METAL MXenes

In mono-transition-metal (mono-M) MXenes (Figure 1.2B), M layers consists of a single type of a transition metal, viz., in Ti_2CT_x, V_2CT_x, $Ti_3C_2T_x$ and $Nb_4C_3T_x$. Though many possible combinations of transition metals and carbon/nitrogen are

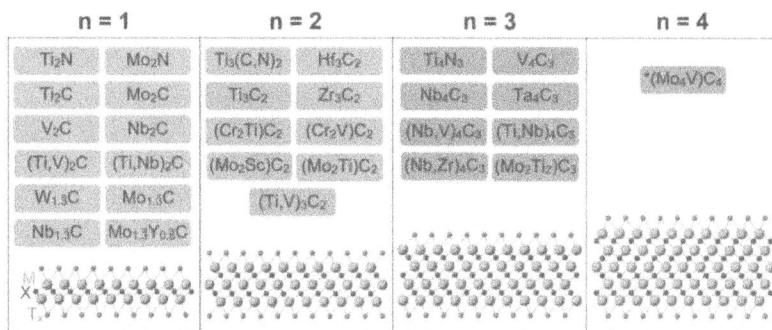

FIGURE 1.4 MXene compositions reported to date. MXenes discovered to date include mono-M MXenes, ordered double-transition-metal MXenes, solid-solution MXenes and ordered divacancy MXenes.

(Image adapted with permission from Ref. [55]. © 2020, American Chemical Society.)

possible, only 14 mono-M MXenes have been synthesized successfully. MXenes generally adopt three structures having one metal on the M site that is inherited from the parent MAX phases (M_2C, M_3C_2 and M_4C_3). They are synthesized by selectively etching out the A element from a MAX-phase or other layered precursor (such as Mo_2Ga_2C) that contains general formula $M_{n+1}AX_n$, wherein M represents an early transition metal, A is an element from group 13 or 14 of the periodic table, X is C and/or N and n = 1–4. The general structure of MXenes is a hexagonal close-packed (hcp) structure. Nevertheless, M atoms change their ordering while going from M_2X to M_3X2 and M_4X_3. In M_2X, M atoms have ABABAB ordering (hcp stacking), while in the cases of M_3C_2 and M_4C_3, M atoms follow ABCABC ordering (face-centred cubic stacking) [16].

Among different types, MXenes having low molecular weights, that is, mono-transition metals, show few interesting applications, including energy storage applications and gravimetric capacity. Theoretical and experimental studies are available on the applicability of these MXenes [16]. For instance, theoretical studies are helpful to define the most promising candidates for energy storage applications. On the basis of theoretical gravimetric capacity (i.e. amount of charge that can be stored per gram of material), it is considered that MXenes having low molecular weights, viz., Ti_2C, Nb_2C, V_2C and Sc_2C, are found to be the most promising ones among other types of MXenes. Thus, M_2X electrodes exhibit higher gravimetric capacities compared to their M_3X_2 and M_4X_3 counterparts. The reason behind this is the existence of a strong bond between M and X that is difficult to break, due to which it is rational to presume that ions can penetrate only via the MXene sheets. The available experimental data also supports this fact [16]. For instance, although Ti_2C and Ti_3C have the same surface chemistry, Ti_3C exhibits higher gravimetric capacity than Ti_2C due to the presence of one inactive layer in Ti_3C. The rich transition metal chemistry of MXenes has led to various computational studies exploring the effect of M, X and T; the number of M layers; and the lattice strain on the electronic, thermal and mechanical characteristics of MXenes. For instance, DFT and molecular dynamics predicted that M_2X MXenes are found to be stiffer and stronger than their M_3X_2 and M_4X_3 counterparts [16].

1.3 DOUBLE-TRANSITION-METAL MXenes

In the second type of MXenes, that is, double-transition-metal (DTM) MXenes, the M atoms can exist either in random (solid solution) or ordered arrangement, wherein the ordered phase is found to be energetically more stable. The popular DTM carbides that have been successfully synthesized include $Mo_2Ti_2C_3$, Cr_2TiC_2, Mo_2TiC_2 and Mo_4VC_4. In a few MXenes, viz., Mo_2TiC_2, $Mo_2Ti_2C_3$ and Cr_2TiC_2, the Mo or Cr atoms lie on the outer edges of the MXene, and the electrochemical properties of the MXenes are controlled by these atoms.

In literature, there is no direct method available to convert mono-M MXenes to DTM MXenes, and the synthesis of DTM MXenes is done only via selective etching of the A-group elements from their DTM MAX-phase precursors. Till now, experimentally synthesized and theoretically predicted DTM MXenes are carbides only; no DTM MXene nitrides or carbonitrides have been investigated. The composition of the MAX phases governs the composition of DTM MXenes, as the layered structures of DTM MXenes have been derived from their parent MAX phases. Although more than 20 DTM MXenes have been produced from their parent MAX, there are many synthesized DTM MAX phases that have not yet been selectively etched to their corresponding DTM MXenes [79].

MXenes can also be synthesized from non-MAX-phase precursors. Mo_2CTx is the first MXene of this category; it was synthesized via etching Ga layers from Mo_2Ga_2C. This phase consists of two A-element layers (Ga) separating the carbide layers. $Zr_3C_2T_X$ has been produced from another non-MAX-phase precursor by selectively etching aluminium carbide (Al_3C_3) layers from $Zr_3Al_3C_5$, not just the Al layers. $Zr_3Al_3C_5$ is the part of a family of layered ternary transition-metal carbides having general formulae of $MnAl_3C_{n+2}$ and $MnAl_4C_{n+3}$, wherein M is a transition metal, typically Hf or Zr, and n = 1–3. Here, Al-C units separate the M_2C or M_3C_2 layers and it is energetically more feasible to etch the Al-C units compared to just Al layers in $Zr_3Al_3C_5$. This finding may contribute to the synthesis of new MXenes from non-MAX precursors [79].

On the basis of the structure, DTM MXenes are further categorized into two types, that is, ordered and solid-solution MXenes (Figure 1.2B).

1.3.1 ORDERED MXenes

In ordered DTM MXenes, two different transition metals (M′ and M″) occupy the M layers in specific positions, defined as in-plane order or out-of-plane order (Figures 1.3B and C). In-plane-ordered MXenes (such as $Mo_{4/3}Y_{2/3}CT_x$) consist of two different transition metals that are ordered with alternating sites in each M-layer atomic plane. However, out-of-plane MXenes (such as $M'_2M''X_2T_x$ or $M'_2M''_2X_3T_x$) consist of ordered transition metal in different atomic planes wherein inner layers of M″ transition metals are sandwiched between outer layers of M′ transition metals [79].

Just like other MXenes, the structural configurations of out-of-plane-ordered DTM MXenes commence with the synthesis of their precursor MAX phases. Experimentally synthesized out-of-plane-ordered DTM MAX phases are Mo_2ScAlC_2, Cr_2VAlC_2, Mo_2TiAlC_2, Cr_2TiAlC_2, Ti_2ZrAlC_2, $Cr_2V_2AlC_3$ and $Mo_2Ti_2AlC_3$. When ordered MAX phases are selectively etched, their derivative MXenes keep the structural ordering of the corresponding MAX phases [79].

In case of in-plane MXenes, the unique atomic ordering is mainly derived from their in-plane ordered MAX-phase precursors, wherein M'' atoms are slightly extended out of the M layers towards the A layers, which results in two different types of MXenes after selective etching. Whilst a milder etching condition, viz., shorter etching time or lower hydrogen fluoride concentration, leads to the removal of the A layer only and forms $M'_{4/3}M''_{2/3}CT_x$, the use of stronger etching conditions, such as longer etching time or higher HF concentration, removes M'' along with the A layers, which leads to the formation of divacancy-ordered MXenes, $Mo_{4/3}CT_x$ [79].

1.3.2 SOLID-SOLUTION MXenes

In contrast to ordered MXenes, solid-solution MXenes consist of random distribution of two different transition metals placed in M-sites with general formulas $(M'_{2-y}M''_y)$ C, $(M'_{3-y}M''_y)C_2$, $(M'_{4-y}M''_y)C_3$ or $(M'_{5-y}M''_y)C_4$ (Figure 1.2B). Whilst ordered DTM structures are unique to MAX phases and MXenes amongst all known materials, solid-solution DTM MXenes are similar to solid solutions in other materials such as bulk nitrides and carbides. The stoichiometric ratio control of transition metals provides continuous control over the MXenes structures and properties [79]. Various MXenes show solid-solution MXenes, such as M_2CT_x, $M_3C_2T_x$, $M_4C_3T_x$ and even $M_5C_4T_x$. Out of these MXenes, $M_5C_4T_x$ is found to be highest order of MXenes till date and has only been observed as solid-solution MXenes. By controlling the stoichiometric ratio of two transition metals in solid-solution MXenes, their properties, such as electrochemical, electrical, chemical and optical, can be tuned between the properties of their two representative mono-M MXenes [79].

Like ordered MXenes, the stoichiometry control in solid-solution MXenes arises from their MAX-phase precursors. Although solid-solution MXenes were discovered in early 2012, with more than thousands of predicted compositions of solid solutions, there are a restricted number of both experimental and theoretical studies available on solid-solution MXenes [79]. In addition, many solid-solution MAX phases, such as $(Cr_\alpha Mn_{1-\alpha})_2AlC$, have not yet been etched to MXene.

1.4 SUMMARY AND CONCLUSION

There has been a drastic improvement in studies based on MXene compounds since 2017, including opening new applications and opportunities with a lot of experimental challenges. The main challenge of existing with MXenes is the scalable, safe and efficient synthesis method, and the development of various surface terminations other than O and OH still in their infant stage. In addition, Gogotsi et al. in a recent editorial from the ACS Nano clearly stated the research challenges that exist in the synthesis and development of MXenes for different applications [14, 58]. Hence,

MXene is still in its initial stage and there is a lot to move forward. Even though there is a lot of review literature about various synthesis methods and applications of MXenes, very little of the literature is focused on in-depth study.

In this book, different synthesis strategies were adopted for the preparation of MXene and structural tuning properties. The topics covered and discussed in this book are as follows: Chapter 2, Synthesis and Processing Strategies of MXenes; Chapter 3, Surface Functionalization and Interfacial Design of MXenes; Chapter 4, Solid-Solution MXenes and Their Properties; Chapter 5, Composites of MXenes; Chapter 6, Electrical Conductivity of MXenes-Based Polymer Composites; Chapter 7, Electromagnetic Interference Shielding Behavior of MXenes: Theoretical and Experimental Perspectives; Chapter 8, Role of Porous MXenes: Foams and Aerogels in EMI Shielding; Chapter 9, Role of MXene-Based Conductive Polymer Composites in EMI Shielding; Chapter 10, MXene-Polymer Nanocomposites for Biomedical Applications; Chapter 11, Role of MXene/Rubber Composites in EMI Shielding; and Chapter 12, Advancement in Nanostructured Carbide/Nitrides MXenes with Different Architecture for Electromagnetic Interference Shielding Application.

This book will cover different aspects of MXenes in a single literature, and this will give the readers an idea about the efficient synthesis methods for producing MXenes.

Moreover, the book discusses the existing applications and the challenges of taking the MXene from lab to industry. Here, in Table 1.1, we summarize the existing review literature that covers the properties and applications of MXenes.

TABLE 1.1

Summarizations of the Existing Review Literature That Covers the Properties and Application of Various MXenes

Sl No.	Title of the Paper	Year Up to Which Literature Was Covered	Reviewed Topics	Ref
1	Recent advances in MXenes: From fundamentals to applications	2019	MAX phase, MXene synthesis, characterization and properties of MXenes, applications in ferroelectricity, piezoelectricity, thermoelectricity, superconductivity catalysis, photocatalysis, batteries, gas sensors and hydrogen storage	[54]
2	Recent advances in 2D MXenes for enhanced cation intercalation in energy harvesting applications: A review	2020	Synthesis of MXene, structure and properties, application in Li-ion batteries, supercapacitors and non-lithium batteries	[59]

(Continued)

TABLE 1.1 (*Continued*)
Summarizations of the Existing Review Literature That Covers the Properties
and Application of Various MXenes

Sl No.	Title of the Paper	Year Up to Which Literature Was Covered	Reviewed Topics	Ref
3	Review of MXene electrochemical microsupercapacitors	2020	MXene synthesis approaches, MXene in aqueous and non-aqueous electrolytes, fabrication method for MXene microsupercapacitors, photolithography, laser engraving, screen printing, inkjet printing and other fabrication methods	[60]
4	Review of MXenes as new nanomaterials for energy storage/delivery and selected environmental applications	2019	Fabrication and property of MXene, energy application, energy conversion, energy storage supercapacitor, batteries, environmental application: adsorption, membrane, photocatalysis and antimicrobial application	[61]
5	Electronic properties and applications of MXenes: A theoretical review	2017	MAX phase, 2D MXene structural and mechanical properties, electronic properties, surface state properties, optical, magnetic and transport properties, applications: low work function electron emitters, catalysis and photocatalysis for hydrogen evolution, energy conversion thermoelectric devices, hydrogen storage, ion-batteries and supercapacitors, nanoribbon, nanotube and heterostructure MXenes	[62]
6	MXene-based fibers, yarns, and fabrics for wearable energy storage devices	2020	MXene synthesis, fabrication method, MXene-based fibers, yarn, and fabrics, electrode properties and device performances	[63]
7	MXene: Are they emerging materials for Analytical Chemistry application? – A review	2020	MXene-based electrochemical sensors, optical sensors, adsorbent for analytical extraction, green synthesis of MXene	[26]

(Continued)

TABLE 1.1 (*Continued*)

Summarizations of the Existing Review Literature That Covers the Properties and Application of Various MXenes

Sl No.	Title of the Paper	Year Up to Which Literature Was Covered	Reviewed Topics	Ref
8	2D transition metal carbides (MXenes): Application as an electrically conducting material	2020	Structural and electrical properties of MXene, application of MXenes in EMI shielding, transparent and flexible electrode, sensors, thermal heaters	[64]
9	Applications of 2D MXenes in energy conversion and storage systems	2019	MXene and MAX phase, properties, synthetic methods, electrochemical energy storage, energy utilization and conversion	[65]
10	2D metal carbides and nitrides (MXenes) for energy storage	2017	Synthesis of MXene, structure and properties, energy storage application of 2D carbides, application other than energy storage	[16]
11	2D MXenes for electromagnetic shielding: A review	2020	Mechanism of EMI shielding, MXene in EMI shielding materials	[66]
12	A review on MXene-based nanomaterials as adsorbents in aqueous solution	2020	Fabrication and characterization of MXene, removal of inorganic contaminant by MXene nanomaterial, removal of organic contaminant by MXene nanomaterial, regeneration of MXene nanomaterials	[67]
13	Adsorptive environmental applications of MXene nanomaterials: A review	2018	Structure and surface terminations of MXenes, progress of MXene in adsorption remediation of pollutants	[68]
14	MXenes: An introduction of their synthesis, selecte properties, and applications	2019	Synthesis and processing of MXene, family of materials with versatile chemical composition, properties from multi-layered to exfoliated MXene, promising applications for MXenes	[69]
15	Computational discovery and design of MXenes for energy applications: Status, successes, and opportunities	2019	MXene synthesis, surface terminations, intrinsic properties of MXenes, MXene materials for energy storage, electrocatalytic properties, high-throughput computation and machine learning to MXenes	[70]

(*Continued*)

TABLE 1.1 (*Continued*)

Summarizations of the Existing Review Literature That Covers the Properties and Application of Various MXenes

Sl No.	Title of the Paper	Year Up to Which Literature Was Covered	Reviewed Topics	Ref
16	Experimental and theoretical advances in MXene-based gas sensors	2021	Experimental perspectives, theoretical perspectives	[71]
17	MXene/polymer membranes: Synthesis, properties, and emerging applications.	2020	Synthesis of MXene, properties, preparation approach for MXene/polymer membrane, other methods, properties of MXene/polymer membrane, applications	[72]
18	MXenes: New horizons in catalysis	2020	MXene in a nutshell, survey of MXene in catalysis, future of MXene in catalysis	[73]
19	Photocatalytic applications of two-dimensional Ti_3C_2 MXenes: A review	2020	Synthesis methods for Ti_3C_2 MXene, Ti_3C_2 MXene as co-catalyst in photocatalyst, Ti_3C_2 MXene derivatives as a photocatalyst, advancement of Ti_3C_2 MXene in photocatalysis	[74]
20	Rational design of flexible two-dimensional MXenes with multiple functionalities	2019	Stability of MAX phase and MXenes, mechanical properties, energy storage, catalytic properties, thermoelectric properties, electronic/magnetic properties, topological properties	[75]
21	Rational design of two-dimensional transition metal carbide/nitride (MXene) hybrids and nanocomposites for catalytic energy storage and conversion	2020	Design strategy for MXene hybrids and composites, MXene hybrids for water splitting reactions, electrocatalysis for metal-air and metal-sulphur batteries, MXene hybrids for emerging EC, PC and PEC, reactions of interest	[76]
22	Recent advances in MXenes for lithium-ion capacitors	2019	Applications to Li-ion capacitors	[77]
23	Titanium carbide (Ti_3C_2) MXene as a promising co-catalysts for photocatalytic CO_2 conversion to energy-efficient fuels: A review	2021	Fundamental of MXene, MXene as a co-catalyst for photocatalytic CO_2 reduction, MXene as a co-catalyst for CO_2 methanation, MXene as a co-catalyst for photocatalytic flared gas reforming	[78]

CONFLICTS OF INTEREST

The authors declare no competing interest.

ACKNOWLEDGEMENTS

All authors would like to thank Indian Institute of Technology Jodhpur – India, Indian Institute of Science – Bengaluru, and Andhra University College of Engineering – Visakhapatnam – India, for resources and technical support. Dr. Srikanth Ponnada would like to thank University Grants Commission-India and Department of Science and Technology – India. Authors would like to convey special thanks to Dr. Nikhitha Joseph and Dr. Poushali Das for key suggestions.

REFERENCES

1. Akinwande, D., Brennan, C.J., Bunch, J.S., Egberts, P., Felts, J.R., Gao, H., Huang, R., Kim, J.S., Li, T., Li, Y. and Liechti, K.M., A review on mechanics and mechanical properties of 2D materials—Graphene and beyond. Extreme Mechanics Letters, (2017), 13, 42–77.
2. Mas-Balleste, R., Gomez-Navarro, C., Gomez-Herrero, J. and Zamora, F., 2D materials: To graphene and beyond. Nanoscale, (2011), 3(1), 20–30.
3. Glavin, N.R., Rao, R., Varshney, V., Bianco, E., Apte, A., Roy, A., Ringe, E. and Ajayan, P.M., Emerging applications of elemental 2D materials. Advanced Materials, (2020), 2(7), 1904302.
4. Zhang, H., Introduction: 2D materials chemistry. Chemical Reviews, (2018), 118(13), 6089–6090.
5. Momeni, K., Ji, Y., Wang, Y., Paul, S., Neshani, S., Yilmaz, D.E., Shin, Y.K., Zhang, D., Jiang, J.W., Park, H.S. and Sinnott, S., Multiscale computational understanding and growth of 2D materials: A review. npj Computational Materials, (2020), 6(1), 1–18.
6. Batmunkh, M., Bat-Erdene, M. and Shapter, J.G., Phosphorene and phosphorene-based materials–prospects for future applications. Advanced Materials, (2016), 28(39), 8586–8617.
7. Chowdhury, T., Sadler, E.C. and Kempa, T.J., Progress and prospects in transition-metal dichalcogenide research beyond 2D. Chemical Reviews, (2020), 120(22), 12563–12591.
8. Zhao, Z., Sun, Y. and Dong, F., Graphitic carbon nitride-based nanocomposites: A review. Nanoscale, (2015), 7(1), 15–37.
9. Caldwell, J.D., Aharonovich, I., Cassabois, G., Edgar, J.H., Gil, B. and Basov, D.N., Photonics with hexagonal boron nitride. Nature Reviews Materials, (2019), 4(8), 552–567.
10. Manzeli, S., Ovchinnikov, D., Pasquier, D., Yazyev, O.V. and Kis, A., 2D transition metal dichalcogenides. Nature Reviews Materials, (2017), 2(8), 1–15.
11. Chhowalla, M., Shin, H.S., Eda, G., Li, L.J., Loh, K.P. and Zhang, H., The chemistry of two-dimensional layered transition metal dichalcogenide nanosheets. Nature Chemistry, (2013), 5(4), 263–275.
12. Naguib, M., Kurtoglu, M., Presser, V., Lu, J., Niu, J., Heon, M., Hultman, L., Gogotsi, Y. and Barsoum, M.W., Two-dimensional nanocrystals produced by exfoliation of Ti_3AlC_2. Advanced Materials, (2011), 23, 4248–4253.
13. Gogotsi, Y. and Anasori, B., The rise of MXenes. ACS Nano, (2019), 13(8), 8491–8494.
14. Gogotsi, Y. and Huang, Q., MXenes: Two-dimensional building blocks for future materials and devices. ACS Nano, (2021), 15(4), 5775–5780.

15. Shekhirev, M., Shuck, C.E., Sarycheva, A. and Gogotsi, Y., Characterization of MXenes at every step, from their precursors to single flakes and assembled films. Progress in Materials Science, (2020), 100757.
16. Anasori, B., Lukatskaya, M.R. and Gogotsi, Y., 2D metal carbides and nitrides (MXenes) for energy storage. Nature Reviews Materials, (2017), 2(2), 1–17.
17. Alhabeb, M., Maleski, K., Anasori, B., Lelyukh, P., Clark, L., Sin, S. and Gogotsi, Y., Guidelines for synthesis and processing of two-dimensional titanium carbide ($Ti_3C_2T_x$ MXene). Chemistry of Materials, (2017), 29(18), 7633–7644.
18. Liu, F., Zhou, A., Chen, J., Jia, J., Zhou, W., Wang, L. and Hu, Q., Preparation of Ti_3C_2 and Ti_2C MXenes by fluoride salts etching and methane adsorptive properties. Applied Surface Science, (2017), 416, 781–789.
19. Naguib, M., Mochalin, V.N., Barsoum, M.W. and Gogotsi, Y., 25th anniversary article: MXenes: A new family of two-dimensional materials. Advanced Materials, (2014), 26(7), 992–1005.
20. Xu, B. and Gogosti, Y., MXenes: From discovery to application. Advanced Functional Materials, (2020), 30, 2007011.
21. Liu, Z. and Alshareef, H.N., MXenes for optoelectronic devices. Advanced Electronic Materials, (2021), 2100295.
22. Yu, H., Wang, Y., Jing, Y., Ma, J., Du, C.F. and Yan, Q., Surface modified MXene-based nanocomposites for electrochemical energy conversion and storage. Small, (2019), 15(25), 1901503.
23. Zhang, A., Liu, R., Tian, J., Huang, W. and Liu, J., MXene-based nanocomposites for energy conversion and storage applications. Chemistry—A European Journal, (2020), 26(29), 6342–6359.
24. Peng, J., Chen, X., Ong, W.J., Zhao, X. and Li, N., Surface and heterointerface engineering of 2D MXenes and their nanocomposites: Insights into electro- and photocatalysis. Chem, (2019), 5(1), 18–50.
25. Xie, X. and Zhang, N., Positioning MXenes in the photocatalysis landscape: Competitiveness, challenges, and future perspectives. Advanced Functional Materials, (2020), 30(36), 2002528.
26. Sajid, M. MXenes: Are they emerging materials for Analytical Chemistry applications?—A review. Analytica Chimica Acta, (2020).
27. Sun, Z.M. Progress in research and development on MAX phases: A family of layered ternary compounds. International Materials Reviews, (2011), 56(3), 143–166.
28. Weng, H., Ranjbar, A., Liang, Y., Song, Z., Khazaei, M., Yunoki, S., Arai, M., Kawazoe, Y., Fang, Z. and Dai, X., Large-gap two-dimensional topological insulator in oxygen functionalized MXene. Physical Review B, (2015), 92(7), 075436.
29. Khazaei, M., Ranjbar, A., Arai, M. and Yunoki, S., Topological insulators in the ordered double transition metals $M_2'M''C_2$ MXenes (M'= Mo, W; M''= Ti, Zr, Hf). Physical Review B, (2016), 94(12), 125152.
30. Si, C., Jin, K.H., Zhou, J., Sun, Z. and Liu, F., Large-gap quantum spin Hall state in MXenes: D-band topological order in a triangular lattice. Nano Letters, (2016), 16(10), 6584–6591.
31. Khazaei, M., Ranjbar, A., Arai, M., Sasaki, T. and Yunoki, S., Electronic properties and applications of MXenes: A theoretical review. Journal of Materials Chemistry C, (2017), 5(10), 2488–2503.
32. Hu, M., Zhang, H., Hu, T., Fan, B., Wang, X. and Li, Z., Emerging 2D MXenes for supercapacitors: Status, challenges and prospects. Chemical Society Reviews, (2020), 49(18), 6666–6693.
33. Aslam, M.K. and Xu, M., A mini-review: MXene composites for sodium/potassium-ion batteries. Nanoscale, (2020), 12(30), 15993–16007.

34. Greaves, M., Barg, S. and Bissett, M.A., MXene-based anodes for metal-ion batteries. Batteries & Supercaps, (2020), 3(3), 214–235.

35. Lei, Y., Zhao, W., Zhang, Y., Jiang, Q., He, J.H., Baeumner, A.J., Wolfbeis, O.S., Wang, Z.L., Salama, K.N. and Alshareef, H.N., A MXene-based wearable biosensor system for high-performance in vitro perspiration analysis. Small, (2019), 15(19), 1901190.

36. Elemike, E.E., Osafile, O.E. and Omugbe, E., New perspectives 2Ds to 3Ds MXenes and graphene functionalized systems as high-performance energy storage materials. Journal of Energy Storage, (2021), 42, 102993.

37. Xie, Y., Zhang, H., Huang, H., Wang, Z., Xu, Z., Zhao, H., Wang, Y., Chen, N. and Yang, W., High-voltage asymmetric MXene-based on-chip micro-supercapacitors. Nano Energy, (2020), 74, 104928.

38. Hasan, M.M., Hossain, M.M. and Chowdhury, H.K., Two-dimensional MXene-based flexible nanostructures for functional nanodevices: A review. Journal of Materials Chemistry A, (2021), 9(6), 3231–3269.

39. Ma, C., Ma, M.G., Si, C., Ji, X.X. and Wan, P., Flexible MXene-based composites for wearable devices. Advanced Functional Materials, (2021), 31(22), 2009524.

40. Sundaram, A., Ponraj, J.S., Wang, C., Peng, W.K., Manavalan, R.K., Dhanabalan, S.C., Zhang, H. and Gaspar, J., Engineering of 2D transition metal carbides and nitrides MXenes for cancer therapeutics and diagnostics. Journal of Materials Chemistry B, (2020), 8(23), 4990–5013.

41. Huang, H., Jiang, R., Feng, Y., Ouyang, H., Zhou, N., Zhang, X. and Wei, Y., Recent development and prospects of surface modification and biomedical applications of MXenes. Nanoscale, (2020), 12(3), 1325–1338.

42. Li, X., Bai, Y., Shi, X., Su, N., Nie, G., Zhang, R., Nie, H. and Ye, L., Applications of MXene $(Ti_3C_2T_x)$ in photocatalysis: A review. Materials Advances, (2021).

43. Zhu, W., Panda, S., Lu, C., Ma, Z., Khan, D., Dong, J., Sun, F., Xu, H., Zhang, Q. and Zou, J., Using a self-assembled two-dimensional MXene-based catalyst (2D-Ni@ Ti_3C_2) to enhance hydrogen storage properties of MgH2. ACS Applied Materials & Interfaces, (2020), 12(45), 50333–50343.

44. Tahir, M., Ali Khan, A., Tasleem, S., Mansoor, R. and Fan, W.K., Titanium carbide (Ti_3C_2) MXene as a promising co-catalyst for photocatalytic CO_2 conversion to energy-efficient fuels: A review. Energy & Fuels, (2021).

45. Wang, H. and Lee, J.M., Recent advances in structural engineering of MXene electrocatalysts. Journal of Materials Chemistry A, (2020), 8(21), 10604–10624.

46. Song, D., Jiang, X., Li, Y., Lu, X., Luan, S., Wang, Y., Li, Y. and Gao, F., Metal–organic frameworks-derived MnO_2/Mn_3O_4 micro cuboids with hierarchically ordered nanosheets and Ti_3C_2 MXene/Au NPs composites for electrochemical pesticide detection. Journal of Hazardous Materials, (2019), 373, 367–376.

47. Barsoum, M.W. The MN+ 1AXN phases: A new class of solids: Thermodynamically stable nanolaminates. Progress in Solid State Chemistry, (2000), 28(1–4), 201–281.

48. Shahzad, F., Iqbal, A., Kim, H. and Koo, C.M., 2D transition metal carbides (MXenes): Applications as an electrically conducting material. Advanced Materials, (2020), 32(51), 2002159.

49. Anasori, B., Xie, Y., Beidaghi, M., Lu, J., Hosler, B.C., Hultman, L., Kent, P.R., Gogotsi, Y. and Barsoum, M.W., Two-dimensional, ordered, double transition metals carbides (MXenes). ACS Nano, (2015), 9507–9516.

50. J. Rosen, M. Dahlqvist, Q. Tao, L. Hultman, in 2D Metal Carbides and Nitrides (MXenes) (Eds: B. Anasori, Y. Gogotsi), Springer, New York (2019), 37.

51. Tao, Q., Dahlqvist, M., Lu, J., Kota, S., Meshkian, R., Halim, J., Palisaitis, J., Hultman, L., Barsoum, M.W., Persson, P.O. and Rosen, J., Two-dimensional $Mo_{1.33}C$ MXene with divacancy ordering prepared from parent 3D laminate with in-plane chemical ordering. Nature Communications, (2017), 8(1), 1–7.

52. Tao, Q., Lu, J., Dahlqvist, M., Mockute, A., Calder, S., Petruhins, A., Meshkian, R., Rivin, O., Potashnikov, D., Caspi, E.A.N. and Shaked, H., Atomically layered and ordered rare-earth i-MAX phases: A new class of magnetic quaternary compounds. Chemistry of Materials, (2019), 31(7), 2476–2485.
53. Thore, A. and Rosén, J., An investigation of the in-plane chemically ordered atomic laminates $(Mo_{2/3}Sc_{1/3})_2AlC$ and $(Mo_{2/3}Y_{1/3})2AlC$ from first principles. Physical Chemistry Chemical Physics, (2017), 19(32), 21595–21603.
54. Khazaei, M., Mishra, A., Venkataramanan, N.S., Singh, A.K. and Yunoki, S., Recent advances in MXenes: From fundamentals to applications. Current Opinion in Solid State and Materials Science, (2019), 23(3), 164–178.
55. Deysher, G., Shuck, C.E., Hantanasirisakul, K., Frey, N.C., Foucher, A.C., Maleski, K., Sarycheva, A., Shenoy, V.B., Stach, E.A., Anasori, B. and Gogotsi, Y., Synthesis of Mo_4VAlC_4 MAX phase and two-dimensional Mo_4VC_4 MXene with five atomic layers of transition metals. ACS Nano, (2019), 14(1), 204–217.
56. Zhang, J., Kong, N., Uzun, S., Levitt, A., Seyedin, S., Lynch, P.A., Qin, S., Han, M., Yang, W., Liu, J. and Wang, X., Scalable manufacturing of free-standing, strong $Ti_3C_2T_x$ MXene films with outstanding conductivity. Advanced Materials, (2020), 32(23), 2001093.
57. Hantanasirisakul, K. and Gogotsi, Y., Electronic and optical properties of 2D transition metal carbides and nitrides (MXenes). Advanced Materials, (2018), 30(52), 1804779.
58. Seredych, M., Shuck, C.E., Pinto, D., Alhabeb, M., Precetti, E., Deysher, G., Anasori, B., Kurra, N. and Gogotsi, Y., High-temperature behavior and surface chemistry of carbide MXenes studied by thermal analysis. Chemistry of Materials, (2019), 31(9), 3324–3332.
59. Hemanth, N.R. and Kandasubramanian, B., Recent advances in 2D MXenes for enhanced cation intercalation in energy harvesting applications: A review. Chemical Engineering Journal, (2020), 392, 123678.
60. Jiang, Q., Lei, Y., Liang, H., Xi, K., Xia, C. and Alshareef, H.N., Review of MXene electrochemical microsupercapacitors. Energy Storage Materials, (2020), 27, 78–95.
61. Jun, B.M., Kim, S., Heo, J., Park, C.M., Her, N., Jang, M., Huang, Y., Han, J. and Yoon, Y., Review of MXenes as new nanomaterials for energy storage/delivery and selected environmental applications. Nano Research, (2019), 12(3), 471–487.
62. Khazaei, M., Ranjbar, A., Arai, M., Sasaki, T. and Yunoki, S., Electronic properties and applications of MXenes: A theoretical review. Journal of Materials Chemistry C, (2017), 5(10), 2488–2503.
63. Levitt, A., Zhang, J., Dion, G., Gogotsi, Y. and Razal, J.M., MXene-based fibers, yarns, and fabrics for wearable energy storage devices. Advanced Functional Materials, (2020), 30(47), 2000739.
64. Shahzad, F., Iqbal, A., Kim, H. and Koo, C.M., 2D transition metal carbides (MXenes): Applications as an electrically conducting material. Advanced Materials, (2020), 32(51), 2002159.
65. Pang, J., Mendes, R.G., Bachmatiuk, A., Zhao, L., Ta, H.Q., Gemming, T., Liu, H., Liu, Z. and Rummeli, M.H., Applications of 2D MXenes in energy conversion and storage systems. Chemical Society Reviews, (2019), 48(1), 72–133.
66. Iqbal, A., Sambyal, P. and Koo, C.M., 2D MXenes for electromagnetic shielding: A review. Advanced Functional Materials, (2020), 30(47), 2000883.
67. Jeon, M., Jun, B.M., Kim, S., Jang, M., Park, C.M., Snyder, S.A. and Yoon, Y., A review on MXene-based nanomaterials as adsorbents in aqueous solution. Chemosphere, (2020), 261, 127781.
68. Zhang, Y., Wang, L., Zhang, N. and Zhou, Z., Adsorptive environmental applications of MXene nanomaterials: A review. RSC Advances, (2018), 8(36), 19895–19905.

69. Verger, L., Natu, V., Carey, M. and Barsoum, M.W., MXenes: An introduction of their synthesis, select properties, and applications. Trends in Chemistry, (2019), 1(7), 656–669.

70. Zhan, C., Sun, W., Xie, Y., Jiang, D.E. and Kent, P.R., Computational discovery and design of MXenes for energy applications: Status, successes, and opportunities. ACS Applied Materials & Interfaces, (2019), 11(28), 24885–24905.

71. Mehdi Aghaei, S., Aasi, A. and Panchapakesan, B., Experimental and theoretical advances in MXene-based gas sensors. ACS Omega, (2021), 6(4), 2450–2461.

72. Gao, L., Li, C., Huang, W., Mei, S., Lin, H., Ou, Q., Zhang, Y., Guo, J., Zhang, F., Xu, S. and Zhang, H., MXene/polymer membranes: Synthesis, properties, and emerging applications. Chemistry of Materials, (2020), 32(5), 1703–1747.

73. Morales-García, A., Calle-Vallejo, F. and Illas, F., MXenes: New horizons in catalysis. ACS Catalysis, (2020), 10(22), 13487–13503.

74. Huang, K., Li, C., Li, H., Ren, G., Wang, L., Wang, W. and Meng, X., Photocatalytic applications of two-dimensional Ti3C2 MXenes: a review. ACS Applied Nano Materials, (2020), 3(10), 9581–9603.

75. Fu, Z., Wang, N., Legut, D., Si, C., Zhang, Q., Du, S., Germann, T.C., Francisco, J.S. and Zhang, R., Rational design of flexible two-dimensional MXenes with multiple functionalities. Chemical Reviews, (2019), 119(23), 11980–12031.

76. Lim, K.R.G., Handoko, A.D., Nemani, S.K., Wyatt, B., Jiang, H.Y., Tang, J., Anasori, B. and Seh, Z.W., Rational design of two-dimensional transition metal carbide/nitride (MXene) hybrids and nanocomposites for catalytic energy storage and conversion. ACS Nano, (2020), 14(9), 10834–10864.

77. Zhang, X., Wang, L., Liu, W., Li, C., Wang, K. and Ma, Y., Recent advances in MXenes for lithium-ion capacitors. ACS Omega, (2019), 5(1), 75–82.

78. Tahir, M., Ali Khan, A., Tasleem, S., Mansoor, R. and Fan, W.K., Titanium carbide (Ti3C2) MXene as a promising co-catalyst for photocatalytic CO_2 conversion to energy-efficient fuels: A review. Energy & Fuels, (2021), 35(13), 10374–10404.

79. Hong, W., Wyatt, B.C., Nemani, S.K. and Anasori, B., Double transition-metal MXenes: Atomistic design of two-dimensional carbides and nitrides. MRS Bulletin, (2020), 45(10), 850–861.

2 Synthesis and Processing Strategies of MXenes

Sayan Ganguly
Bar-Ilan Institute for Nanotechnology and
Advanced Materials, Department of Chemistry
Bar-Ilan University
Ramat-Gan, Israel

CONTENTS

2.1 INTRODUCTION

The emergence of 2D materials, which are characterised by extremely large aspect ratios and thicknesses of only a few atomic layers, has generated significant interest in the field of materials research in recent years. These materials, which typically have a layered structure held together by weak van der Waals forces, are exfoliated physically or chemically from the bulk 3D predecessors. They have exceptional chemical, mechanical, electrical, thermal, and optical characteristics, thanks to the decreased dimensionality and the quantum confinement effect from their bulk 3D counterpart. They are a viable contender in the fields of electronics, energy storage, biotechnology, and photovoltaics, because their characteristics may be further changed to a desired optimal level. Since the discovery of graphene [1] in 2004, quite a few novel and notable classes of 2D materials have been introduced. These include transition metal dichalcogenides (TMDs) and oxides [2]; hexagonal boron nitrides [3]; and other single-element 2D materials such as silicene [4], borophene [5], phosphorene [6], and so on. Among them, graphene and 2D TMDs seem to be the two most prevalent types of 2D materials in the field of study. Graphene, which is an atomically thin sheet of graphite, has been pushing the boundaries of study further and farther each year because to the exceptional and one-of-a-kind qualities it possesses, such as ultra-high carrier mobility [7–9]. It is also possible for it to be assisted in the form of membranes, which can be used for the purification of water and the separation of gases, or in the form of aerogels and hydrogels, which can be used for the storage of energy, dye adsorption, and so on [10]. In 2011, Naguib et al. [11] conducted research on the exfoliation of 2D transition metal carbides by selectively etching 'A' elements from the so-called MAX phases. This was done in order to study the exfoliation. The suffix 'ene' added to the end of the names of these 2D-layered MXenes indicates that they are comparable to graphene [8, 12]. Since that time, more than a thousand patent applications and hundreds of research articles have been generated, and one book [13] has been written about the subject. More important than the number of patents that have been applied for and granted is the extremely diverse range of applications that may be used, which includes not only electronics but also medical, sensing, communication, optoelectronics, and tribology, amongst a great number of other fields.

In this chapter, a viewpoint from the primary researchers who discovered MXenes a decade ago has been accounted. This perspective not only discusses the most significant recent advances in synthesis, the expansion of the MXene family, and the understanding of the structure and chemistry of these compounds, along with their properties and applications, but it also suggests potential future research avenues in this rapidly expanding field.

2.2 MXenes SYNTHESIS BY DIFFERENT TYPES OF ETCHING METHODS

As was mentioned earlier, the primary focus of our initial study on 2D transition metal carbide was on titanium carbide (Ti_3C_2), which was produced by utilising aqueous hydrofluoric acid (HF) as an etchant for aluminium at room temperature (RT). MXenes are typically created by removing the A layers from the parent MAX phases using selective etching as the production method [9]. Etchanting has been done to a large extent using acids that contain aqueous fluoride for the reason previously stated. Figure 2.1 presents a diagrammatic representation of the process that takes place during the synthesis of MXenes from MAX phases.

Different types of etching methods have been depicted as classifications in Figure 2.2. Within the scope of this method, stacked MAX-phase powders are mixed with aqueous HF acid at room temperature for a predetermined amount of time. As a consequence of this, the 'A' layers of the MAX phase are etched in a selective manner, and the metallic bonds that were present between the MAX layers are replaced with the more fragile bonds that are formed by surface terminations, such as hydroxyl, fluoride, or oxygen, on the surface of MXene. After this, the mixture is washed with deionised water and subjected to further centrifugation and filtration in order to separate the supernatant from the solid. The pH of the mixture is then kept within the range of 4 to 6 at all times. This results in the formation of few-layer (FL) MXene. When there are fewer than five layers in an MXene, it is referred to as an FL [14]. It should be noted that a further fall in the pH of the MXene solution, for example, around one in the case of Ti_3C_2Tx, might lead to the crumpling of MXene flakes. This is something that should be taken into consideration [15]. The first example of MXene synthesis from a non-MAX-phase precursor, Mo_2Ga_2C, was reported; in this case, Ga layers were etched to generate Mo_2CT_x MXenes [16]. The presence of two 'A' layers of Ga in Mo_2Ga_2C is what sets it apart from a MAX

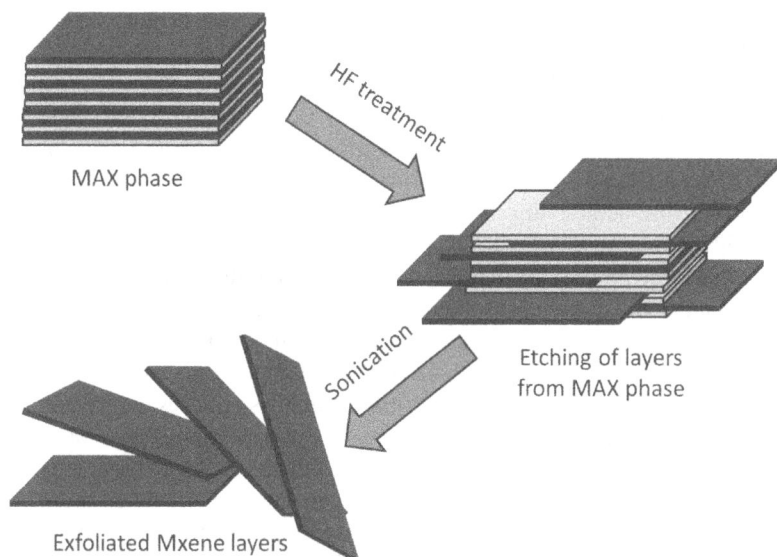

FIGURE 2.1 Schematic illustration of MXenes synthesis via HF etching method.

phase, which only has one. The $Zr_3C_2T_x$ MXene [17] generated from $Zr_3Al_3C_5$ is yet another instance where the aluminium carbide (Al_3C_3) layer was etched in addition to the aluminium (Al) layer. Etching parameters (acid concentration, temperature, and duration) were altered for each composition in order to accomplish the conversion. This was necessary because the binding strengths between the various M components and Al layers were distinct [18]. Tuning the etching conditions is important for a number of reasons, one of which is to maintain high yields and minimise the degradation of 2D flakes in the acid. This is necessary because the chemical stability of 2D flakes differs depending on the structure and composition of the material.

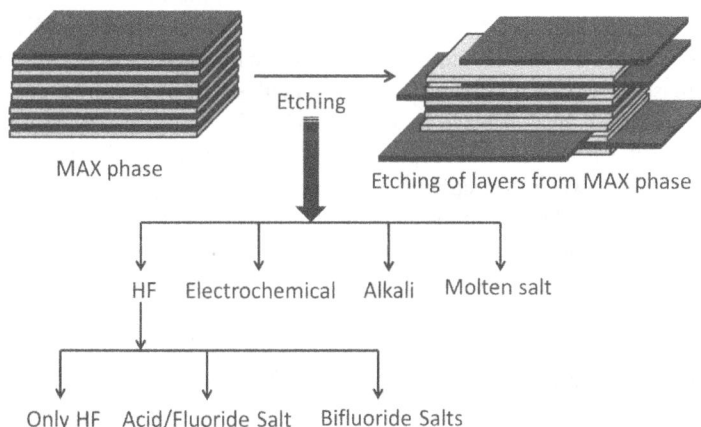

FIGURE 2.2 Classifications of different types of etching methods for preparing the MXenes.

Etching at high temperatures has also been used to produce MXenes starting from MAX phase starting material. In 2016. a molten fluoride salt mixture consisting of 59% KF, 29% LiF, and 12% NaF was used in an inert atmosphere of argon gas at 550°C to etch the Al layer part from Ti_4AlN_3 powder. This powder was then further delaminated by tetrabutylammonium hydroxide (TBAOH) to produce monolayers of $Ti_4N_3T_x$ MXene. The first experimental synthesis [19]. There are also reports of In layers from Ti_2InC sublimating at 800°C to form TiC_x in a vacuum [20] and Si layers from Ti_3SiC_2 being removed with molten cryolite at 960°C [21]. In spite of this, the carbides that were produced had a 3D cubic structure rather than a 2D structure, as a direct result of the treatment circumstances that included gas and temperature. In addition, it was discovered that ordered non-stoichiometric transition metal carbides are only stable below 800°C, which is in agreement with the phase diagram for these compounds [22]. In order to avoid the formation of 3D structures, the heat treatment and synthesis methods involving MXenes must be carried out at temperatures lower than those mentioned. Various approaches have been developed in order to manufacture 2D metal nitrides, such as the migration-enhanced encapsulated growth technique for GaN and the scalable salt-templated synthesis method for MoN [23]. Both of these methods were used to produce GaN. The final method, in which 2D hexagonal oxides are ammoniated to make 2D nitrides, was also used to produce other types of 2D nitrides, such as V_2N and W_2N [24]. Later, in 2017, Urbankowski et al. reported on the synthesis of 2D V_2N and Mo_2N from their respective 2D carbide MXenes V_2CT_x and Mo_2CT_x by ammoniation at 600°C, such that the C atoms could be replaced by N atoms. This was accomplished by removing the C atoms from the carbide MXenes first. In all instances, the 2D nitrides in question were generated via ammoniation; nevertheless, the processes led to their crystal forms being quite different from one another. During the second approach, Mo_2N maintained the MXene structure, whereas the V_2CT_x MXene transformed into a mixed phase consisting of cubic VN and trigonal V_2N [25]. In a manner analogous to that of the HF etching process, the in situ HF etching procedure also imparts surface functionalities, like OH, F, and O. The MXenes that may be produced using the latter technique are frequently found to be accompanied by intercalations of water molecules, which results in an extended period of time required for drying. When the MXenes that were formed by the in situ HF etching are dried, the interlayer spacing will dramatically decrease because the interlayer water molecules will no longer be present. The interlayer spacing is also affected by the sort of surface terminations that are used. Because F groups have a high hydrophobicity, they exhibit a clear inverse connection with the number of water molecules and the interlayer spacing. This is because water molecules are repelled by F groups. That is, the decrease in the amount of interlayer water molecules along with the interlayer spacing occurs as a direct result of the increasing concentration of the F groups.

The majority of the aforementioned processes include the use of acids to erode away the A atom layers. In point of fact, it is anticipated that the alkali will likewise be able to achieve the selective etching of the MAX phase. Xie et al. described a two-step etching technique that involved soaking the Ti_3AlC_2 in a 1 (M) NaOH solution for 100 h, followed by soaking it in a 1 (M) H_2SO_4 solution for 2 hours at 80°C. This enabled for the surface etching of the MAX phase into $Ti_3C_2T_x$. The MXene-covered

MAX phase has been reported elsewhere [26]. During this procedure, the alkali was responsible for removing the Al atoms from the MAX phase layers. On the other hand, H_2SO_4 was in charge of removing the Al atoms that were surface-exposed. The approach made it possible to effectively etch MAX phase using a low concentration of alkali as an etchant; nevertheless, it was only possible to etch the surface layer of MAX phase, which resulted in an exceedingly poor yield of MXene. In addition to this, it proved difficult to separate the etched MXenes from their MAX phase precursor. One more thing that slowed down the alkali etching process was the creation of oxide and hydroxide layers on top of the MAX phase. In one instance, a core–shell MAX@$K_2Ti_8O_{17}$ composite was produced by etching Ti_3SiC_2 at 200°C using a hydrothermal reaction. On the other hand, the use of NaOH resulted in the development of $Na_2Ti_7O_{15}$ onto the surface of the MAX phase, which prevented the procedure from yielding pure MXenes. When both the temperature and the concentration of the alkali are raised to a certain point, the reaction that takes place between the alkali and the MAX phase will go through a series of qualitative shifts. For instance, the Al layer may be effectively removed from Ti_3AlC_2 by applying 27.5 millimetres of sodium hydroxide at a temperature of 270°C in order to get $Ti_3C_2T_x$ with a yield of 92% [27].

The aqueous etching techniques which result in hydrophilic MXenes are applicable to the vast majority of Al-containing MAX phases when the temperature of operation is kept low. The aqueous system is ineffective, however, when dealing with the non-Al MAX or nitrides MAX phases. Based on the theoretical estimation, it was established that the transformation of $Ti_{n+1}AlN_n$ to $Ti_{n+1}N_n$ had a greater energy barrier than that of the $Ti_{n+1}AlC_n$ to $Ti_{n+1}C_n$. This reflects the thermodynamic limitation to etch the $Ti_{n+1}AlN_n$ owing to the strong bonding between Ti and Al atoms. Additionally, the cohesive energy of $Ti_{n+1}N_n$ was found to be lower than that of its comparable carbides. This is an indication that $Ti_{n+1}N_n$ has poor structural stability, which would make its dissolution in the fluorine-containing aqueous solution very simple [28]. Since transition metal halides are electron acceptors, they are capable of reacting with the A layer of the MAX phase while they are in a molten state. In this particular setting, Huang et al. etched Ti_3AlC_2, Ti_2AlC, Ti_2AlN, and V_2AlC MAX phases in a mixed $ZnCl_2$/NaCl/KCl molten salt system while the surrounding atmosphere was shielded by nitrogen [29]. In this molten salt system, $ZnCl_2$ was utilised as an etchant for the etching of the MAX phase, while NaCl and KCl with a molar ratio of 1:1 were used to generate the molten salt bath and decrease the melting point of the eutectic system. The molten salt bath was used to etch the MAX phase. During the etching process, the Zn^{2+} interacted with the A atoms of the MAX phase, which resulted in the transformation of the Al atoms that were only weakly bound into Al^{3+}. After this, the Zn atoms that were reduced move into the A-layer locations, which results in the formation of a new Zn-MAX phase, also known as Ti_3ZnC_2. In spite of the fact that the nonaqueous molten salt etching method has a wider etching range and is chemically safe, it is still in the early stages of development. As a result, in-depth research into the physical and chemical characteristics of produced MXenes, such as electrical conductivity, hydrophilicity, or mechanical property, is required. In addition, the created MXenes have structures that are similar to accordions, which render them inappropriate for the formation of nanoscale complexes.

It is possible to develop theoretical models of MXene that are practical, and these models may anticipate the behaviour of MXenes for specific applications on the basis of their surface-termination and electrical properties. In addition to the two types of in situ HF generating systems described earlier, several additional publications are employing an approach that is conceptually analogous to the etch MAX phase. Wu et al. reported a novel hybrid etchant that was created by combining NH_4F with the low eutectic mixed solvent of choline chloride and oxalic acid. This etchant was used to etch the MAX phase using a hydrothermal process at temperatures ranging from 100 to 180°C for a period of 24 hours [30]. In this etching system, the oxalic acid can react with NH_4F to create HF and break the Ti–Al bonds in the Ti_3AlC_2 to yield multilayered $Ti_3C_2T_x$ MXene. This process takes place when the oxalic acid comes into contact with the NH_4F. While chlorine ions have the ability to intercalate between the layers of MXenes, this has the effect of increasing the interlayer gap. In lithium-ion batteries, the dual-functional capability of this approach can increase both the lithium-ion kinetics and the reversible capacity of the MXenes-based anodes that are created. In a similar fashion, some fluoride-containing ionic liquids, such as 1-ethyl-3-methylimidazolium tetrafluoroborate ($EMIMBF_4$) and 1-butyl-3-methyl-imidazolium hexafluorophosphate ($BMIMPF_6$), can also be used to etch Ti_3AlC_2 and Ti2AlC at a temperature of 80°C for a period of 20 hours. The overall kinetics of the process is dependent on the acidity of dissociated organic anions in water and the interactions with F dissociated from ionic liquids [31]. The proposed method can circumvent the need to use an acidic solution for etching; however, this does not mean that the process is completely acid-free.

In order to create Ti_2C MXenes, Mei et al. suggested using a thermal reduction approach that used Ar/H_2 [32]. In this study, advancement was made over the MAX phase by using Ti, TiS_2, and commercial graphene to synthesise a novel Ti_2SC MAX phase with S-based A-layers. This work was an innovation over the MAX phase. After some time, the Ti_2C MXenes were formed by the dissociation of sulphur atoms with weak bonds at a very high temperature. The approach, on the other hand, places a heavy reliance on the MAX phase, which contains sulphur, and frequently produces an incomplete etching of the MAX phase. In addition, it was discovered that the TiO_2 nucleates at temperatures higher than 700°C, which has an effect on the amount of MXene produced. It has been demonstrated that certain species of algae are capable of removing MXene nanosheets from layered ternary precursors. In order to induce delamination and cleavage of MAX phases, Zada et al. employed organic acids made from algae to etch the Al atoms in MAX phase where the bioactive compounds may work as intercalators [33]. The V_2CT_x MXenes that were produced after an entire day of etching at RT had a lateral size of between 50 and 100 nm and an average thickness of less than 1.8 nm.

2.3 DELAMINATION AND INTERCALATION OF MXenes

When etching 'A' element from a layered ternary precursor using top-down procedures, the typical outcome is accordion-like MXenes that are piled on top of one another. To get single-layer MXene nanosheets, the processes of intercalation and delamination are necessary. Single-layer MXene nanosheets, as opposed to stacked MXenes, offer superior chemical properties. These nanosheets have a high specific surface area, are effective at repelling water, and have a diverse chemical composition on their surfaces. The research community's focus has shifted toward developing effective delamination techniques as a result of the appealing applications of 2D MXenes. In point of fact, an ultrasonic treatment was utilised in the first report of MXenes in order to delaminate the accordion-like MXenes into a few-layer thick structure. The strong contact that exists between the MXene layers, on the other hand, caused the delaminated layers to have a poor yield and rendered them incapable of being application-productive. Therefore, breaking the predominant interlayer pressures is the most important step in the process of separating the stacked nanosheets of MXenes. It has been demonstrated that injecting organic molecules or inorganic ions into these layers is a feasible alternative for reducing the strength of the interlayer contacts and increasing the interlayer separation. Because of their rheological property, hydrophilicity, and plasticity, MXenes are regarded to be comparable to clays such as kaolinite. Therefore, the organic species, which are frequently used as intercalators for clay, would have an advantageous effect in the case of MXenes. By intercalating dimethyl sulfoxide between the layers of multilayer $Ti_3C_2T_x$, Mashtalir et al. observed that the multilayer $Ti_3C_2T_x$ could be delaminated into single-layer MXene nanosheets. This discovery was published in 2013 [34]. The increased interlayer spacing has the potential to dramatically lower the van der Waals forces that are present between the MXene layers. This will make it easier to exfoliate the multilayer $Ti_3C_2T_x$ using the straightforward method of ultrasonication. Through the successful exfoliation of the dimethyl sulfoxide (DMSO)-intercalated multilayer $Ti_3C_2T_x$ to the single or few-layer $Ti_3C_2T_x$, the shape of the delaminated $Ti_3C_2T_x$ was further validated. In addition, the delaminated $Ti_3C_2T_x$ nanosheets were provided with an outstanding hydrophilicity and a negatively charged surface as a result of the abundant oxygen-containing surface terminations. As a result, the delaminated $Ti_3C_2T_x$ nanosheets may be equally disseminated in the deionised water, which results in the formation of a stable colloidal suspension in the absence of any surfactant. MXene films stacked by delaminated sheets showed much greater interplanar spacing as compared to multilayer $Ti_3C_2T_x$ films. This was advantageous for exposing more surface-active sites, which resulted in improved electrochemical performance. When put through the same tests as their multilayer counterparts, delaminated $Ti_3C_2T_x$ nanosheets were shown to have a capacity that was four times higher than that of their multilayer counterparts when used as anodes in lithium-ion batteries. In addition to DMSO, many additional organic solvents, such as hydrazine monohydrate (HM), N,N-dimethylformamide (DMF), and urea, have been investigated for their potential use as intercalators in the process of exfoliating multilayer $Ti_3C_2T_x$. Intercalation and delamination of multilayer MXenes have also been accomplished with the help of tetramethylammonium hydroxide (TMAOH). In addition to the effects of van der Waals forces, Han et al. argued that the Ti–Ti bonds and the Ti–Al bonds that exist inside the $Ti_3C_2T_x$ layers are the fundamental hurdles. Because of this, it was unable to achieve full intercalation by only employing an ultrasonic treatment [35]. TMAOH was used in a hydrothermal method, which allowed it to infiltrate and intercalate into the multilayer MXenes, hence making the following delamination of the film easier to accomplish. Similar to how the MXenes could be intercalated and delaminate with the use of microwave treatment, the yield of monolayer MXene nanosheets produced by the TMAOH was quite a bit low, which limited its usage for preparative applications [36]. TMAOH has intercalated itself into the MAX phase's interlayers, where it then interacts with the Al atoms. Because of this, it was possible to produce delaminated MXenes with surface-covered $Al(OH)_4$- groups without using sonication at any point in the process. The resulting MXene solution exhibited clear Tyndall effects, which is corroborative of the fact that the delaminated monolayer MXene flakes had a high degree of hydrophilicity. The transmission electron microscope (TEM) picture was used to evaluate how such an approach affected the delamination of the material. The MXene dispersion is typically recovered using a centrifugal technique after the intercalation stage of multilayer MXenes. In most cases, the required centrifugal speed is 3,500 revolutions per minute, and the required time period is around one hour. A low temperature must be maintained at all times in order to exclude the possibility of MXenes becoming oxidised. The undelaminated multilayer MXenes or the stacked nanosheets are precipitated after the centrifugation, and the supernatant, which is the stable single-layer MXene colloidal solution, may be collected afterwards. At this time, the delamination of multilayer MXenes is typically performed for products that have been produced using HF etching or some other type of aqueous etching process. Therefore, further research is required to determine whether or not these organic intercalators may be used in systems that do not include water.

LiCl may be employed as an intercalator for multilayer MXenes, which allows for the insertion of Li^+ and results in larger interlayer spacing. After ultrasonic treatment, the multilayer MXene may be delaminated into monolayer MXene nanosheets because the interlayer van der Waals force has been reduced as a result of the decreased interlayer van der Waals force. On the other hand, this approach can only be used with multilayer MXenes that have been etched using an HF/HCl combination as the etchant [37].

Ghidiu et al. etched the $Ti_3C_2T_x$ using a solution that included both HF and LiCl, which allowed for the formation of an accordion-like $Ti_3C_2T_x$ structure by the intercalation of Li^+. However, the subsequent ultrasonic treatment was unable to successfully delaminate the material [38]. As a result, it is not unreasonable to hypothesise that the incorporation of HCl into the delamination process plays a significant part in the overall outcome. However, the particular process of delamination has not yet been established at this time. It is important to note that further ion exchange can be performed in order to embed macromolecular compounds into the interlayers after the Li^+ ions from the HF/LiCl etchants have inserted themselves into the $Ti_3C_2T_x$ interlayer. Ghidiu and colleagues first synthesised Li-$Ti_3C_2T_x$, and then they substituted big $[(CH_3)_3NR]^+$ cations for the interlayer Li^+. R in this context refers to an alkyl chain. By selecting a different alkyl chain, it is possible to adjust the interlayer spacing of MXenes within the range of 5–28 Å, which ultimately leads to the fine-tuning of the conductivity of MXenes [39].

2.4 DELAMINATION BY MECHANICAL FORCES

Another method for separating the nanosheets is called mechanical delamination, and it works by applying either longitudinal or transverse stress on the surface of the multilayer structure. Exfoliation of graphene, molybdenum disulfide, and a variety of other layered materials has demonstrated the efficacy of this approach [40]. On the other hand, MXenes have greater interlayer contacts, which are difficult to totally overcome when subjected to a stress of this mechanical kind. Through the use of an adhesive tape delamination method, Xu et al. and Lai et al. were able to delaminate Ti_2CT_x nanosheets onto a Si wafer [41], in spite of the fact that effective delamination was performed to some extent and the MXene nanosheets obtained were those with a few layers. In addition, the mechanical technique has a limited yield of delamination, which prevents it from being used for manufacturing on a wide scale [42]. In more recent times, it has also been claimed that the delamination of MXenes may be accomplished with the assistance of cyclic freezing and thawing. The molecules of frozen water were able to increase the interlayer spacing of multilayer MXenes, which resulted in a weakening of the strong van der Waals forces and made it possible for MXenes to easily delaminate without the need of any intercalators [43].

2.5 LARGE SCALE SYNTHESIS

There is never a scarcity of innovative materials with distinctive features that merit further investigation and development in the published scientific literature. It is essential to have a fundamental understanding of the link between the material and its qualities in order to rationally develop products and customise them to specific applications. This relationship may be understood in terms of the material's chemistry, structure, and processing. However, for materials to be used on a global scale, a different approach needs to be taken, and that is to assess the practicability of their application in the market. When it comes to certain applications, such as energy storage, catalysis, water desalination/purification, or structural composites, it is very necessary that the materials be able to be manufactured in large quantities and at a low cost. The fact that the components utilised are readily available, low in cost, and risk-free to work with is one of the fundamental tenets of these applications. Take, for example, the field of catalysis. For the past few decades, researchers have been looking for ways to reduce the use of noble metals (such as platinum, palladium, and gold) by discovering alternative catalysts, utilising nano-sized catalysts, single-atom catalysts, or finding clever solutions for catalyst regeneration. In order for materials to be used in industry, an additional set of considerations must first take place. These include the cost and availability of precursors and the materials that can be produced from them, along with the materials' toxicity, impact on the environment, and the ability to be manufactured. In many situations, it is preferable to utilise catalysts that are less expensive and more readily available, even if these catalysts do not function as well as their noble metal analogues. The same is true in the field of energy storage; both the cathode and the anode in batteries are made from abundant and inexpensive materials (graphite and $LiFePO_4$ or $LiNiCoAlO_2$, for example, in electric cars). This stands in stark contrast to the thousands of research articles that are published on energy storage each year that use new materials, many of which are quite unusual. On the other hand, in order to use these materials effectively in applications such as biosensors or electronics, it is not necessary to be able to mass manufacture large quantities of them at once. Because each device only requires extremely small quantities of the materials, it is less critical that they are readily available and relatively inexpensive. When this is taken into consideration, it becomes abundantly evident that we, as scientists, have a moral obligation to take into account the practicability of our investigations.

In particular, the synthesis of 2D materials may be carried out using one of two main strategies: either a bottom-up or a top-down strategy. Approaches that start from the bottom and work their way up basically construct the materials atom by atom. This indicates that it is feasible to make essentially defect-free materials, with specified orientations exposed, or to develop materials or structures that won't naturally form.. The processes of physical vapour deposition and chemical vapour deposition are the two basic methods that are utilised in bottom-up synthesis. In spite of the fact that there have been numerous recent developments in these methods, which have resulted in lower costs and bigger crystals, it is abundantly clear that these methods are pricey, and the scalability is restricted [44]. Bottom-up procedures are great for applications that need only a tiny amount of materials, such those found in electronics, or highly precise structures; but, these technologies are less viable for the manufacture of large quantities of inexpensive materials in bulk. There are many different methods of production that may be used for top-down synthesis. Some of these methods include selectively etching a bulk material or exfoliating bulk materials by the use of mechanical force, solution processing, or sonication. When top-down methods are used, the materials that are created will, in general, be generated in a manner that is less uniform. There will be a dispersion in the particle size, defect density, and characteristics as a result of this. On the other hand, top-down techniques make use of conventional chemical engineering procedures. They can be easily scaled using the technologies that are now available, and there are only a few new barriers that need to be overcome. Because of this, methods that include top-down synthesis are more cost-effective, making them more suitable for applications involving mass utilisation.

One of the most pressing concerns for people working in the field of materials science is getting MXenes out of the lab and into practical use on an industrial scale. There are just a few technologies that are now available that might be used to create 2D materials such as graphene on a preparative scale. This is the case as of right now. The fact that the majority of 2D materials are generated using a bottom-up technique, which limits the substrate size's preparation scale, is often the source of the difficulty that is associated with the manufacture of large quantities of 2D materials. In the case of alternative pathways, such as the hydrothermal one, the morphologies and characteristics of the products would differ depending on the magnitude of the reaction [45]. On the other hand, MXenes are created by etching the precursors using top-down techniques, which makes it simple to scale up production of these compounds. When the synthesis method is scaled up to a larger scale, there are significant issues associated with maintaining the uniformity of the morphology and the intrinsic features of the generated MXene nanosheets. The HF etching method, which was the first etchant to be researched and developed, has demonstrated that it has the potential to become a pathway that may realise the large-scale creation of MXenes, which has encouraging implications for industrial production. Shuck et al. demonstrated that the etching of MAX phase may be amplified using a combined HF/HCl etchant to the level of 50 g per time [46].

The vacuum filtration of the single-layer MXene colloidal solution results in the production of an important derivative product of MXenes called the flexible free-standing membrane. MXene membranes are able to have good mechanical properties, flexibility, and electrical conductivity thanks to the large lateral size and dense stacking of MXene nanosheets. This enables the membranes to recognise their potential in applications such as energy storage, water treatment, electromagnetic interference shielding, sensors, and other similar applications. Producing an MXene membrane that has a high surface area continues to be difficult due to the fact that the diameter of the MXene is restricted by the size of the filter. Because of this, preparation techniques for large-size MXene membranes that do not use vacuum filtration have been investigated. It is anticipated that a production technique for large-area MXene membranes would have both high efficiency and low energy consumption as its optimum characteristics. After screening the precursor MAX phase to make the particle size consistent, Zhang et al. used an etching technique to manufacture large-scale MXene nanosheets [47]. To create a free-standing large-size MXene membrane with good mechanical strength and electrical conductivity, a scraping blade approach was applied. In a separate experiment, Deng et al. were successful in mass producing a flexible MXene membrane by employing an electrochemical deposition method with a stainless steel mesh serving as the anode [48]. In a matter of minutes, this technique can manufacture a flexible membrane with an area of 500 cm^2, and it also allows for size selectiveness by varying the amount of time spent depositing the material. The membranes that were created have high ion rejection capacity against metal ions of tiny size and have the potential to be utilised in applications involving the treatment of water or the environment. When preparing MXenes on a large scale, it is important to take into consideration the effects of precursor particle size, because this might have an effect on the etching kinetics along with the quality of the end products. Sieves were used by Mashtalir et al. to separate the Ti_3AlC_2 particles into three distinct portions [49].

Those Ti_3AlC_2 particles with a size between 38 and 53 μm were able to be totally etched into $Ti_3C_2T_x$ after being submerged in HF for two hours at room temperature. The resulting $Ti_3C_2T_x$ had the greatest c-Lattice parameter of 19.64 Å. Furthermore, Naguib et al. discovered that after the V_2AlC precursor was ground in advance, its etching time in 50 weight percent HF at room temperature was reduced from 90 hours (yield greater than 60%) to 8 hours (yield greater than 55%), and the interlayer spacing of the MXene that was obtained increased from 19.73 to 23.96 Å [50]. It is possible to draw the conclusion that the particle size of the precursor has an immediate and decisive impact on the size of the MXenes. The MAX phases that are big lead to incomplete etching, whereas the MAX phases that are tiny (100–200 nm) make the corresponding MXenes have a small lateral size and can even be over etched or dissolved directly in etchants. When the kinetics of etching and the lateral size of MXene nanosheets are taken into consideration, a MAX phase with a particle size on the micron scale should be considered comparatively acceptable. At the moment, the large-scale preparation procedures are only used for the $Ti_3C_2T_x$ MXene, and the scalable manufacture of high-order structure MXenes has not been documented. However, this is expected to change in the near future. Because the existing aqueous etching of higher-order structured MXenes often requires critical conditions such as highly concentrated HF (50 wt%) and extended etching time (>69 h), it is more challenging to expand this method to large-scale manufacturing [19].

2.6 STORAGE OF MXenes

MXenes provide a significant challenge in terms of their stability since they are susceptible to oxidation in moist settings and, as a result, their characteristics degrade with time. In terms of their chemical make-up, MXenes have a thermodynamically unstable structure. In contrast to the surface, which is densely populated with functional terminations and preserves the integrity of the structure, edge-based M atoms are typically unstable due to the incomplete bonding and make it simple to nucleate transition metal oxide particles when they interact with oxygen or water molecules. The surface is saturated with functional terminations and preserves the integrity of the structure. The initial oxidation then extends to the entire surface, which causes the structural breaking with increasing edges, which in turn accelerates the domino effect of the oxidative deterioration process. The electrical conductivity and hydrophilic properties that are intrinsic to the colloidal dispersion of monolayer $Ti_3C_2T_x$ are gradually lost when it undergoes the slow transformation to TiO_2 in an environment consisting of water and oxygen [51]. In order to find a solution to this bottleneck problem, several research projects have been carried out to get a comprehensive understanding of the fundamental oxidation mechanism, the elements that might influence it, and the innovative methods that can create oxidation-resistant MXenes. Zhang et al. investigated how the oxidation process of MXenes was affected by the surrounding environment and the temperature. According to the findings of this research, the structural stability of $Ti_3C_2T_x$ is superior in an argon-saturated environment at low temperatures than it is in the air/Ar condition at room temperature [52]. According to the findings, oxidation happened quite fast for the $Ti_3C_2T_x$ when it was kept in an environment of storage that contained air at RT; on the other hand, its characteristics were able to be well retained even after storage for 24 days under an argon atmosphere at a low temperature. It was discovered that smaller MXene flakes in solution had a greater propensity to oxidise than their bigger counterparts, showing that the process of deterioration was edge-dependent. This was discovered by further choosing the size of MXene flakes that were present in solution. Thus, an inert environment may be ascribed to the isolation of $Ti_3C_2T_x$ with oxygen, while a low temperature can constrain the nucleation of metal oxides begun from the flake's edges. Both of these factors contribute to the process of isolating $Ti_3C_2T_x$ with oxygen. Through the use of molecular dynamics modelling, Lotfi et al. computed the oxidation rate of MXenes in H_2O_2, humid air, and dry air [53]. The researchers found that the oxidation rate of MXenes fell in turn under each of the aforementioned three conditions. Through the use of comparative experimental analysis, Huang et al. were able to further differentiate between the functions played by water molecules and oxygen in the oxidation process of MXenes [54]. When exposed to aqueous circumstances, Ti2CTx oxidised after a week, regardless of whether it was saturated with oxygen or argon gas. However, when the Ti2CTx was moved to a solution containing isopropanol, the oxidation rate decreased dramatically even in an atmosphere that was oxygen saturated. After being stored for one week, it was determined to be stable. According to the findings of the study, the oxidation of MXenes is more likely caused by the presence of water in the system than by the presence of dissolved oxygen gas.

The presence of oxygen-containing terminations on MXenes' surfaces makes it possible for these surfaces to have a negative charge in the aqueous system. On the other hand, it has been suggested that the charge of edges might be positive [55]. Therefore, the absorption of OH over a positively charged MXene-edge under alkaline circumstances may result in the deprotonation of surface groups and an acceleration of the process of surface oxidation [56]. Lamellar stacking with a less-exposed surface and a delayed oxidation process was created while the Zeta potential was lowered under acidic circumstances. However, the very acidic environment may eventually create corrosion damage and may permit more exposed surface edges, both of which may make the MXene oxidation process worse.

2.7 SUMMARY AND OUTLOOK

MXenes are a relatively new type of material that have just developed to compete with traditional 2D materials in a variety of applications. Only about 70 MAX phases are thought to exist at this time, despite the fact that there are approximately 792 possible combinations that can be interpreted to form MAX phases (11 M, 12 A, 2X, and 3n values). Excluding solid solutions and ordered-M layers from consideration, only about 20 of these 70 phases have been etched into a 2D MXene material. In addition, prior to 2018, only Al has been successfully etched out of the 12 'A' elements that constitute MAX phases. In that year, Si was successfully etched out of Ti_3SiC_2 to make Ti_3C_2 MXene. As a result, the identification of novel MAX phases and the 2D MXenes that correspond to them becomes an important research avenue.

The chapter describes how far along MXenes are in the process of synthesis. This includes the fundamentals of the etching, intercalation, and delamination processes, along with the conditional effect on the morphology, surface groups, and physicochemical characteristics of MXenes. In addition to this, the subject is expanded to include the production of MXenes on a big scale. In this key portion, we also cover how stable MXenes are, as well as the various storage techniques that are currently in use. The development of MXenes for use in industrial applications is dependent on the large-scale preparation techniques being discovered. The HF etching approach was the only one that could etch 50 g of MXene precursor without affecting the purity of the MXene, showing that it had the potential to satisfy the large-scale requirement. Despite the environmental impact of this process, it was the only one that could do so. In contrast to MXene powders, derivatives like flexible membranes and fibres have recently attracted a lot of attention on preparative scales due to their potential application in flexible electronics and wearables. Intercalation is an important part in the processing of MXenes, as well as their characteristics and performance in a variety of different applications. However, there is still a lack of understanding regarding the nature of intercalants in confinement, their interaction with the catalytically active MXene nanosheets, and their interaction with solvents (e.g., confined water). Intercalation in $Ti_3C_2T_x$ has been the primary focus of the majority of the effort. The use of MXenes in a wide variety of applications, including, but not limited to, hydrogen storage, supercapacitors, batteries, and desalination, are still in the early phases of development. There should be a substantial amount of work done on all of its uses, both in the realm of business and in the realm of research. Last but not least, the majority of recent publications on the subject of the characteristics of MXenes, such as electrical, electronic, mechanical, and so on, are based on theoretical studies such as DFT and MD. A comprehensive experimental investigation into these discoveries is necessary in order to validate those theoretical estimates.

REFERENCES

1. Novoselov KS, Geim AK, Morozov SV, Jiang D-e, Zhang Y, Dubonos SV, et al. Electric field effect in atomically thin carbon films. Science. 2004;306:666–9.
2. Ataca C, Sahin H, Ciraci S. Stable, single-layer MX2 transition-metal oxides and dichalcogenides in a honeycomb-like structure. The Journal of Physical Chemistry C. 2012;116:8983–99.
3. Pacile D, Meyer J, Girit Ç, Zettl A. The two-dimensional phase of boron nitride: few-atomic-layer sheets and suspended membranes. Applied Physics Letters. 2008;92:133107.
4. Lalmi B, Oughaddou H, Enriquez H, Kara A, Vizzini S, Ealet B, et al. Epitaxial growth of a silicene sheet. Applied Physics Letters. 2010;97:223109.
5. Mannix AJ, Zhou X-F, Kiraly B, Wood JD, Alducin D, Myers BD, et al. Synthesis of borophenes: anisotropic, two-dimensional boron polymorphs. Science. 2015;350:1513–6.
6. Liu H, Neal AT, Zhu Z, Luo Z, Xu X, Tománek D, et al. Phosphorene: an unexplored 2D semiconductor with a high hole mobility. ACS Nano. 2014;8:4033–41.
7. Banszerus L, Schmitz M, Engels S, Goldsche M, Watanabe K, Taniguchi T, et al. Ballistic transport exceeding 28 µm in CVD grown graphene. Nano letters. 2016;16:1387–91.
8. Das P, Ganguly S, Saha A, Noked M, Margel S, Gedanken A. Carbon-dots-initiated photopolymerization: an in situ synthetic approach for MXene/poly (norepinephrine)/copper hybrid and its application for mitigating water pollution. ACS Applied Materials & Interfaces. 2021;13:31038–50.
9. Ganguly S, Das P, Saha A, Noked M, Gedanken A, Margel S. Mussel-inspired poly-norepinephrine/MXene-based magnetic nanohybrid for electromagnetic interference shielding in X-band and strain-sensing performance. Langmuir. 2022;38:3936–50.
10. Gogotsi Y, Anasori B. The rise of MXenes. ACS Nano. 2019;13:8491–4.
11. Naguib M, Kurtoglu M, Presser V, Lu J, Niu J, Heon M, et al. Two-dimensional nano-crystals produced by exfoliation of Ti_3AlC_2. Advanced Materials. 2011;23:4248–53.
12. Naguib M, Gogotsi Y. Synthesis of two-dimensional materials by selective extraction. Accounts of Chemical Research. 2015;48:128–35.
13. Anasori B, Lukatskaya MR, Gogotsi Y. 2D metal carbides and nitrides (MXenes) for energy storage. Nature Reviews Materials. 2017;2:1–17.
14. Naguib M, Mochalin VN, Barsoum MW, Gogotsi Y. 25th anniversary article: MXenes: a new family of two-dimensional materials. Advanced Materials. 2014;26:992–1005.
15. Natu V, Clites M, Pomerantseva E, Barsoum MW. Mesoporous MXene powders synthesized by acid induced crumpling and their use as Na-ion battery anodes. Materials Research Letters. 2018;6:230–5.

16. Halim J, Kota S, Lukatskaya MR, Naguib M, Zhao MQ, Moon EJ, et al. Synthesis and characterization of 2D molybdenum carbide (MXene). Advanced Functional Materials. 2016;26:3118–27.

17. Zhou J, Zha X, Chen FY, Ye Q, Eklund P, Du S, et al. A two-dimensional zirconium carbide by selective etching of Al_3C_3 from nanolaminated $Zr_3Al_3C_5$. Angewandte Chemie International Edition. 2016;55:5008–13.

18. Khazaei M, Arai M, Sasaki T, Estili M, Sakka Y. Trends in electronic structures and structural properties of MAX phases: a first-principles study on M_2AlC (M= Sc, Ti, Cr, Zr, Nb, Mo, Hf, or Ta), M_2AlN, and hypothetical M_2AlB phases. Journal of Physics: Condensed Matter. 2014;26:505503.

19. Urbankowski P, Anasori B, Makaryan T, Er D, Kota S, Walsh PL, et al. Synthesis of two-dimensional titanium nitride Ti_4N_3 (MXene). Nanoscale. 2016;8:11385–91.

20. Barsoum M, Golczewski J, Seifert H, Aldinger F. Fabrication and electrical and thermal properties of Ti2InC, Hf2InC and (Ti, Hf) 2InC. Journal of Alloys and Compounds. 2002;340:173–9.

21. Barsoum MW, El-Raghy T, Farber L, Amer M, Christini R, Adams A. The topotactic transformation of Ti_3SiC_2 into a partially ordered cubic Ti ($C_{0.67}Si_{0.06}$) phase by the diffusion of Si into molten cryolite. Journal of the Electrochemical Society. 1999;146:3919.

22. Gusev A, Rempel A. Atomic ordering and phase equilibria in strongly nonstoichiometric carbides and nitrides. Materials Science of Carbides, Nitrides and Borides. Springer; 1999. p. 47–64.

23. Al Balushi ZY, Wang K, Ghosh RK, Vilá RA, Eichfeld SM, Caldwell JD, et al. Two-dimensional gallium nitride realized via graphene encapsulation. Nature Materials. 2016;15:1166–71.

24. Xiao X, Yu H, Jin H, Wu M, Fang Y, Sun J, et al. Salt-templated synthesis of 2D metallic MoN and other nitrides. ACS Nano. 2017;11:2180–6.

25. Urbankowski P, Anasori B, Hantanasirisakul K, Yang L, Zhang L, Haines B, et al. 2D molybdenum and vanadium nitrides synthesized by ammoniation of 2D transition metal carbides (MXenes). Nanoscale. 2017;9:17722–30.

26. Xie X, Xue Y, Li L, Chen S, Nie Y, Ding W, et al. Surface Al leached Ti_3AlC_2 as a substitute for carbon for use as a catalyst support in a harsh corrosive electrochemical system. Nanoscale. 2014;6:11035–40.

27. Li T, Yao L, Liu Q, Gu J, Luo R, Li J, et al. Fluorine-free synthesis of high-purity $Ti_3C_2T_x$ (T= OH, O) via alkali treatment. Angewandte Chemie International Edition. 2018;57:6115–9.

28. Zhang N, Hong Y, Yazdanparast S, Zaeem M. Superior structural, elastic and electronic properties of 2D titanium nitride MXenes over carbide MXenes: a comprehensive first principles study. 2D Mater. 2018;5:045004.

29. Li M, Lu J, Luo K, Li Y, Chang K, Chen K, et al. Element replacement approach by reaction with Lewis acidic molten salts to synthesize nanolaminated MAX phases and MXenes. Journal of the American Chemical Society. 2019;141:4730–7.

30. Wu J, Wang Y, Zhang Y, Meng H, Xu Y, Han Y, et al. Highly safe and ionothermal synthesis of Ti_3C_2 MXene with expanded interlayer spacing for enhanced lithium storage. Journal of Energy Chemistry. 2020;47:203–9.

31. Husmann S, Budak Ö, Shim H, Liang K, Aslan M, Kruth A, et al. Ionic liquid-based synthesis of MXene. Chemical Communications. 2020;56:11082–5.

32. Mei J, Ayoko GA, Hu C, Sun Z. Thermal reduction of sulfur-containing MAX phase for MXene production. Chemical Engineering Journal. 2020;395:125111.

33. Zada S, Dai W, Kai Z, Lu H, Meng X, Zhang Y, et al. Algae extraction controllable delamination of vanadium carbide nanosheets with enhanced near-infrared photothermal performance. Angewandte Chemie International Edition. 2020;59:6601–6.

34. Mashtalir O, Naguib M, Mochalin VN, Dall'Agnese Y, Heon M, Barsoum MW, et al. Intercalation and delamination of layered carbides and carbonitrides. Nature Communications. 2013;4:1–7.
35. Han F, Luo S, Xie L, Zhu J, Wei W, Chen X, et al. Boosting the yield of MXene 2D sheets via a facile hydrothermal-assisted intercalation. ACS Applied Materials & Interfaces. 2019;11:8443–52.
36. Wu W, Xu J, Tang X, Xie P, Liu X, Xu J, et al. Two-dimensional nanosheets by rapid and efficient microwave exfoliation of layered materials. Chemistry of Materials. 2018;30:5932–40.
37. Driscoll N, Richardson AG, Maleski K, Anasori B, Adewole O, Lelyukh P, et al. Two-dimensional Ti_3C_2 MXene for high-resolution neural interfaces. ACS Nano. 2018;12:10419–29.
38. Ghidiu M, Halim J, Kota S, Bish D, Gogotsi Y, Barsoum MW. Ion-exchange and cation solvation reactions in Ti_3C_2 MXene. Chemistry of Materials. 2016;28:3507–14.
39. Ghidiu M, Kota S, Halim J, Sherwood AW, Nedfors N, Rosen J, et al. Alkylammonium cation intercalation into Ti_3C_2 (MXene): effects on properties and ion-exchange capacity estimation. Chemistry of Materials. 2017;29:1099–106.
40. DiCamillo K, Krylyuk S, Shi W, Davydov A, Paranjape M. Automated mechanical exfoliation of MoS_2 and $MoTe_2$ layers for two-dimensional materials applications. IEEE Transactions on Nanotechnology. 2018;18:144–8.
41. Xu J, Shim J, Park JH, Lee S. MXene electrode for the integration of WSe_2 and MoS_2 field effect transistors. Advanced Functional Materials. 2016;26:5328–34.
42. Hu T, Hu M, Li Z, Zhang H, Zhang C, Wang J, et al. Interlayer coupling in two-dimensional titanium carbide MXenes. Physical Chemistry Chemical Physics. 2016;18:20256–60.
43. Huang X, Wu P. A facile, high-yield, and freeze-and-thaw-assisted approach to fabricate MXene with plentiful wrinkles and its application in on-chip micro-supercapacitors. Advanced Functional Materials. 2020;30:1910048.
44. Long M, Wang P, Fang H, Hu W. Progress, challenges, and opportunities for 2D material based photodetectors. Advanced Functional Materials. 2019;29:1803807.
45. Sun D, Feng S, Terrones M, Schaak RE. Formation and interlayer decoupling of colloidal $MoSe_2$ nanoflowers. Chemistry of Materials. 2015;27:3167–75.
46. Shuck CE, Gogotsi Y. Taking MXenes from the lab to commercial products. Chemical Engineering Journal. 2020;401:125786.
47. Zhang J, Kong N, Uzun S, Levitt A, Seyedin S, Lynch PA, et al. Scalable manufacturing of free-standing, strong $Ti_3C_2T_x$ MXene films with outstanding conductivity. Advanced Materials. 2020;32:2001093.
48. Deng J, Lu Z, Ding L, Li Z-K, Wei Y, Caro J, et al. Fast electrophoretic preparation of large-area two-dimensional titanium carbide membranes for ion sieving. Chemical Engineering Journal. 2021;408:127806.
49. Mashtalir O, Naguib M, Dyatkin B, Gogotsi Y, Barsoum MW. Kinetics of aluminum extraction from Ti_3AlC_2 in hydrofluoric acid. Materials Chemistry and Physics. 2013;139:147–52.
50. Naguib M, Halim J, Lu J, Cook KM, Hultman L, Gogotsi Y, et al. New two-dimensional niobium and vanadium carbides as promising materials for Li-ion batteries. Journal of the American Chemical Society. 2013;135:15966–9.
51. Mashtalir O, Cook KM, Mochalin VN, Crowe M, Barsoum MW, Gogotsi Y. Dye adsorption and decomposition on two-dimensional titanium carbide in aqueous media. Journal of Materials Chemistry A. 2014;2:14334–8.
52. Zhang CJ, Pinilla S, McEvoy N, Cullen CP, Anasori B, Long E, et al. Oxidation stability of colloidal two-dimensional titanium carbides (MXenes). Chemistry of Materials. 2017;29:4848–56.

53. Lotfi R, Naguib M, Yilmaz DE, Nanda J, Van Duin AC. A comparative study on the oxidation of two-dimensional Ti_3C_2 MXene structures in different environments. Journal of Materials Chemistry A. 2018;6:12733–43.
54. Huang S, Mochalin VN. Hydrolysis of 2D transition-metal carbides (MXenes) in colloidal solutions. Inorganic Chemistry. 2019;58:1958–66.
55. Natu V, Sokol M, Verger L, Barsoum MW. Effect of edge charges on stability and aggregation of $Ti_3C_2T_z$ MXene colloidal Suspensions. The Journal of Physical Chemistry C. 2018;122:27745–53.
56. Zhao X, Vashisth A, Blivin JW, Tan Z, Holta DE, Kotasthane V, et al. pH, nanosheet concentration, and antioxidant affect the oxidation of $Ti_3C_2T_x$ and Ti_2CT_x MXene dispersions. Advanced Materials Interfaces. 2020;7:2000845.

3 Surface Functionalization and Interfacial Design of MXenes

Anita Wojciechowska, Madhurya Chandel, and Agnieszka Maria Jastrzębska
Faculty of Materials Science and Engineering
Warsaw University of Technology
Warsaw, Poland

CONTENTS

3.1   INTRODUCTION

A large interest can be seen in solving the emerging problem of electromagnetic radiation. Electromagnetic interference (EMI) is a consequence of wireless communication and massive use of various electronic devices that produce electromagnetic radiation and surround people at their every step. Such electromagnetic radiation is unfavorable both for human health and for the operation quality of other electronic devices [1, 2]. Unfortunately, the EMI cannot be fully eliminated. The only ways of protection consider reduction or shielding with specially designed materials. The design aims at both limiting a radiative emission of a device and protecting it against radiation coming from the outside.

Currently, metal shields are a popular material used in EMI but their considerable weight and corrosion tendency encourages the search for an alternative material. The perfect candidate should be light, cheap, durable, and easy to manufacture. Such

DOI: 10.1201/9781003281511-3

features can be handled using polymer matrix-based composites. At first, carbon-based fillers such as nanotubes or graphene [3] were investigated, but no significant breakthroughs were reported [4]. Therefore, attention was shifted to other members of the two-dimensional (2D) family, including MXenes.

MXene phases are collectively named as early transition metal carbides, nitrides and carbonitrides. The stoichiometry-related chemical formula of this novel family of 2D materials is $M_{n+1}X_nT_x$, where M stands for an early transition metal (e.g., Ti); X is carbon or nitrogen, n = 1, 2, 3 [5], 4 [6], or 5 [7]; while T_x corresponds to surface functional groups. MXene phases are obtained by chemical etching of the starting MAX phase in which 'A element' (a metal belonging to group 13 or 14 of the periodic table) is removed between the M and X layers *via* chemical reactions. Importantly, A-empty places in the crystal lattice are immediately replaced with chemical functional groups, as collectively denoted by T_x.

Understanding the role of T_x is important for EMI shielding since it determines the quality of interactions of MXenes with EMI, and together with M and X composition, creates the so-called unique MXene chemistry. Elucidating the nature and exact composition of T_x in MXenes already requires a lot of effort, since different synthesis methods deliver different T_x compositions [8]. The problems with surface composition of MXene may influence processing pathways, making them difficult to control. More complexity comes from variable surface terminations, when considering MXene as a functional additive to polymer matrix. This complexity can be, however, mastered in order to tune MXenes by choosing the most relevant surface modification, as presented in Figure 3.1.

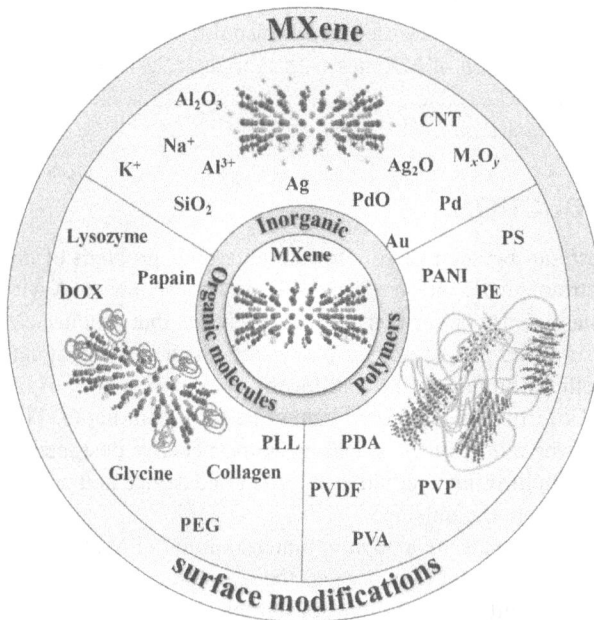

FIGURE 3.1 A schematic presenting a plethora of possibilities to modify MXenes for EMI shielding applications.

Synthetic polymers have been extensively used to prepare MXene-based composites due to their low cost, ease of fabrication, and tunable functionalities. In the case of MXene, the surface functional groups that arise during the etching process make MXene hydrophilic. The same functional groups can interact with hydrophilic polymers such as poly-diallyl dimethylammonium chloride (PDDA), polyacrylic acid (PAA), and so on [9–12] and produce MXene/polymer composites for EMI shielding. MXene can be homogeneously dispersed and mixed into a polymer, depending on the application. The first MXene/polymer structure was reported in 2014 using $Ti_3C_2T_x$ [12]. After the first report, the MXene polymer research increased, but it is still smaller than MXene and other derivatives. Therefore, a lot of possibilities can be explored.

3.2   SURFACE CHEMISTRY OF THE MULTILAYERED MXenes

The first method ever described for MAX phase etching, involved the addition of 48% hydrofluoric acid (HF) [13]. The chemical reactions that stand behind developing the native (pristine or bare) surface are as follows:

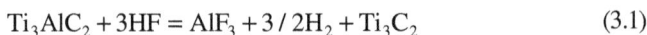

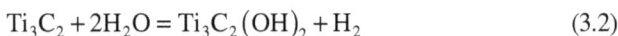

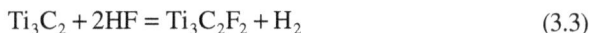

$$Ti_3AlC_2 + 3HF = AlF_3 + 3/2H_2 + Ti_3C_2 \tag{3.1}$$

$$Ti_3C_2 + 2H_2O = Ti_3C_2(OH)_2 + H_2 \tag{3.2}$$

$$Ti_3C_2 + 2HF = Ti_3C_2F_2 + H_2 \tag{3.3}$$

The preceding chemical reactions, associated with HF etching, deliver $=O$, -F and -OH functional groups to MXene's surface [14]. The HF etching also results in MXenes having a greater abundance of the oxygen-containing surface groups [15, 16].

HF-etched MXenes can be collected as a sediment and form a clay-like paste. Since the clay is made of T_x-terminated multilayered MXene sheets, it swells when hydrated and can be shaped into a highly conductive solid, or it can be rolled into films of tens of micrometers thickness. These films are well-stacked and characterized by a high volumetric capacity of up to 900 F cm^{-3} [17, 18], which is important for creating MXene-based devices and also EMI shielding.

Apart from the concentrated HF-based etching, several other approaches are available (see Figure 3.2). For instance, the A-element can be etched out with ammonium bifluoride (NH_4HF_2) salt or $NaHF_2$ solution. The advantage of this method in terms of EMI shielding is being more sustainable so that the whole approach is more environmentally friendly. The approach leads to simultaneous intercalation of NH_3 and NH_4^+ between M and X layers as well as adds Na to the MXene surface [19].

In 2020, the NH_4HF_2 was combined with polar organic solvents such as propylene carbonate. The resulting $Ti_3C_2T_x$ was rich in fluorine termination, thus showing different optical, electronic and catalytic properties compared to oxygen-functionalized

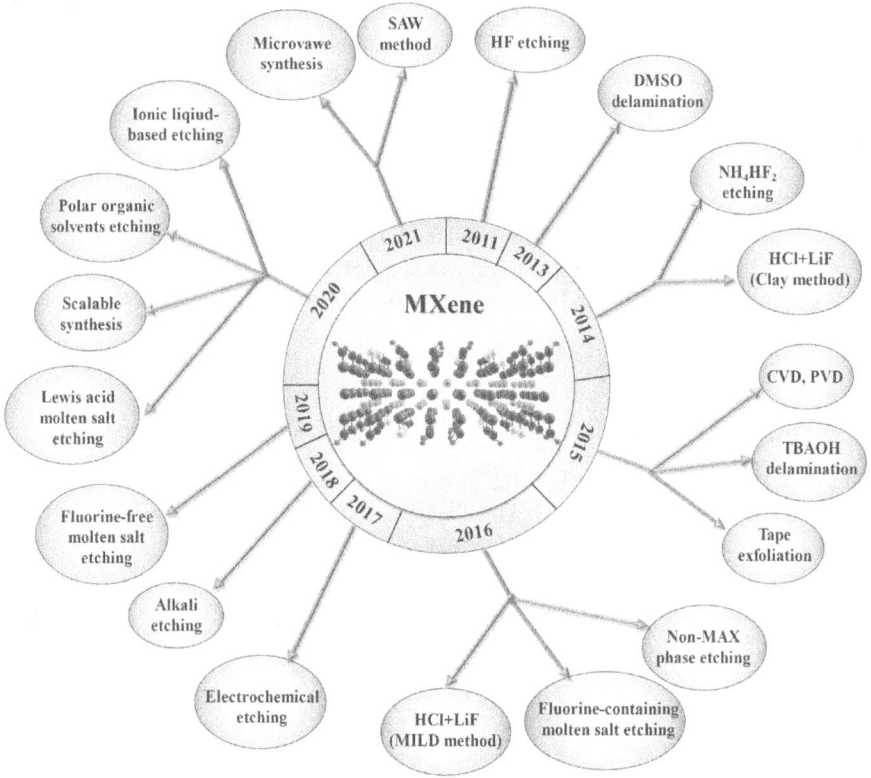

FIGURE 3.2 A schematic presenting an outstanding effort to develop the universal synthesis and delamination technique for the most recognized Ti_3C_2Tx MXene phase.

MXene [20]. As in the case of etching with concentrated HF, the process is carried out at room temperature and the following reactions take place [21]:

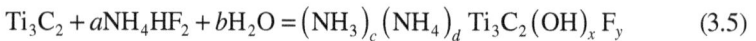

$$Ti_3AlC_2 + 3NH_4HF_2 = (NH_4)_3 AlF_6 + Ti_3C_2 + 3/2H_2 \qquad (3.4)$$

$$Ti_3C_2 + aNH_4HF_2 + bH_2O = (NH_3)_c (NH_4)_d Ti_3C_2(OH)_x F_y \qquad (3.5)$$

As a result of the etching, spacing between Ti_3C_2 layers increases, resulting in a multilayered MXene structure. Nevertheless, it is also possible to etch Ti_3AlC_2 MAX phase using the so-called molten salt technique [22]. At an elevated temperature, the $ZnCl_2$ and several other Lewis salts melt down into a liquid, thus forming an acidic environment that enables MXene etching. In addition, there is no need for using water and such obtained MXenes are not terminated with oxygen-bearing functional groups [23].

Moreover, it is possible to tune MXene properties by choosing a suitable surface termination. For instance, when etched with $ZnCl_2$ or $CdBr_2$, MXenes are surface-functionalized with -Cl or -Br, respectively [24]. Both M-Cl and M-Br are much weaker than M-F or M-OH [25]. Thanks to this, -Cl and -Br groups can be easily replaced by sulfur, selenium, tellurium, oxygen or NH groups [22]. Importantly, such ability to tune MXene surface has not been previously demonstrated for other 2D materials like graphene or transition metal dichalcogenides.

Furthermore, MAX-phase can be etched with NaOH or KOH bases, which leads to F-free MXenes [26–28]. However, this process was not fully confirmed in terms of full etching of A element and requires the use of elevated temperature in combination with a concentrated alkali environment. The resulting multilayered MXene requires further processing to obtain 2D flakes more compatible with a potential matrix for EMI shielding applications. Therefore, the alkali-based etching is not used, so far, on a larger scale.

3.3 SURFACE CHEMISTRY OF THE DELAMINATED MXenes

The EMI shielding is a high-tech application, which may require the use of single 2D sheets, apart from the multilayered MXenes. Single MXene sheets have more active surface, due to size reduction, and interact efficiently with other materials. Therefore, a delamination is further needed to obtain a colloidal dispersion of single MXene sheets that can be further processed.

MXene delamination is based on using large organic cations and molecules right after etching out the A element. In particular, the dimethyl sulfoxide (DMSO) was first used as the intercalating compound for MXene. Later on, researchers verified the efficiency of urea, hydrazine monohydrate ($N_2H_2 \cdot H_2O$) or N, N-dimethylformamide (DMF) [29]. It was also noted that DMSO is partially effective for $Ti_3C_2T_x$ but the delamination was not that effective and was not suitable for other types of MXenes [29] such as Ti_3CNT_x [30].

Therefore, intensive research was carried out on searching for a universal compound for MXene delamination. It was found, for instance, that hydrazine monohydrate ($N_2H_4 \cdot H_2O$) works fine with other types of MXenes beyond only Ti_3C_2 [29]. Furthermore, the isopropyl amine (i-PrA) was used for the intercalation and delamination of Nb_2CT_x, $Ti_3C_2T_x$ and $Nb_4C_3T_x$ [31]. In this approach, active ammonium cations R-NH_3^+ are formed in a water-based environment and further intercalate between MXene layers due to electrostatic interactions. In addition, the i-PrA molecules have alkyl chains, thanks to which they can overcome the steric obstacle after intercalation, thus separating the MXene layers from each other. Apart from i-PrA, other large organic compounds such as n-butylamine (n-BA), choline hydroxide (ChOH) and tetrabutylammonium hydroxide (TBAOH) had a similar effect on MXene delamination. These large molecules are organic bases, thus playing the role of a universal intercalating agent. Similarly to i-PrA, they increase material swelling, which, in combination with mild sonication, leads to efficient intercalation and delamination of Ti_3CNT_x and V_2CT_x [30, 32].

The $Ti_3C_2T_x$ delamination can be also achieved by intercalating with tetramethylammonium hydroxide (TMAOH). The TMA$^+$ cations act both as an etchant and intercalating agent, thus allowing the concentration of HF used in the etching process to be reduced. In the first step, TMA$^+$ removes Al layer from the MAX phase structure *via* chemical reaction. In the next step, the TMA$^+$ cations intercalate between the $Ti_3C_2T_x$ layers, leading to the formation of one or few-layered MXene sheets. However, one must bear in mind that organic compounds used for the intercalation often remain as a residue on MXene surface or stay deeply intercalated within ultrathin spaces formed between the layers. This is underpinned by strong chemical interaction between TMA^+ and the MXene, which does not allow fully washing out the intercalated residue. Since TMA^+ is a highly reactive organic agent, its presence in MXene affects both chemistry and properties of the final MXene and can affect the EMI shielding efficiency.

The delamination can be carried out by lithium or sodium fluoride salts (LiF or NaF) mixed with hydrochloric acid (HCl). This approach was first introduced in 2014 [18] and was further called a clay method or minimally intensive delamination (MILD). In the MILD approach, the HF is formed *in situ* during the reaction of HCl and fluoride salt near the MXene surface. Herein, apart from the previously mentioned ordinary terminating species (=O, -F, -OH), the additional species, such as -Cl, -Li or -Na, readily appear on the MXene surface, while the presence of Li and Na strictly depends on chosen Li-salt. Their presence may help in achieving better SE values since the MILD method improves the general quality of MXene flakes.

To date, obtaining single MXene sheets without any surface termination is a great challenge, but could be interesting in terms of testing them in EMI applications. On the one hand, it has been proven that the presence of T_x positively impact the physicochemical properties of MXenes and helps in processability [33] due to a hydrophilic character coming from the presence of surface functionalization [34]. On the other hand, finding a way to control the T_x composition is in no way easier. For instance, changing the specific arrangement or concentration of chosen terminating groups on the MXene surface can be achieved by adjusting the etchant concentration or etching time, or by using various types of post-treatments [35, 36]. This could further influence the physical properties of the MXenes [37].

Such a possibility was tested so far with theoretical calculations [38]. The electronic properties of various MXene phases such as M_2C (M = Sc, Ti, V, Cr, Zr, Nb or Ta) and M_2N (M = Ti, Cr or Zr) changed upon chemical functionalization of the surface with -F, -OH or =O. The surface functionalized MXenes in theory become either conductors or ferromagnets. This property may be used for designing EMI tunability of MXenes in terms of on-demand enhancing the SE of the protective layer.

3.4   SURFACE CHEMISTRY OF OTHER MXene-LIKE STRUCTURES

MXene phases can be obtained by selective etching of precursors other than only MAX phases. In particular, layered compounds such as $Zr_3Al_3C_5$ [39], $Hf_3(AlSi)_4C_5$ [40] or Mo_2Ga_2C [41] have other composition than M_nAX_{n-1} and cannot be labelled as MAX phases. Importantly, $Zr_3C_2T_x$, $Hf_3C_2T_x$ and Mo_2CT_x MXenes were successfully obtained while using them as parental phases. These were, however, not yet tested for EMI shielding applications.

Another way for obtaining MXenes without using any MAX phase is the chemical vapor deposition (CVD) technique. CVD allowed so far to successfully obtain MXene-type layered Mo_2C compound without many defects and a thickness of several nanometers [42]. Notably, CVD is a promising approach for obtaining thin MXene films, applicable in electronics, optoelectronics and photovoltaics. Therefore, it could be useful also for obtaining thin films for EMI shielding.

Since CVD is carried out in an argon-based environment, and the oxygen and water are not involved in the process, the resulting MXenes should not have surface functional groups. Unfortunately, the CVD has low efficiency and is of high cost. Nevertheless, since there is a high demand for developing perfect thin MXene films for electronic applications, more progress in CVD-MXenes is expected in the nearest future so that EMI shielding applications can be also tackled.

3.5 MXene/POLYMER COMPOSITES FOR USE IN ELECTROMAGNETIC INTERFERENCE (EMI) SHIELDING

In the context of EMI requirements, MXene phases are of interest since they combine conductivity with excellent water dispersibility. Therefore, the EMI challenge can be faced by conductive MXenes, which collect charge carriers directly interacting with an electromagnetic field [43]. Being hydrophilic material, they facilitate many approaches to develop EMI shielding composites that are based on a polymer matrix.

Several review papers have been published about MXene/polymer assemblies and different polymer composites. The synthesis, properties and applications of MXene and polymer-based membranes were discussed [44]. The synthesis and applications of MXene-functionalized polymer composites has been explained nicely [45]. The focused was on MXene and polymers based electrochemical sensors [46]. Collectively, they explained in detail various applications. However, in this chapter, we focus on MXene/polymer assemblies that provide or could provide advancements to EMI shielding application. We cover useful synthetic methods for their preparation and discuss the important functionalities of MXenes for EMI.

MXenes are easily processable, thus allowing them to be applied in structures having both simple and complicated architectures. For instance, a thin film composed of $Ti_3C_2T_x$ and polyvinylidene fluoride (PVDF) can be obtained using a compaction method [47]. The $Ti_3C_2T_x$/PVP composite of 2 mm thickness showed a good shielding effectiveness (SE) of 48.47 ± 3.5 dB at 22.55 vol.% $Ti_3C_2T_x$ content. It was shown that it promoted multifaceted electronic utilization, thus proving to be a promising EMI shielding composite.

Thermal properties are another crucial criterion that are required for various applications. These are mainly related to materials processing, but essential thermal transport property is fairly dominated by charge carrier mobility [48–50]. The thermal conductivity of an MXene/polymer hybrid is primarily affected by the loading, distribution, phase structure, and interfacial thermal resistance [51]. According to the density functional theory (DFT) first-principles calculations, the calculated thermal conductivities of MXene are higher than metals and semiconducting low-dimensional materials, which could facilitate better thermal management of EMI shielding coatings [48–50]. For instance, MXene can be well dispersed in the hydrophilic epoxy resin, which is a thermosetting plastic of outstanding physical properties and high performance. Fabrication of MXene/epoxy resin can be carried out by a simple casting and curing process due to the strong interaction between the MXene surface and the polymer [52–54].

Polymer resins composites for EMI shielding can be fabricated by another simple compaction method. Such obtained $Ti_3C_2T_x$/epoxy resin composite showed a 41 dB of SE [55]. The composite also had excellent mechanical properties and an attractive price. The practical application can be facilitated by the already wide use of polymer resins in electronic equipment enclosures.

Apart from simple compaction, other reported methods of producing MXene/polymer composites for EMI shielding were vacuum-assisted filtration and freeze-drying [56]. The vacuum-assisted filtration is a simple and controllable process that produces composite films varied by thickness and layers composition. The MXene/polymer films prepared in this way form homogeneous and electrically conductive networks so that better conductivity and SE are obtained. While the polymer network in such composite is formed by traditional chemical bonding, MXene and polymer are connected mainly through hydrogen bonds [56].

The vacuum-assisted filtration technique was used to obtain thin films of MXene/sodium alginate composites [4]. The composites showed an extraordinary ability to shield electromagnetic interference with SE values reaching 57 dB for a film thickness of 8 μm. These films, apart from having excellent EMI properties, also had high conductivity, relatively low density, mechanical flexibility and were generally easy to handle. Within the smallest thickness, they exhibited similar EMI shielding efficiency as pure metals.

The vacuum-assisted filtration can also be used to prepare MXene/polymer composites. If MXene flakes are mixed with polymer solution and then vacuum filtered, the polymer chains alternate MXene flakes so that the multilayered MXene/polymer composite can be achieved. Using this type of filtration, translucent MXene/carbon nanotubes (CNT) were composed with a high molecular weight PVA or sodium polystyrene sulfonate (PSS) [57]. The MXene/CNT/polymer films of 0.17 μm thickness showed EMI shielding SE of 58.2 dB. A similar approach was used [58] to obtain an MXene/PVA composite having 19.5 wt.% MXene content and a thickness of 27 μm. The composite had the ability to shield EMI by 44.4 dB.

Another commonly used method for producing MXenes/polymer composites is freeze-drying. The advantage in relation to vacuum-assisted filtration relates to forming a 3D architecture in the form of a foam. Such light-weight foams could efficiently protect the devices against EMI. The freeze-drying process consists of mixing MXene with a polymer solution to self-assembly, then freezing it and subjecting it to ice sublimation at a temperature and pressure below a water's triple point. Thanks to freeze-drying, it is possible to obtain a 3D architecture without changing the material chemical composition.

By using a freeze-drying, a porous Ti_2CT_x/PVA foam was obtained for EMI shielding. With the addition of MXenes 0.15 vol.% and the film thickness of 2 mm, EMI shielding was 28 dB, while the specific shielding effectiveness (SSE) was 5136 dB cm^2 g^{-1}. The EMI shielding performance of the composite was enhanced because of the porous structure of the foam and the layered structure of f-Ti_2CT_x [59].The MXene/sodium alginate (SA) aerogel was obtained using the lyophilization method, which was then coated with a thin layer of polydimethylsiloxane (PDMS), thus obtaining compressible and electrically conductive composite EMI shielding foams. At 95 wt.% of MXenes use, the composite reached an EMI SE of 70.5 dB along with an outstanding conductivity of 2211 S m^{-1} [60].

MXene/polymer composite foams protecting against electromagnetic interference obtained by freeze-drying have a high porosity with regular distribution. Moreover, by creating complete electrically conductive networks, an ideal EMI SE can be obtained with less MXenes. As mentioned before, the MXenes phases show a significant tendency to oxidation, which is a big problem in foam material, which has a large specific surface area. Therefore, a big problem with this type of composite is the poor air stability of the MXenes/polymer composite foam materials protecting against electromagnetic interference.

3.6 MODIFICATION OF THE MXenes WITH POLYMERS

MXene phases have exceptional electrical conductivity, hydrophilicity and chemical activity [16, 61]. Moreover, unlike traditional metal-based materials, they are resistant to unfavorable external conditions that cause corrosion. However, it is impossible to create free-standing 3D structures with excellent mechanical properties by the self-assembly of pristine MXene flakes [10, 62, 63]. Therefore, scientists are combining MXenes with polymers to enable satisfying mechanical properties [64, 65]. MXene polymer composites can also be made by thermal pressing or stacking laminates [66]. In these composites, MXene acts as a filler and the polymer as a matrix, which serves both thermoplastic [67] and thermosetting [52] functions. There are many other methods for obtaining MXene/polymer composites, which are modified in various ways depending on the application needs. Such hybrids are possible to obtain thanks to the hydrophilic nature of MXenes and their compatibility with organic molecules.

The first report on MXene/polymer composites appeared in 2014, in which the $Ti_3C_2T_x$ phase was combined with poly(diallyldimethylammonium chloride) (PDDA) and PVA. The process of producing MXene/polymer films was carried out *via* vacuum-assisted filtration. Since then, many reports have appeared concerning in majority the hydrophilic polymers such as PVA [12], poli acrylic acid (PAA) [68], polyethylene (PE) [67], polypropylene (PP) [69], polystyrene (PS) [70], polyamide (PA) [68], acrylamide [71], silicones [72] and many others. Much of the current research on MXenes/polymer composites is based on conductive polymers, among of which poly(aniline) [73], polyvinylpyrrolidone (PVP) [74] and polypyrrole (PPy) [75] can be distinguished.

The process of bonding the polymer to MXene structure is a promising strategy that could lead to obtaining new hybrids with unique physicochemical properties. The most frequently used methods for constructing more complex polymer-based matrices for embedding MXenes are based on *ex situ* mixing of polymers [12, 71, 76, 77], or *in situ* polymerization [4, 78, 79]. The polymer blending strategy is a simple method leading to well-defined polymer structures. In this case, there is no chemical cross-linking between the MXene phase and the polymer matrix [56]. Using this method of preparing MXene/polymer composites, the interactions between these two compounds can be improved by hydrogen bonding [80] and electrostatic interactions.

The use of the polymer blending method to obtain MXene/polymer composites is possible due to the hydrophilic nature of MXenes. However, only polymers that hydrolyze or are soluble can be used for this method, for example, polyethylene oxide (PEO) [81] or polyvinylidene fluoride (PVDF) [51]. On the other hand, a highly conductive MXene/polystyrene nanocomposite was obtained using the direct mixing method. This composite exploits the interaction between the negative surface charge of MXene and the positively charged polystyrene particles. In the present composition, there were two types of interactions between MXene and the polymer that depended on the concentration of the filler in the polymer matrix. At a low concentration of MXenes, electrostatic interactions played a major role, whereas at a high concentration of MXene, van der Waals interactions and hydrogen bonds played a major role [82].

Due to the oxygen functional groups on the MXene surface, the $Ti_3C_2T_x$ polymer composites were obtained using the direct mixing method. The composite was obtained by directly mixing a delaminated titanium carbide solution with cationic poly(diallyldimethylammonium chloride) or with electrically neutral polyvinyl alcohol (PVA). In this way, flexible highly conductive composites were obtained, the electrical conductivity of which reached 2.2×10^4 S m^{-1}. These composites were also characterized by a much higher tensile strength compared to the starting materials. Therefore, this material can be successfully used, among others, in energy storage devices or wearable electronics [12].

Creating $Ti_3C_2T_x$/PVA gave even fourfold better mechanical properties, compared to films made of only MXenes [83]. The excellent MXene compatibility with PVA creates a unique opportunity to add it more to the matrix. For instance, $Ti_3C_2T_x$/PVA composite film was fabricated and filled with an incredible MXene amount of 87.29 wt.%. Thermal properties of both MXene and PVA increased due to the formation of Ti–O bonding. However, when the amount of MXene in the composite was more than 40 wt.%, it affected scaling-up [84]. In another report, $Ti_3C_2T_x$/PVA heterostructures have been studied with only 10 wt.% of $Ti_3C_2T_x$, showing good dispersion of MXene flakes in the polymer and enhanced dielectric properties due to the nacre-like structure [85].

In the process of direct mixing of polymers with MXenes, dispersants are used. In the case of MXenes, these are polar compounds such as water [59], N,N-dimethylformamide (DMF) [51], DMSO [86] or dimethylacetamide (DMAc) [87]. If the same dispersant also dissolves the polymer, both compounds can be mixed in the same dispersant. The MXenes and polymer solutions can also be mixed in different dispersants, but they must dissolve each other. This approach was used to obtain the MXene/polyacrylamide (PAM) composite. MXene was dispersed in DMSO and PAM disperion where delamination was feasible quite easily. In this way, a composite in the form of a film, characterized by flexibility and high electrical conductivity, was obtained [71].

Another method, very frequently used to obtain MXene/polymer composites, is *in situ* polymerization in which monomers, polymerization initiators and curing agents are mixed with MXenes. Then, the polymerization reaction takes place, as a result of which a composite material with well-distributed MXenes in a polymer matrix is obtained [88]. The polymers used so far include epoxy resin [52], polyaniline (PANI) [89] or PPy [90].

Modified $Ti_3C_2T_x$ MXene and polyaniline (PANI) were used to prepare MXene/PANI hybrid films with different thicknesses (4–90 μm). The *in situ* polymerization method was used, which helps in the uniform distribution of the polymer in the hybrid film. It was also observed that the restacking of MXene flakes affected Braymer chains intercalated in the hybrid film. It was concluded that the good and selective functionalization on the surface MXene would help combine MXene with various polymers [73]. The $Ti_3C_2T_x$/PANI composite could also add a supercapacitor functionality to EMI shielding protection. The effect comes from amino groups present in PANI that increase the electrical conductivity, making the transport of ions faster, and also improve the wettability of the MXenes' surface. The specific capacity of the obtained composite was 164 Fg^{-1}, which is 1.26 times more compared to $Ti_3C_2T_x$ [91].

Using the one-step *in situ* polymerization method, it is possible to produce the MXene/polydopamine (PDA) composite. The obtained composite was characterized by a superior areal capacitance of 715 mF cm^{-2} at a scan rate of 2 mV s^{-1} and maintained 95.5% of initial capacitance after 10,000 cycles, exhibiting long-term cycling stability [92].

In another approach, the electroactive polypyrrole (PPy) was combined with MXene [93]. To prepare the composite, MXene and pyrrole monomer reaction mixture was vigorously stirred. Due to the alignment and intercalation, the polymer chain created spaces between MXene flakes and allow the 3D architecture to expand. This leads to high volumetric capacitances (~1000 F cm^{-3}) and excellent cycling stability (capacitance retention of ~92% after 25,000 cycles). The $Ti_3C_2T_x$/PPy composites could be also prepared by *in situ* electrochemical polymerization for use as capacitor electrodes. The obtained condensate had a capacity of 69.5 mF cm^{-2} and an energy density of 250 mWh cm^{-3} [94].

On the surface of $Ti_3C_2T_x$ MXene *in situ* polymerization of 3,4-ethylene dioxythiophene (EDOT) was achieved without using any oxidant [95]. The thickness of MXene/polymer membrane was controlled by the solution concentration and volume. It was observed that better hybrid formation can be achieved if EDOT monomer is adjacent to the MXene layer for better electron transfer, which further additionally controls its polymerization. Based on its theoretical calculation and analysis, the polymerization reaction pathways were suggested which proceeded through charge-transfer-induced polymerization. Thanks to this, the final hybrid showed an enhanced Li-ion uptake performance with a reversible capacity of up to 300 mAh g^{-1} [95].

Polydopamine (PDA) is another important and interesting polymer that has been used due to inherently strong adhesion to different nanomaterials, without causing any damage to their structure [96]. To test this assumption, $Ti_3C_2T_x$ was combined with PDA into hybrid by using one-step *in situ* polymerization. They have shown that PDA chains were intercalated and aligned into Ti layers thus achieving an outstanding electrochemical performance. They have also shown the dopamine polymerization mechanism, which helped in better understanding the process [92].

A hybrid composed of poly(ethylene oxide) (PEO) and $Ti_3C_2T_x$ was fabricated in the form of a membrane. The membrane showed enhanced ionic conductivity and stability at room temperature and can be used in all-solid-state lithium metal batteries [97]. A similar membrane was developed by casting the chitosan and $Ti_3C_2T_x$ mixture onto a glass plate using the Nafion-based membranes method. It was observed that uniformly distributed and stable MXene strengthens the thermal and mechanical stabilities of the composite membranes. Also it enhances the proton conductivity of composite membrane under various conditions which improved the hydrogen fuel cell performance. The reason behind this improvement is the -OH groups on the surface of MXenes, which easily connected by hydrogen-bonding interaction with the acidic/basic groups of polymer matrix and providing efficient hopping sites [98]. While in another study, MXene/chitosan films were fabricated using a spin coating method, and used for different solvent dehydration study *via* pervaporation. It was observed that incorporation of only 3 wt.% of MXene ($Ti_3C_2T_x$) can enhance its solvent dehydration capacity [99]. Such kinds of studies are further helpful to design a better film/membrane for EMI applications.

3.7 MODIFICATION OF MXenes WITH MACROMOLECULES

Due to the negative surface charge, MXenes can be successfully modified organic molecules. This property can positively influence designing the EMI shielding composite in which MXene lakes should be well-dispersed. The negatively charged surface allows positively charged molecules to attach, and is predominantly driven by electrostatic interactions. This, in turn, can be tracked in-operando. By applying an in-operando zeta potential measurement coupled with particle size measurements with dynamic light scattering (DLS), it becomes possible to conduct studies on stepwise adsorption of cationic organic macromolecules such as lysozyme, collagen or poly-L-lysine on the surface of $Ti_3C_2T_x$ and Ti_2CT_x [100–102]. What is particularly tracked are changes that occur on MXene's surface, starting from a highly negative charge, going through the sign change into positive at point of zero charge (PZC), and finally achieving a highly positive value, after fully covering MXene's surface by the molecules, as presented in Figure 3.3.

It was further noticed that -OH groups present on MXene's surface can be used as excellent active sights for immobilizing enzymes while maintaining the bioactivity and stability of the MXene [103]. Therefore, it becomes possible to immobilize a model enzyme (tyrosine) on the surface of $Ti_3C_2T_x$ to add a functionality of biosensing to the EMI shielding layer. In such composite, a surface-controlled electrochemical process allows an electron transfer between tyrosine and the electrode. The phenol-sensors obtained in this way showed satisfying analytical performance with a low detection limit and high sensitivity, thanks to the large specific surface area and good electrical conductivity of MXene. Moreover, they showed reproducibility, long-term stability and high phenol recovery in real water samples.

Furthermore, glycine was adsorbed on $Ti_3C_2T_x$ surface to prevent restacking of MXenes' sheets, thereby increasing the lifespan of the electrodes for energy storage applications [104]. Adsorption of glycine was followed by the appearance of a chemical bonding between the Ti atoms of $Ti_3C_2T_x$ and the N atoms of glycine. The Ti–N

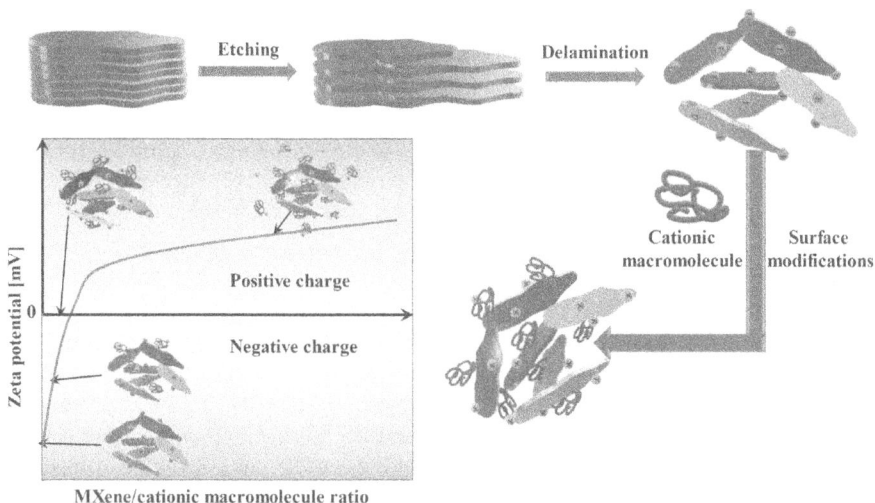

FIGURE 3.3 Schematic image presenting the synthesis and surface modification of MXenes with cationic molecules into MXene/molecule hybrids, as well as in-operando tracking approach, developed for the adsorption of macromolecules on MXene's surface.

bonding led to increased spacing between MXene layers by 1.22 Å, as confirmed by XRD analysis.

There are also other non-covalent and covalent techniques to modify the MXene's surface. For instance, a papain can be immobilized on the surface of $Ti_3C_2T_x$ MXene [105]. It was first physically adsorbed on MXene, and then covalently cross-linked by glutaraldehyde. The theoretical calculations confirmed that the mechanism of enzyme adsorption on MXene is related to the presence of surface functional groups, which leads to higher enzymatic activity and a higher thermal and pH stability. The obtained MXene-organic hybrid showed promise as a biocatalyst with more than 60% of enzymatic activity maintained after the ninth reuse.

Hemoglobin can also be immobilized on $Ti_3C_2T_x$ to obtain a mediator-free biosensor for detecting $NaNO_2$ with a wide linear feature (ranging between 0.5 and 11800 µM) and an exceptionally low detection limit (0.12 µM) [106]. Therefore, $Ti_3C_2T_x$ MXene is an excellent candidate for effective enzyme immobilization due to its biocompatibility for redox protein, which ensures its good bioactivity and stability.

Other enzyme such as poly-L-lysine (PLL) can be also used for this purpose along with collagen [102]. These two compounds are of great biological importance [101]. Since the adsorption may impact MXene bioactivity, it is important to test the biocompatibility of the resulting MXene/bio-macromolecule hybrid. In the case of $Ti_3C_2T_x$/collagen, Ti_2CT_x/collagen and $Ti_3C_2T_x$/PLL, the MMT *in vitro* test didn't show acute toxicity in relation to A375 cells (melanoma cells), HaCaT (benign human keratinocytes), MCF-7 (breast cancer cells) and MCF-10A (benign breast cells). Although a majority of the research on MXene surface modification with macromolecules concerns the phases from Ti-C system, there are some reports available for other MXenes as well. The research on Nb-based MXenes such as

Nb_2CT_x and $Nb_4C_3T_x$ also considered using PLL for surface modification [107]. The *in vitro* research on $Nb_{n+1}C_nT_x$/PLL hybrids has shown promising results on targeting cancer cells and inducing the cell cycle arrest in the apoptotic (G_0/G_1) phase, that is, programmed cell death. This effect is highly desirable for modern cancer treatment strategies [108]. Importantly, the $Nb_{n+1}C_nT_x$/PLL hybrids showed good biocompatibility in relation to benign skin cell lines. The reduced level of reactive oxygen species (ROS) was also shown and verified by non-specific fluorescent dye (DCF-DA) tests and flow cytometry.

While ROS species involve oxygen atoms with an unpaired electron, such as O_2^-, H_2O_2, O_3 or singlet oxygen, they may be dangerous to humans and the environment [109]. They are formed as a natural by-product of normal oxygen metabolism in the cells and play an important role in the production of adenosine-5'-triphosphate (ATP) [110]. Any changes in their natural physiological level are the main marker of disturbances in cell functioning [111]. Therefore, while tackling the safety use of EMI shielding coatings, one must also consider the ROS parameter since it may be disturbed due to changes in MXene's surface parameters. The results of this study indicate no significant differences in the ability to generate ROS in normal HaCaT and MCF-10A cells. A different effect was observed in the case of A375 and MCF-7 cancer cells. The modification of the surface of the nanoflakes resulted in a significant increase in the material's ability to generate reactive oxygen species. This mainly affects melanoma cells, but the trend is also observed in breast cancer cells. Similar results were obtained for the modification of the PLL. A significant influence of the studied nanostructures on the generation of ROS selectively in cancer cells was observed. Therefore, surface-modified MXenes showed no toxicity, thus enabling designing the EMI shielding in a form of coatings or textiles that could have direct contact with human skin.

Another macromolecule of importance is polyethylene glycol (PEG). PEG is fully biocompatible, thus frequently used for stabilizing various nanomaterials for nano-medicinal applications. Therefore, there is no surprise that researchers could use PEG to stabilize MXenes for EMI shielding applications [112]. The obtained Ti_2CT_x/PEG is characterized by good photothermal conversion efficiency and good biocompatibility in a wide range of tested concentrations. The increased affinity of Ti_2CT_x/PEG can also help in interfacial management of MXenes for designing the EMI shielding films or porous 3D architectures of the same.

Interestingly, modification of Nb_2CT_x MXene with polydopamine (PDA) made it possible to obtain a nanoplatform with the addition of Imiquimod R837, being a modifier of the immune response and having a strong indirect antiviral effect [113]. The obtained hybrid showed generally low cytotoxicity, but within 1064 nm NIR-II laser irradiation, its cytotoxicity, both *in vitro* and *in vivo*, increased significantly. The increase in cytotoxicity after irradiation is related to the efficient ability to photothermal conversion, which could be utilized in multifunctional EMI shielding protective coatings.

Other researchers modified Nb_2CT_x MXene with polyvinylpyrrolidone (PVP) [114] and tested their photothermal conversion efficiency. The obtained Nb_2CT_x/PVP hybrid showed excellent biocompatibility and physiological stability, both *in vitro* and *in vivo*. It also showed effective photothermal ablation against tumor xenografts for both NIR-I and NIR-II windows and the possibility of biodegradation *via* human myeloperoxidase.

To advance the multifunctionality of EMI shielding, researchers can use various active macromolecules to modify MXenes. Modification with doxorubicin (DOX) enhanced the luminescence of MXenes [115] since they are characterized by low luminescence in aqueous solutions [24]. The binary hybrid systems are also possible by adding hyaluronic acid (HA) to the MXene surface, apart from DOX, thus enabling bioimaging [115]. The bioimaging can also be facilitated by modifying MXene with contrast agents such as iodine-containing compounds [116, 117]. Moreover, the surface modification with soybean phospholipids (SP) increases the colloidal stability of $Ta_4C_3T_x/SP$ in various media.

Surface modification with macromolecules is equally important for enabling the antibacterial action of MXenes, which, if of high purity, are not antimicrobial [118]. Adding new functionality to the protective layer, apart from only EMI shielding, would give it an innovative impact. As previously described for graphene and graphene oxide, the antimicrobial properties are determined by direct interaction with bacteria on the surface [119]. Since bacterial cells are characterized by a negative surface charge [120, 121], they can interact more efficiently with positively-charged surface-modified MXene, thus ensuring direct bacteria-MXene contact and inactivation. Therefore, the positive surface charge, antimicrobial additive and particle size are key factors in determining the antimicrobial activity of the MXene [122].

Most recent research on MXene's biological properties has confirmed that pure MXenes are not antimicrobial [102, 118, 123]. However, contrary to unmodified $Ti_3C_2T_x$, the $Ti_3C_2T_x/PLL$ hybrid showed antibacterial action against *E. coli* bacteria [102]. It reduced the number of viable bacteria cells by two orders of magnitude at a concentration of 200 mg L^{-1}, but did not completely eliminate *E. coli*, even at higher concentrations. Consequently, MXene 2D flakes being in nanocolloidal solution, even if surface-modified, are far more biostatic than biocidal.

The unique surface chemistry plays a key role in many properties of MXene phases, but to date, there are no universal methods available for controlling the specific composition of surface functionalization [124, 125]. These materials show a significant tendency to oxidation, which causes their chemical degradation, and hence the loss of properties. Researchers found so far the universal method for MXenes' oxidative stabilization by modifying their surface with antioxidants such as L-ascorbic acid [124], sodium ascorbate [126] or polyphosphates [125], which significantly slows down the oxidation process. The effectiveness of this method was confirmed for $Ti_3C_2T_x$ [124], Nb_2CT_x and $Nb_4C_3T_x$ [127] MXenes. While not protected, the $Ti_3C_2T_x$ MXene is oxidized in water and decomposed into TiO_2 and carbon. As depicted in Figure 3.4, the use of an antioxidant such as L-ascorbic acid protects MXenes against oxidation. This is because L-ascorbic acid molecules, acting as reducing agents, protect titanium atoms from oxidation [124].

In summary, there is insufficient research on the interactions of MXenes with biomacromolecules, which limits utilization of this promising strategy in practice. This has been emphasized by Lin, Chen and Si [15] and also in our previous works [100–102, 107, 112]. Consequently, there are many challenges related to the application of 2D MXenes in biotechnology and related fields. In terms of surface

$Ti_3C_2T_x$ in water environment

$Ti_3C_2T_x$ in vitamin C environment

Vitamin C TiO_2

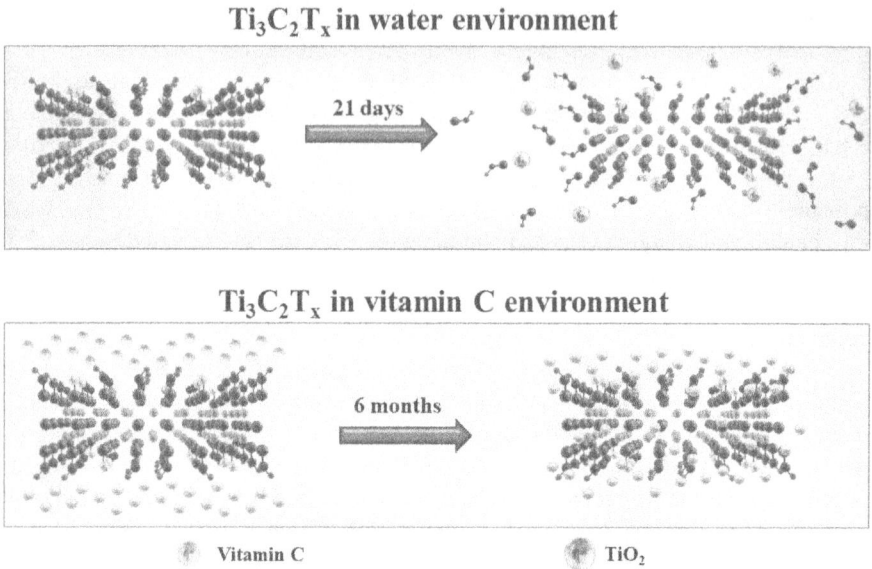

FIGURE 3.4 The mechanisms of Ti_3C_2Tx MXene protection with L-ascorbic acid antioxidant.

interactions with biomacromolecules, the current focus is on MXenes' surface engineering for molecular sensing along with comprehensive biocompatibility and fate assessment, including the protein corona formation and its role in bio-recognition of MXene materials.

3.8 SURFACE MODIFICATION OF MXenes WITH INORGANIC PARTICLES

MXenes have a highly hydrophilic nature, which is associated with the presence of terminal functional groups such as -OH, =O or -F [14, 128]. They allow to functionalize MXene with organic molecules, but also inorganic particles or their precursors. These surface-nucleation sites allow ionic seeding and further growth of amorphous nanoparticles and nanocrystals.

MXenes can also be modified *in situ*, reversibly to a plethora of conventional surface modification techniques. The *in situ* technique uses natural susceptibility of MXenes to surface oxidation due to their low stability [124]. Just ten days after delamination, significant amounts of amorphous titanium superoxide and hydroxide are observed on the surface of $Ti_3C_2T_x$ stored at room temperature, without any enclosure. These kinds of oxide-bearing species on $Ti_3C_2T_x$ surface change to crystalline TiO_2 anatase after passing 25 days, suggesting that any intermediate Ti_xO_y phases transform to TiO_2, which is tightly bonded to the MXenes' surface [129]. The oxidation process occurs in both single and multilayered MXenes exposed to water [8] or air [130]. It can be called *in situ* since it does not require adding any chemical reagents or precursors to MXene dispersion.

The *in situ* oxidation can be controlled and used to achieve metal oxide nanoparticles, having composition M_xO_y. Here, M origins from MXene and oxygen is incorporated from the air or water environment, in which MXene is dispersed [129]. Moreover, the morphology and quality of M_xO_y can be controlled by using various techniques. For instance, the $Ti_3C_2T_x$ MXene can be mildly oxidized in water at elevated temperature or with ultrasounds assistance to obtain smaller or larger TiO_2 nanoparticles on MXene surface [131]. The formation of Ti_3C_2/TiO_2 begins at 2D flake edges or surface defects and continues to fully cover the MXene flake with metal oxide nanoparticles. The *in situ* MXene oxidation can be also tracked by applying UV-V is spectroscopy [129]. The intensity of the characteristic peak present at about 800 nm gives the answer on oxidation rate. In particular, the higher it is, the less oxidized is the MXene.

The $Ti_3C_2T_x$ and Nb_2CT_x MXenes' surfaces can be also *in situ* modified with Ti_xO_y and Nb_xO_y, respectively, by rapidly heating them in air up to 1150°C in 5 seconds and holding at that temperature for 30 seconds. Interestingly, the TiO_2 were present at the edges and in the interlayer space of $Ti_3C_2T_x$ [132]. A similar effect was achieved by annealing the MXene in CO_2, while annealing the $Nb_4C_3T_x$ MXene in CO_2 achieved Nb_4C_3/Nb_2O_5 [133]. Later, various oxidizing agents were used to control the surface oxidation of MXenes. The oxidation degree with oxidizing agents can be controlled by changing the reaction time and the oxidant concentration. The H_2O_2 oxidant was used to oxidize MXenes at room temperature [62] along with NaOH [134] or KOH [135]. Notably, the latter two led to oxidized $Ti_3C_2T_x$, additionally intercalated with Na^+ and K^+. Such an option is desirable in energy applications to increase the energy storage capacity of lithium-ion batteries.

In another approach, the $Ti_3C_2T_x$ was modified by Al-oxoanion, which originated from the MAX phase and was utilized at the etching stage, thus reducing the photothermal effect of the obtained $Ti_3C_2T_x$ phase. Modification of the surface by Al^{3+} led to the recovery of the oxoanion lost during the etching MAX phase to MXene, leading to an increase in photothermal conversion at the level of 58.3% [115].

The M_xO_y can be obtained on the surface of MXene also from external sources. The hydrothermal process was used to obtain rutile TiO_2 on the surface of $Ti_3C_2T_x$. In this process, the $TiCl_4$ was used as a precursor for TiO_2 and the resulting composite was applied to produce hydrogen *via* water splitting [136]. Other methods involve the sol-gel techniques, which additionally allow the incorporation of noble metal nanoparticles to the crystalline TiO_2 layer.

For instance, the Ti_2C was surface-modified TiO_2 and co-modified with Ag_2O, Ag, PdO, Pd and Au [137]. The research was based on the analysis of the activity of the photodegradation reaction of salicylic acid (SA). All produced composites had a high photocatalytic activity, with the SA degradation rate within the range of 86.1–97.1%.

A colloidal Ag, Au and $Pd@Ti_3C_2$ hybrids were obtained by mixing a colloidal solution of $Ti_3C_2T_x$ with aqueous solutions of metal salts [138]. Thus, obtained composites can be used as a substrate for Surface Enhanced Raman Spectroscopy (SERS). Further works used the sol-gel method to prepare Ti_3C_2/TiO_2 composites by adding the titanium oxide precursor in the form of $(tBuO)_4Ti$ to the MXene phase suspension [139]. The obtained Ti_3C_2/TiO_2 composite was used to produce a mediator-free biosensor for the detection of H_2O_2 by additionally immobilizing hemoglobin on material surface. The obtained biosensors showed a low detection limit of 14 nM and a wide linear range of 0.1–380 μm for H_2O_2. A similar approach was used to produce membranes made of silver-modified MXenes [140]. The membranes had excellent antibacterial properties against *E. coli* bacteria thanks to the presence of silver nanoparticles and the tendency of the MXenes phases to oxidize to TiO_2, which has excellent antibacterial properties as well. A similar type of membrane prepared from gold-modified MXene also showed the antibacterial properties [141].

An additional advantage of non-*in situ* methods is that they make possible various metals and M-based oxides that do not necessarily have the same M as in MXene. In particular, there are reports in which Ti_3C_2 was surface modified with Rh [142], Ru [143], bimetallic Ru/Ni [144] or Co/Ni [145], SnO_2 [146], Cu_2O [147], Mn_3O_4 [148], MnO_2 [149], Sb_2O_3 [150] or NiO_2 [151]. The $Ti_3C_2T_x$ can be thus modified by Al_2O_3 and SiO_2, and co-modified by noble metal nanoparticles such as Ag and Pd [152]. As a result, $Ti_3C_2/Al_2O_3/Ag$, $Ti_3C_2/SiO_2/Ag$ and $Ti_3C_2/SiO_2/$ Pd composited were obtained. The modification of Ti_3C_2 with ceramic oxides and noble metal nanoparticles allowed the antibacterial properties against both gram-negative and gram-positive bacterial strains. Additional ecotoxicity studies concluded that the MXene ecotoxicity can be adjusted by modifying it with various nanoparticles.

Since metal oxides are semiconducting, they can be combined with MXenes to facilitate the multifunctionality of EMI shielding coatings by adding a solar-harvesting and conversion functionality. The ultrathin heterojunction of $MXene/Bi_2WO_6$ was produced by *in situ* growth of ultra-thin nanoflakes of Bi_2WO_6 on the surface of Ti_3C_2 nanoflakes [153]. It was used for the photocatalytic reduction of CO_2. The yield of CH_4 and CH_3OH was 4.6 times higher than that of the pure Bi_2WO_6. It was attributed to the increased CO_2 adsorption due to high surface area of MXene and improved pore structure of the final heterojunction. Another type of composite $Ti_2C/g-C_3N_4$ was used for water splitting to produce hydrogen [154]. The efficiency of the catalyst was 47.5 μmol h^{-1} and was 14.5 times higher than that of the pure $g-C_3N_4$. It is anticipated that this catalyst may find application on a larger scale due to its low substrate cost, availability, non-toxicity and high yield.

An electrochemical aptasensor, consisting of gold-modified $Nb_4C_3T_x$ is effective for DNA sensing. This material was used for the selective and sensitive detection of Pb^{2+} ions in water samples. The developed sensor achieved a detection limit of 4 nM with a linear range from 10 nM to 5 μM along with high selectivity and stability. This work showed that there is a possibility of a robust $Nb_4C_3T_x$-based platform for the immobilization of DNA oligonucleotides in a variety of sensing applications [155].

Another compound from the MXenes phase family that has found application in light conversion application is Ta_4C_3 [156]. It was used to build a superparamagnetic theranostic nanoplatform to effectively combat breast cancer but could be also used for developing EMI. The surface of the $Ta_4C_3T_x$ was covered with superparamagnetic iron oxide to achieve Ta_4C_3/Fe_2O_3 composite. Thanks to its high atomic number and high X-ray attenuation coefficient, the composite showed high efficiency in CT imaging with contrast enhancement. In contrast, Fe_2O_3 acted as an excellent contrast agent. This material showed high photothermal conversion (32.5%) which could be further useful for enhancing the SE of EMI shielding devices.

3.9 SUMMARY

MXene phases have unique chemistry with numerous functional groups on their surface. This can be grouped into halide-type terminations (-F, -Cl, -Br), chalcogenide-type (-OH, -O-, -S, -Se, -Te) and imine-type (-NH). Their presence on MXene surfaces is strictly related to chosen synthesis method, and significantly influences their optical, electrical, magnetic, piezoelectric and thermoelectric properties, along with the ability to shield the EMI radiation [157]. According to the state of knowledge, MXenes and their composites are placed among the best-known EMI shielding materials [124]. To achieve EMI shielding of above 30 dB, ordinary metallic materials with a thickness above 1 mm are needed. However, the ultra-thin MXenes far outweigh all known synthetic materials.

MXene phases can be surface modified with organic bio-macromolecules, metals, metal oxides, or polymers. The aim is to improve the physical properties of MXenes phases and could allow new functionalities, for instance, in EMI shielding. The modification of MXenes' surface can be carried out according to two strategies [15], by involving physical (e.g., van der Waals) or chemical (e.g., chemical bonding) forces [42, 158]. The first strategy is based on non-covalent interactions between MXene surfaces and other chemical molecules, biological macromolecules or polymers. While being charged, they attach to MXenes through electrostatic-driven physical interactions, other van der Waals forces or only by a simple physical attraction.

The second method of MXene-surface modification involves the covalent-type bonding formed between MXene surface terminating species and other chemical compounds, both organic and inorganic. Numerous functional groups on MXene surfaces can be used as active sites to form a covalent bonding [159–161]. Nowadays, there is a huge increase in interest in hybrid materials that are formed from the combination of organic and inorganic compounds. Therefore, understanding their interactions is very important when considering, for instance, biotechnological applications. The hybrid materials can be created by designing various types of physico-chemical interactions, which can be divided into two groups [104]. The first of them involves the physical forces, among which we can distinguish weak physical attraction via van der Waals forces, hydrogen bonding [162], π-π interactions [163] or electrostatic interactions [100]. The second group includes chemical interactions that usually result in forming the chemical bonding [164].

Polymer and MXene-based composites have become a potential solution to increase the performance of many materials, due to the different forces and interactions, such as van der Waals (vdW) forces, electrostatic interactions and hydrogen bonds, which also helps to increase the interaction between polymer chains and MXene flakes, while synthesis of the intercalation of polymer chains will help in restacking of MXene flakes, which reduces the oxidation and increases the stability by preventing their contact with oxygen [80, 82, 165]. Altogether, MXene and polymer hybrids have attracted interest because of their improved properties and interfacial design for EMI shielding. The results so far are outstanding, but only preliminarily explored. There is still a lot to investigate in terms of managing challenges such as the problem of MXene homogenous distribution, stress transfer mechanism, MXene size effect and oxidative instability, just to mention a few.

Since MXenes are already well-known for their excellent surface activity, it is reasonable to conclude that many types of bio-macromolecules will readily adsorb to MXenes' surface thus causing changes in their surface charge. If the molecules are biocompatible, this may adjust material biocompatibility and reduce the cytotoxicity *in vitro* by scavenging ROS, thus reducing the oxidative stress in cells [101, 102]. Also, the adsorption of bio-macromolecules can adjust MXene's antimicrobial properties.

To conclude, MXenes are promising in EMI shielding not only due to having rich surface chemistry but also due to large surface-to-volume ratio, high biocompatibility, tunable electronic structure, mechanical properties and many other physical features [166]. However, even such excellent properties may not meet the requirements for more specific applications [167]. This problem can be solved by additional surface modifications and functionalization to improve the compatibility in a chosen matrix, to enable the desired EMI shielding functionality.

REFERENCES

1. Y. Zhang, K. Ruan, X. Shi, H. Qiu, Y. Pan, Y. Yan, J. Gu, $Ti_3C_2T_x$/rGO porous composite films with superior electromagnetic interference shielding performances, Carbon, 175 (2021) 271–280.
2. N.F. Colaneri, L. Schacklette, EMI shielding measurements of conductive polymer blends, IEEE Transactions on Instrumentation and Measurement, 41 (1992) 291–297.
3. M.-S. Cao, X.-X. Wang, W.-Q. Cao, J. Yuan, Ultrathin graphene: electrical properties and highly efficient electromagnetic interference shielding, Journal of Materials Chemistry C, 3 (2015) 6589–6599.
4. F. Shahzad, M. Alhabeb, C.B. Hatter, B. Anasori, S. Man Hong, C.M. Koo, Y. Gogotsi, Electromagnetic interference shielding with 2D transition metal carbides (MXenes), Science, 353 (2016) 1137–1140.
5. M.W. Barsoum, The $M_{(N+1)}AX_{(N)}$ phases: A new class of solids; Thermodynamically stable nanolaminates, Progress in Solid State Chemistry, 28 (2000) 201–281.
6. M. Naguib, O. Mashtalir, J. Carle, V. Presser, J. Lu, L. Hultman, Y. Gogotsi, M.W. Barsoum, Two-dimensional transition metal carbides, American Chemical Society, 6 (2012) 1322–1331, DOI: 10.1021/nn204153h.
7. G. Deysher, C.E. Shuck, K. Hantanasirisakul, N.C. Frey, A.C. Foucher, K. Maleski, A. Sarycheva, V.B. Shenoy, E.A. Stach, B. Anasori, Y. Gogotsi, Synthesis of Mo_4VAlC_4 MAX phase and two-dimensional Mo_4VC_4 MXene with five atomic layers of transition metals, ACS Nano, 14 (2020) 204–217, DOI: 10.1021/acsnano.9b07708.
8. O. Mashtalir, K.M. Cook, V.N. Mochalin, M. Crowe, M.W. Barsoum, Y. Gogotsi, Dye adsorption and decomposition on two-dimensional titanium carbide in aqueous media, Journal of Materials Chemistry A, 2 (2014) 14334–14338.
9. M. Carey, Z. Hinton, M. Sokol, N.J. Alvarez, M.W. Barsoum, Nylon-6/$Ti_3C_2T_z$ MXene nanocomposites synthesized by in situ ring opening polymerization of ε-caprolactam and their water transport properties, ACS Applied Materials & Interfaces, 11 (2019) 20425–20436.

10. J. Liu, H.B. Zhang, X. Xie, R. Yang, Z. Liu, Y. Liu, Z.Z. Yu, Multifunctional, superelastic, and lightweight MXene/polyimide aerogels, Small, 14 (2018) 1802479.
11. Y. Li, X. Zhou, J. Wang, Q. Deng, M. Li, S. Du, Y.-H. Han, J. Lee, Q. Huang, Facile preparation of in situ coated $Ti_3C_2T_x/Ni_{0.5}Zn_{0.5}Fe_2O_4$ composites and their electromagnetic performance, RSC Advances, 7 (2017) 24698–24708.
12. Z. Ling, C.E. Ren, M.-Q. Zhao, J. Yang, J.M. Giammarco, J. Qiu, M.W. Barsoum, Y. Gogotsi, Flexible and conductive MXene films and nanocomposites with high capacitance, Proceedings of the National Academy of Sciences, 111 (2014) 16676–16681.
13. M. Naguib, M. Kurtoglu, V. Presser, J. Lu, J. Niu, M. Heon, L. Hultman, Y. Gogotsi, M.W. Barsoum, Two-dimensional nanocrystals produced by exfoliation of Ti_3AlC_2, Advanced Materials, 23 (2011) 4248–4253, DOI: 10.1002/adma.201102306.
14. Q. Peng, J. Guo, Q. Zhang, J. Xiang, B. Liu, A. Zhou, R. Liu, Y. Tian, Unique lead adsorption behavior of activated hydroxyl group in two-dimensional titanium carbide, Journal of the American Chemical Society, 136 (2014) 4113–4116.
15. H. Lin, Y. Chen, J. Shi, Insights into 2D MXenes for versatile biomedical applications: current advances and challenges ahead, Advanced Science, 5 (2018) 1800518.
16. A. Lipatov, M. Alhabeb, M.R. Lukatskaya, A. Boson, Y. Gogotsi, A. Sinitskii, Effect of synthesis on quality, electronic properties and environmental stability of individual monolayer Ti_3C_2 MXene flakes, Advanced Electronic Materials, 2 (2016) 1600255.
17. M.R. Lukatskaya, O. Mashtalir, C.E. Ren, Y. Dall'Agnese, P. Rozier, P.L. Taberna, M. Naguib, P. Simon, M.W. Barsoum, Y. Gogotsi, Cation intercalation and high volumetric capacitance of two-dimensional titanium carbide, Science, 341 (2013) 1502–1505.
18. M. Ghidiu, M.R. Lukatskaya, M.Q. Zhao, Y. Gogotsi, M.W. Barsoum, Conductive two-dimensional titanium carbide 'clay' with high volumetric capacitance, Nature, 516 (2014) 78–81, DOI: 10.1038/nature13970.
19. Y. Zhao, M. Zhang, H. Yan, Y. Feng, X. Zhang, R. Guo, Few-layer large $Ti_3C_2T_x$ sheets exfoliated by $NaHF_2$ and applied to the sodium-ion battery, Journal of Materials Chemistry A, 9 (2021) 9593–9601.
20. V. Natu, R. Pai, M. Sokol, M. Carey, V. Kalra, M.W. Barsoum, 2D $Ti_3C_2T_z$ MXene synthesized by water-free etching of Ti_3AlC_2 in polar organic solvents, Chem, 6 (2020) 616–630.
21. J. Halim, M.R. Lukatskaya, K.M. Cook, J. Lu, C.R. Smith, L.-Å. Näslund, S.J. May, L. Hultman, Y. Gogotsi, P. Eklund, Transparent conductive two-dimensional titanium carbide epitaxial thin films, Chemistry of Materials, 26 (2014) 2374–2381.
22. V. Kamysbayev, A.S. Filatov, H. Hu, X. Rui, F. Lagunas, D. Wang, R.F. Klie, D.V. Talapin, Covalent surface modifications and superconductivity of two-dimensional metal carbide MXenes, Science, 369 (2020) 979–983.
23. S.-Y. Pang, Y.-T. Wong, S. Yuan, Y. Liu, M.-K. Tsang, Z. Yang, H. Huang, W.-T. Wong, J. Hao, Universal strategy for HF-free facile and rapid synthesis of two-dimensional MXenes as multifunctional energy materials, Journal of the American Chemical Society, 141 (2019) 9610–9616.
24. N. Ahmed, S.-E. Oh, Toxicity assessment of selected heavy metals in water using a seven-chambered sulfur-oxidizing bacterial (SOB) bioassay reactor, Sensors and Actuators B: Chemical, 258 (2018) 1008–1014.
25. D. Kim, T.Y. Ko, H. Kim, G.H. Lee, S. Cho, C.M. Koo, Nonpolar organic dispersion of 2D $Ti_3C_2T_x$ MXene flakes *via* simultaneous interfacial chemical grafting and phase transfer method, ACS Nano, 13 (2019) 13818–13828.
26. X. Xie, Y. Xue, L. Li, S. Chen, Y. Nie, W. Ding, Z. Wei, Surface Al leached Ti_3AlC_2 as a substitute for carbon for use as a catalyst support in a harsh corrosive electrochemical system, Nanoscale, 6 (2014) 11035–11040.

27. G. Zou, Q. Zhang, C. Fernandez, G. Huang, J. Huang, Q. Peng, Heterogeneous Ti_3SiC_2@ C-containing $Na_2Ti_7O_{15}$ architecture for high-performance sodium storage at elevated temperatures, ACS Nano, 11 (2017) 12219–12229.

28. G. Li, L. Tan, Y. Zhang, B. Wu, L. Li, Highly efficiently delaminated single-layered MXene nanosheets with large lateral size, Langmuir, 33 (2017) 9000–9006.

29. O. Mashtalir, M. Naguib, V.N. Mochalin, Y. Dall'Agnese, M. Heon, M.W. Barsoum, Y. Gogotsi, Intercalation and delamination of layered carbides and carbonitrides, Nature Communications, 4 (2013) 1716, DOI: 10.1038/ncomms2664.

30. M. Naguib, R.R. Unocic, B.L. Armstrong, J. Nanda, Large-scale delamination of multi-layers transition metal carbides and carbonitrides "MXenes", Dalton Transactions, 44 (2015) 9353–9358, DOI: 10.1039/C5DT01247C.

31. O. Mashtalir, M.R. Lukatskaya, M.Q. Zhao, M.W. Barsoum, Y. Gogotsi, Amine-assisted delamination of Nb_2C MXene for Li-ion energy storage devices, Advanced Materials, 27 (2015) 3501–3506.

32. X. Yu, X. Cai, H. Cui, S.-W. Lee, X.-F. Yu, B. Liu, Fluorine-free preparation of titanium carbide MXene quantum dots with high near-infrared photothermal performances for cancer therapy, Nanoscale, 9 (2017) 17859–17864.

33. I. Shein, A. Ivanovskii, Planar nano-block structures $Ti_{n+1}Al_{0.5}C_n$ and $Ti_{n+1}C_n$ (n = 1, and 2) from MAX phases: Structural, electronic properties and relative stability from first principles calculations, Superlattices and Microstructures, 52 (2012) 147–157.

34. H. Kim, Z. Wang, H.N. Alshareef, MXetronics: Electronic and photonic applications of MXenes, Nano Energy, 60 (2019) 179–197.

35. R. Thakur, A. VahidMohammadi, J. Moncada, W.R. Adams, M. Chi, B. Tatarchuk, M. Beidaghi, C.A. Carrero, Insights into the thermal and chemical stability of multilayered V_2CT_x MXene, Nanoscale, 11 (2019) 10716–10726.

36. O. Mashtalir, M. Naguib, B. Dyatkin, Y. Gogotsi, M.W. Barsoum, Kinetics of aluminum extraction from Ti_3AlC_2 in hydrofluoric acid, Materials Chemistry and Physics, 139 (2013) 147–152.

37. T. Schultz, N.C. Frey, K. Hantanasirisakul, S. Park, S.J. May, V.B. Shenoy, Y. Gogotsi, N. Koch, Surface termination dependent work function and electronic properties of $Ti_3C_2T_x$ MXene, Chemistry of Materials, 31 (2019) 6590–6597.

38. M. Khazaei, M. Arai, T. Sasaki, C.-Y. Chung, N.S. Venkataramanan, M. Estili, Y. Sakka, Y. Kawazoe, Novel electronic and magnetic properties of two-dimensional transition metal carbides and nitrides, Advanced Functional Materials, 23 (2013) 2185–2192, DOI: 10.1002/adfm.201202502.

39. J. Zhou, X. Zha, F.Y. Chen, Q. Ye, P. Eklund, S. Du, Q. Huang, A two-dimensional zirconium carbide by selective etching of Al_3C_3 from nanolaminated $Zr_3Al_3C_5$, Angewandte Chemie International Edition, 55 (2016) 5008–5013.

40. J. Zhou, X. Zha, X. Zhou, F. Chen, G. Gao, S. Wang, C. Shen, T. Chen, C. Zhi, P. Eklund, Synthesis and electrochemical properties of two-dimensional hafnium carbide, ACS Nano, 11 (2017) 3841–3850.

41. R. Meshkian, L.-Å. Näslund, J. Halim, J. Lu, M.W. Barsoum, J. Rosen, Synthesis of two-dimensional molybdenum carbide, Mo_2C, from the gallium based atomic laminate Mo_2Ga_2C, Scripta Materialia, 108 (2015) 147–150.

42. B. Anasori, M.R. Lukatskaya, Y. Gogotsi, 2D metal carbides and nitrides (MXenes) for energy storage, Nature Review Materials, 2 (2017) 16098, DOI: 10.1038/natrevmats.2016.98.

43. D. Chung, Electromagnetic interference shielding effectiveness of carbon materials, Carbon, 39 (2001) 279–285.

44. G. Gao, A.P. O'Mullane, A. Du, 2D MXenes: a new family of promising catalysts for the hydrogen evolution reaction, ACS Catalysis, 7 (2017) 494–500.

45. J. Jimmy, B. Kandasubramanian, Mxene functionalized polymer composites: Synthesis and applications, European Polymer Journal, 122 (2020) 109367.
46. J.-C. Gui, L. Han, W.-Y. Cao, Lamellar MXene: A novel 2D nanomaterial for electrochemical sensors, Journal of Applied Electrochemistry, 51 (2021) 1509–1522.
47. K. Rajavel, S. Luo, Y. Wan, X. Yu, Y. Hu, P. Zhu, R. Sun, C. Wong, 2D $Ti_3C_2T_x$ MXene/ polyvinylidene fluoride (PVDF) nanocomposites for attenuation of electromagnetic radiation with excellent heat dissipation, Composites Part A: Applied Science and Manufacturing, 129 (2020) 105693.
48. X.-H. Zha, J. Yin, Y. Zhou, Q. Huang, K. Luo, J. Lang, J.S. Francisco, J. He, S. Du, Intrinsic structural, electrical, thermal, and mechanical properties of the promising conductor Mo_2C MXene, The Journal of Physical Chemistry C, 120 (2016) 15082–15088.
49. X.-H. Zha, J. Zhou, Y. Zhou, Q. Huang, J. He, J.S. Francisco, K. Luo, S. Du, Promising electron mobility and high thermal conductivity in Sc_2CT_2 (T = F, OH) MXenes, Nanoscale, 8 (2016) 6110–6117.
50. X.-H. Zha, Q. Huang, J. He, H. He, J. Zhai, J.S. Francisco, S. Du, The thermal and electrical properties of the promising semiconductor MXene Hf_2CO_2, Scientific Reports, 6 (2016) 1–10.
51. Y. Cao, Q. Deng, Z. Liu, D. Shen, T. Wang, Q. Huang, S. Du, N. Jiang, C.-T. Lin, J. Yu, Enhanced thermal properties of poly (vinylidene fluoride) composites with ultrathin nanosheets of MXene, RSC Advances, 7 (2017) 20494–20501.
52. L. Wang, L. Chen, P. Song, C. Liang, Y. Lu, H. Qiu, Y. Zhang, J. Kong, J. Gu, Fabrication on the annealed $Ti_3C_2T_x$ MXene/Epoxy nanocomposites for electromagnetic interference shielding application, Composites Part B: Engineering, 171 (2019) 111–118.
53. Y. Zou, L. Fang, T. Chen, M. Sun, C. Lu, Z. Xu, Near-infrared light and solar light activated self-healing epoxy coating having enhanced properties using MXene flakes as multifunctional fillers, Polymers, 10 (2018) 474.
54. R. Kang, Z. Zhang, L. Guo, J. Cui, Y. Chen, X. Hou, B. Wang, C.-T. Lin, N. Jiang, J. Yu, Enhanced thermal conductivity of epoxy composites filled with 2D transition metal carbides (MXenes) with ultralow loading, Scientific Reports, 9 (2019) 1–14.
55. J.-W. Gu, Q.-Y. Zhang, H.-C. Li, Y.-S. Tang, J. Kong, J. Dang, Study on preparation of SiO2/epoxy resin hybrid materials by means of sol-gel, Polymer-Plastics Technology and Engineering, 46 (2007) 1129–1134.
56. P. Song, B. Liu, H. Qiu, X. Shi, D. Cao, J. Gu, MXenes for polymer matrix electromagnetic interference shielding composites: A review, Composites Communications, 24 (2021) 100653.
57. G.M. Weng, J. Li, M. Alhabeb, C. Karpovich, H. Wang, J. Lipton, K. Maleski, J. Kong, E. Shaulsky, M. Elimelech, Layer-by-layer assembly of cross-functional semi-transparent MXene-carbon nanotubes composite films for next-generation electromagnetic interference shielding, Advanced Functional Materials, 28 (2018) 1803360.
58. X. Jin, J. Wang, L. Dai, X. Liu, L. Li, Y. Yang, Y. Cao, W. Wang, H. Wu, S. Guo, Flame-retardant poly (vinyl alcohol)/MXene multilayered films with outstanding electromagnetic interference shielding and thermal conductive performances, Chemical Engineering Journal, 380 (2020) 122475.
59. H. Xu, X. Yin, X. Li, M. Li, S. Liang, L. Zhang, L. Cheng, Lightweight Ti_2CT_x MXene/ poly (vinyl alcohol) composite foams for electromagnetic wave shielding with absorption-dominated feature, ACS Applied Materials & Interfaces, 11 (2019) 10198–10207.
60. X. Wu, B. Han, H.-B. Zhang, X. Xie, T. Tu, Y. Zhang, Y. Dai, R. Yang, Z.-Z. Yu, Compressible, durable and conductive polydimethylsiloxane-coated MXene foams for high-performance electromagnetic interference shielding, Chemical Engineering Journal, 381 (2020) 122622.

61. M. Naguib, V.N. Mochalin, M.W. Barsoum, Y. Gogotsi, 25th anniversary article: MXenes: a new family of two-dimensional materials, Advanced Materials, 26 (2014) 992–1005.

62. B. Ahmed, D.H. Anjum, M.N. Hedhili, Y. Gogotsi, H.N. Alshareef, H_2O_2 assisted room temperature oxidation of Ti_2C MXene for Li-ion battery anodes, Nanoscale, 8 (2016) 7580–7587.

63. T. Shang, Z. Lin, C. Qi, X. Liu, P. Li, Y. Tao, Z. Wu, D. Li, P. Simon, Q.H. Yang, 3D macroscopic architectures from self-assembled MXene hydrogels, Advanced Functional Materials, 29 (2019) 1903960.

64. M.-s. Cao, Y.-Z. Cai, P. He, J.-C. Shu, W.-Q. Cao, J. Yuan, 2D MXenes: electromagnetic property for microwave absorption and electromagnetic interference shielding, Chemical Engineering Journal, 359 (2019) 1265–1302.

65. M. Han, C.E. Shuck, R. Rakhmanov, D. Parchment, B. Anasori, C.M. Koo, G. Friedman, Y. Gogotsi, Beyond $Ti_3C_2T_x$: MXenes for electromagnetic interference shielding, ACS Nano, 14 (2020) 5008–5016.

66. H. Yang, J. Dai, X. Liu, Y. Lin, J. Wang, L. Wang, F. Wang, Layered PVB/$Ba_3Co_2Fe_{24}O_{41}$/Ti_3C_2 Mxene composite: enhanced electromagnetic wave absorption properties with high impedance match in a wide frequency range, Materials Chemistry and Physics, 200 (2017) 179–186.

67. H. Zhang, L. Wang, Q. Chen, P. Li, A. Zhou, X. Cao, Q. Hu, Preparation, mechanical and anti-friction performance of MXene/polymer composites, Materials & Design, 92 (2016) 682–689.

68. M.S. Carey, On the Synthesis & Characterization of $Ti_3C_2T_x$ MXene Polymer Composites, Drexel University, 2017.

69. Y. Shi, C. Liu, L. Liu, L. Fu, B. Yu, Y. Lv, F. Yang, P. Song, Strengthening, toughing and thermally stable ultra-thin MXene nanosheets/polypropylene nanocomposites via nanoconfinement, Chemical Engineering Journal, 378 (2019) 122267.

70. J.-Y. Si, B. Tawiah, W.-L. Sun, B. Lin, C. Wang, A.C.Y. Yuen, B. Yu, A. Li, W. Yang, H.-D. Lu, Functionalization of MXene nanosheets for polystyrene towards high thermal stability and flame retardant properties, Polymers, 11 (2019) 976.

71. T.S.M. Naguib, S. Lai, M. S. Rager, T. Aytug, M. P. Paranthaman, M.-Q. Zhao, Y. Gogotsi, $Ti_3C_2T_x$ (MXene)–polyacrylamide nanocomposite films, RSC Advances, 102 (2016) 72069–72073.

72. L. Hao, H. Zhang, X. Wu, J. Zhang, J. Wang, Y. Li, Novel thin-film nanocomposite membranes filled with multi-functional $Ti_3C_2T_x$ nanosheets for task-specific solvent transport, Composites Part A: Applied Science and Manufacturing, 100 (2017) 139–149.

73. A. VahidMohammadi, J. Moncada, H. Chen, E. Kayali, J. Orangi, C.A. Carrero, M. Beidaghi, Thick and freestanding MXene/PANI pseudocapacitive electrodes with ultrahigh specific capacitance, Journal of Materials Chemistry A, 6 (2018) 22123–22133.

74. N.S. Seroka, M.A. Mamo, Application of functionalised MXene-carbon nanoparticle-polymer composites in resistive hydrostatic pressure sensors, SN Applied Sciences, 2 (2020) 1–11.

75. W. Zhang, J. Ma, W. Zhang, P. Zhang, W. He, J. Chen, Z. Sun, A multidimensional nanostructural design towards electrochemically stable and mechanically strong hydrogel electrodes, Nanoscale, 12 (2020) 6637–6643.

76. M. Boota, M. Pasini, F. Galeotti, W. Porzio, M.-Q. Zhao, J. Halim, Y. Gogotsi, Interaction of polar and nonpolar polyfluorenes with layers of two-dimensional titanium carbide (MXene): intercalation and pseudocapacitance, Chemistry of Materials, 29 (2017) 2731–2738.

77. E.A. Mayerberger, O. Urbanek, R.M. McDaniel, R.M. Street, M.W. Barsoum, C.L. Schauer, Preparation and characterization of polymer-$Ti_3C_2T_x$ (MXene) composite nanofibers produced via electrospinning, Journal of Applied Polymer Science, 134 (2017).

78. M. Boota, B. Anasori, C. Voigt, M.Q. Zhao, M.W. Barsoum, Y. Gogotsi, Pseudocapacitive electrodes produced by oxidant-free polymerization of pyrrole between the layers of 2D titanium carbide (MXene), Advanced Materials, 28 (2016) 1517–1522.

79. M. Zhu, Y. Huang, Q. Deng, J. Zhou, Z. Pei, Q. Xue, Y. Huang, Z. Wang, H. Li, Q. Huang, Highly flexible, freestanding supercapacitor electrode with enhanced performance obtained by hybridizing polypyrrole chains with MXene, Advanced Energy Materials, 6 (2016) 1600969.

80. S. Tu, Q. Jiang, X. Zhang, H.N. Alshareef, Large dielectric constant enhancement in MXene percolative polymer composites, ACS Nano, 12 (2018) 3369–3377.

81. Z. Huang, S. Wang, S. Kota, Q. Pan, M.W. Barsoum, C.Y. Li, Structure and crystallization behavior of poly (ethylene oxide)/$Ti_3C_2T_x$ MXene nanocomposites, Polymer, 102 (2016) 119–126.

82. R. Sun, H.B. Zhang, J. Liu, X. Xie, R. Yang, Y. Li, S. Hong, Z.Z. Yu, Highly conductive transition metal carbide/carbonitride (MXene)@ polystyrene nanocomposites fabricated by electrostatic assembly for highly efficient electromagnetic interference shielding, Advanced Functional Materials, 27 (2017) 1702807.

83. Z. Li, L. Wang, D. Sun, Y. Zhang, B. Liu, Q. Hu, A. Zhou, Synthesis and thermal stability of two-dimensional carbide MXene Ti_3C_2, Materials Science and Engineering: B, 191 (2015) 33–40.

84. S. Sarikurt, D. Çakır, M. Keçeli, C. Sevik, The influence of surface functionalization on thermal transport and thermoelectric properties of MXene monolayers, Nanoscale, 10 (2018) 8859–8868.

85. S.A. Mirkhani, A. Shayesteh Zeraati, E. Aliabadian, M. Naguib, U. Sundararaj, High dielectric constant and low dielectric loss *via* poly (vinyl alcohol)/$Ti_3C_2T_x$ MXene nanocomposites, ACS Applied Materials & Interfaces, 11 (2019) 18599–18608.

86. R. Han, X. Ma, Y. Xie, D. Teng, S. Zhang, Preparation of a new 2D MXene/PES composite membrane with excellent hydrophilicity and high flux, RSC Advances, 7 (2017) 56204–56210.

87. M. Fei, R. Lin, Y. Deng, H. Xian, R. Bian, X. Zhang, J. Cheng, C. Xu, D. Cai, Polybenzimidazole/Mxene composite membranes for intermediate temperature polymer electrolyte membrane fuel cells, Nanotechnology, 29 (2017) 035403.

88. X. Chen, Y. Zhao, L. Li, Y. Wang, J. Wang, J. Xiong, S. Du, P. Zhang, X. Shi, J. Yu, MXene/polymer nanocomposites: preparation, properties, and applications, Polymer Reviews, 61 (2021) 80–115.

89. H. Wei, J. Dong, X. Fang, W. Zheng, Y. Sun, Y. Qian, Z. Jiang, Y. Huang, $Ti_3C_2T_x$ MXene/polyaniline (PANI) sandwich intercalation structure composites constructed for microwave absorption, Composites Science and Technology, 169 (2019) 52–59.

90. Y. Tong, M. He, Y. Zhou, X. Zhong, L. Fan, T. Huang, Q. Liao, Y. Wang, Hybridizing polypyrrole chains with laminated and two-dimensional $Ti_3C_2T_x$ toward high-performance electromagnetic wave absorption, Applied Surface Science, 434 (2018) 283–293.

91. Y. Ren, J. Zhu, L. Wang, H. Liu, Y. Liu, W. Wu, F. Wang, Synthesis of polyaniline nanoparticles deposited on two-dimensional titanium carbide for high-performance supercapacitors, Materials Letters, 214 (2018) 84–87.

92. H. Wang, L. Li, C. Zhu, S. Lin, J. Wen, Q. Jin, X. Zhang, In situ polymerized $Ti_3C_2T_x$/ PDA electrode with superior areal capacitance for supercapacitors, Journal of Alloys and Compounds, 778 (2019) 858–865.

93. M.Q. Zhao, C.E. Ren, Z. Ling, M.R. Lukatskaya, C. Zhang, K.L. Van Aken, M.W. Barsoum, Y. Gogotsi, Flexible MXene/carbon nanotube composite paper with high volumetric capacitance, Advanced Materials, 27 (2015) 339–345.

94. X. Wu, B. Huang, R. Lv, Q. Wang, Y. Wang, Highly flexible and low capacitance loss supercapacitor electrode based on hybridizing decentralized conjugated polymer chains with MXene, Chemical Engineering Journal, 378 (2019) 122246.

95. C. Chen, M. Boota, X. Xie, M. Zhao, B. Anasori, C.E. Ren, L. Miao, J. Jiang, Y. Gogotsi, Charge transfer induced polymerization of EDOT confined between 2D titanium carbide layers, Journal of Materials Chemistry A, 5 (2017) 5260–5265.

96. H. Lee, S.M. Dellatore, W.M. Miller, P.B. Messersmith, Mussel-inspired surface chemistry for multifunctional coatings, Science, 318 (2007) 426–430.

97. Q. Pan, Y. Zheng, S. Kota, W. Huang, S. Wang, H. Qi, S. Kim, Y. Tu, M.W. Barsoum, C.Y. Li, 2D MXene-containing polymer electrolytes for all-solid-state lithium metal batteries, Nanoscale Advances, 1 (2019) 395–402.

98. Y. Liu, J. Zhang, X. Zhang, Y. Li, J. Wang, $Ti_3C_2T_x$ filler effect on the proton conduction property of polymer electrolyte membrane, ACS Applied Materials & Interfaces, 8 (2016) 20352–20363.

99. Z. Xu, G. Liu, H. Ye, W. Jin, Z. Cui, Two-dimensional MXene incorporated chitosan mixed-matrix membranes for efficient solvent dehydration, Journal of Membrane Science, 563 (2018) 625–632.

100. A. Rozmysłowska-Wojciechowska, T. Wojciechowski, W. Ziemkowska, L. Chlubny, A. Olszyna, A.M. Jastrzębska, Surface interactions between 2D Ti_3C_2/Ti_2C MXenes and lysozyme, Applied Surface Science, 473 (2019) 409–418, DOI: 10.1016/j.apsusc.2018.12.081.

101. A. Rozmyslowska-Wojciechowska, A. Szuplewska, T. Wojciechowski, S. Pozniak, J. Mitrzak, M. Chudy, W. Ziemkowska, L. Chlubny, A. Olszyna, A.M. Jastrzebska, A simple, low-cost and green method for controlling the cytotoxicity of MXenes, Materials Science & Engineering C: Materials for Biological Applications, 111 (2020) 110790, DOI: 10.1016/j.msec.2020.110790.

102. A. Rozmysłowska-Wojciechowska, J. Mitrzak, A. Szuplewska, M. Chudy, J. Woźniak, M. Petrus, T. Wojciechowski, A.S. Vasilchenko, A.M. Jastrzębska, Engineering of 2D Ti_3C_2 MXene surface charge and its influence on biological properties, Materials, 13 (2020) 2347.

103. L. Wu, X. Lu, Z.-S. Wu, Y. Dong, X. Wang, S. Zheng, J. Chen, 2D transition metal carbide MXene as a robust biosensing platform for enzyme immobilization and ultrasensitive detection of phenol, Biosensors and Bioelectronics, 107 (2018) 69–75.

104. C. Chen, M. Boota, P. Urbankowski, B. Anasori, L. Miao, J. Jiang, Y. Gogotsi, Effect of glycine functionalization of 2D titanium carbide (MXene) on charge storage, Journal of Materials Chemistry A, 6 (2018) 4617–4622.

105. C. Li, X. Feng, L. Sun, L. Zhou, J. Sun, Z. Wang, D. Liao, P. Lan, X. Lan, Non-covalent and covalent immobilization of papain onto Ti3C2 MXene nanosheets, Enzyme and Microbial Technology, 148 (2021) 109817.

106. H. Liu, C. Duan, C. Yang, W. Shen, F. Wang, Z. Zhu, A novel nitrite biosensor based on the direct electrochemistry of hemoglobin immobilized on MXene-Ti_3C_2, Sensors and Actuators B: Chemical, 218 (2015) 60–66.

107. A. Jastrzebska, A. Szuplewska, A. Rozmysłowska-Wojciechowska, J. Mitrzak, T. Wojciechowski, M. Chudy, D. Moszczyńska, A. Wójcik, K. Prenger, M. Naguib, Juggling surface charges of 2D niobium carbide MXenes for a reactive oxygen species scavenging and effective targeting of the malignant melanoma cell cycle into programmed cell death, ACS Sustainable Chemistry & Engineering, 8 (2020) 7942–7951, DOI: 10.1021/acssuschemeng.0c01609.

108. M. Zabadaj, A. Szuplewska, D. Kalinowska, M. Chudy, P. Ciosek-Skibińska, Studying pharmacodynamic effects in cell cultures by chemical fingerprinting – SIA electronic tongue versus 2D fluorescence soft sensor, Sensors and Actuators B: Chemical, 272 (2018) 264–273.

109. Y. Wang, L. Xiong, M. Tang, Toxicity of inhaled particulate matter on the central nervous system: neuroinflammation, neuropsychological effects and neurodegenerative disease, Journal of Applied Toxicology, 37 (2017) 644–667.

110. Y. Wang, M. Tang, Review of in vitro toxicological research of quantum dot and potentially involved mechanisms, Science of the Total Environment, 625 (2018) 940–962.

111. W. Wang, C. Zeng, Y. Feng, F. Zhou, F. Liao, Y. Liu, S. Feng, X. Wang, The size-dependent effects of silica nanoparticles on endothelial cell apoptosis through activating the p53-caspase pathway, Environmental Pollution, 233 (2018) 218–225.

112. A. Szuplewska, D. Kulpinska, A. Dybko, A.M. Jastrzebska, T. Wojciechowski, A. Rozmyslowska, M. Chudy, I. Grabowska-Jadach, W. Ziemkowska, Z. Brzozka, A. Olszyna, 2D Ti_2C (MXene) as a novel highly efficient and selective agent for photothermal therapy, Materials Science & Engineering C: Materials for Biological Applications, 98 (2019) 874–886, DOI: 10.1016/j.msec.2019.01.021.

113. Y. Lu, X. Zhang, X. Hou, M. Feng, Z. Cao, J. Liu, Functionalized 2D Nb_2C nanosheets for primary and recurrent cancer photothermal/immune-therapy in the NIR-II biowindow, Nanoscale, 13 (2021) 17822–17836.

114. H. Lin, S. Gao, C. Dai, Y. Chen, J. Shi, A two-dimensional biodegradable niobium carbide (MXene) for photothermal tumor eradication in NIR-I and NIR-II biowindows, Journal of the American Chemical Society, 139 (2017) 16235–16247, DOI: 10.1021/jacs.7b07818.

115. G. Liu, J. Zou, Q. Tang, X. Yang, Y. Zhang, Q. Zhang, W. Huang, P. Chen, J. Shao, X. Dong, Surface Modified Ti_3C_2 MXene nanosheets for tumor targeting photothermal/photodynamic/chemo synergistic therapy, ACS Applied Materials & Interfaces, 9 (2017) 40077–40086, DOI: 10.1021/acsami.7b13421.

116. H. Lusic, M.W. Grinstaff, X-ray-computed tomography contrast agents, Chemical Reviews, 113 (2013) 1641–1666.

117. C. Caro, M. Dalmases, A. Figuerola, M.L. García-Martín, M.P. Leal, Highly water-stable rare ternary Ag-Au–Se nanocomposites as long blood circulation time X-ray computed tomography contrast agents, Nanoscale, 9 (2017) 7242–7251.

118. T. Ozulumba, G. Ingavle, Y. Gogotsi, S. Sandeman, Moderating cellular inflammation using 2-dimensional titanium carbide MXene and graphene variants, Biomaterials Science, 9 (2021) 1805–1815.

119. L.V. Efremova, A.S. Vasilchenko, E.G. Rakov, D.G. Deryabin, Toxicity of graphene shells, graphene oxide, and graphene oxide paper evaluated with *Escherichia coli* biotests, BioMed Research International, 2015 (2015).

120. W.W. Wilson, M.M. Wade, S.C. Holman, F.R. Champlin, Status of methods for assessing bacterial cell surface charge properties based on zeta potential measurements, Journal of Microbiological Methods, 43 (2001) 153–164.

121. A.P.F. Maillard, J.C. Espeche, P. Maturana, A.C. Cutro, A. Hollmann, Zeta potential beyond materials science: applications to bacterial systems and to the development of novel antimicrobials, Biochimica et Biophysica Acta (BBA)-Biomembranes, 1863 (2021) 183597.

122. D.G. Deryabin, L.V. Efremova, A.S. Vasilchenko, E.V. Saidakova, E.A. Sizova, P.A. Troshin, A.V. Zhilenkov, E.A. Khakina, A zeta potential value determines the aggregate's size of penta-substituted [60] fullerene derivatives in aqueous suspension whereas positive charge is required for toxicity against bacterial cells, Journal of Nanobiotechnology, 13 (2015) 1–13.

123. A.M. Jastrzębska, E. Karwowska, T. Wojciechowski, W. Ziemkowska, A. Rozmysłowska, L. Chlubny, A. Olszyna, The atomic structure of Ti_2C and Ti_3C_2 MXenes is responsible for their antibacterial activity toward *E. coli* bacteria, Journal of Materials Engineering and Performance, 28 (2019) 1272–1277.

124. X. Zhao, A. Vashisth, E. Prehn, W. Sun, S.A. Shah, T. Habib, Y. Chen, Z. Tan, J.L. Lutkenhaus, M. Radovic, Antioxidants unlock shelf-stable $Ti_3C_2T_x$ (MXene) nanosheet dispersions, Matter, 1 (2019) 513–526.

125. V. Natu, J.L. Hart, M. Sokol, H. Chiang, M.L. Taheri, M.W. Barsoum, Edge capping of 2D-MXene sheets with polyanionic salts to mitigate oxidation in aqueous colloidal suspensions, Angewandte Chemie, 131 (2019) 12785–12790.

126. C.-W. Wu, B. Unnikrishnan, I.-W.P. Chen, S.G. Harroun, H.-T. Chang, C.-C. Huang, Excellent oxidation resistive MXene aqueous ink for micro-supercapacitor application, Energy Storage Materials, 25 (2020) 563–571.

127. I.J. Echols, D.E. Holta, V.S. Kotasthane, Z. Tan, M. Radovic, J.L. Lutkenhaus, M.J. Green, Oxidative Stability of $Nb_{n+1}C_nT_z$ MXenes, The Journal of Physical Chemistry C, 125 (2021) 13990–13996.

128. H. Zhang, F.-Z. Dai, H. Xiang, X. Wang, Z. Zhang, Y. Zhou, Phase pure and well crystalline Cr_2AlB_2: A key precursor for two-dimensional CrB, Journal of Materials Science & Technology, 35 (2019) 1593–1600, DOI: 10.1016/j.jmst.2019.03.031.

129. C.J. Zhang, S. Pinilla, N. McEvoy, C.P. Cullen, B. Anasori, E. Long, S.-H. Park, A. Seral-Ascaso, A. Shmeliov, D. Krishnan, C. Morant, X. Liu, G.S. Duesberg, Y. Gogotsi, V. Nicolosi, Oxidation stability of colloidal two-dimensional titanium carbides (MXenes), Chemistry of Materials, 29 (2017) 4848–4856.

130. C. Zhang, Y. Xie, G. Sun, A. Pentecost, J. Wang, W. Qiao, L. Ling, D. Long, Y. Gogotsi, Ion intercalation into graphitic carbon with a low surface area for high energy density supercapacitors, Journal of The Electrochemical Society, 161 (2014) A1486.

131. A. Jastrzębska, A. Szuplewska, A. Rozmysłowska-Wojciechowska, M. Chudy, A. Olszyna, M. Birowska, M. Popielski, J. Majewski, B. Scheibe, V. Natu, On tuning the cytotoxicity of Ti_3C_2 (MXene) flakes to cancerous and benign cells by post-delamination surface modifications, 2D Materials, 7 (2020) 025018.

132. M. Naguib, O. Mashtalir, M.R. Lukatskaya, B. Dyatkin, C. Zhang, V. Presser, Y. Gogotsi, M.W. Barsoum, One-step synthesis of nanocrystalline transition metal oxides on thin sheets of disordered graphitic carbon by oxidation of MXenes, Chemical Communications, 50 (2014) 7420–7423.

133. C. Zhang, S.J. Kim, M. Ghidiu, M.Q. Zhao, M.W. Barsoum, V. Nicolosi, Y. Gogotsi, Layered orthorhombic $Nb_2O_5@Nb_4C_3T_x$ and $TiO_2@Ti_3C_2T_x$ hierarchical composites for high performance Li-ion batteries, Advanced Functional Materials, 26 (2016) 4143–4151.

134. J. Huang, R. Meng, L. Zu, Z. Wang, N. Feng, Z. Yang, Y. Yu, J. Yang, Sandwich-like $Na_{0.23}TiO_2$ nanobelt/Ti_3C_2 MXene composites from a scalable in situ transformation reaction for long-life high-rate lithium/sodium-ion batteries, Nano Energy, 46 (2018) 20–28.

135. Y. Dong, Z.-S. Wu, S. Zheng, X. Wang, J. Qin, S. Wang, X. Shi, X. Bao, Ti_3C_2 MXene-derived sodium/potassium titanate nanoribbons for high-performance sodium/potassium ion batteries with enhanced capacities, ACS Nano, 11 (2017) 4792–4800.

136. H. Wang, R. Peng, Z.D. Hood, M. Naguib, S.P. Adhikari, Z. Wu, Titania composites with 2 D transition metal carbides as photocatalysts for hydrogen production under visible-light irradiation, ChemSusChem, 9 (2016) 1490–1497.

137. T. Wojciechowski, A. Rozmyslowska-Wojciechowska, G. Matyszczak, M. Wrzecionek, A. Olszyna, A. Peter, A. Mihaly-Cozmuta, C. Nicula, L. Mihaly-Cozmuta, S. Podsiadlo, D. Basiak, W. Ziemkowska, A. Jastrzebska, Ti_2C MXene modified with ceramic oxide and noble metal nanoparticles: synthesis, morphostructural properties, and high photocatalytic activity, Inorganic Chemistry, 58 (2019) 7602–7614.

138. E. Satheeshkumar, T. Makaryan, A. Melikyan, H. Minassian, Y. Gogotsi, M. Yoshimura, One-step solution processing of Ag, Au and Pd@MXene Hybrids for SERS, Science Reports, 6 (2016) 32049.

139. F. Wang, C. Yang, M. Duan, Y. Tang, J. Zhu, TiO_2 nanoparticle modified organ-like Ti_3C_2 MXene nanocomposite encapsulating hemoglobin for a mediator-free biosensor with excellent performances, Biosensors and Bioelectronics, 74 (2015) 1022–1028.

140. K. Rasool, R.P. Pandey, P.A. Rasheed, G.R. Berdiyorov, K.A. Mahmoud, MXenes for environmental and water treatment applications, 2D Metal Carbides and Nitrides (MXenes), Springer, 2019, pp. 417–444.

141. W. Qu, H. Zhao, Q. Zhang, D. Xia, Z. Tang, Q. Chen, C. He, D. Shu, Multifunctional Au/Ti_3C_2 photothermal membrane with antibacterial ability for stable and efficient solar water purification under the full spectrum, ACS Sustainable Chemistry & Engineering, 9 (2021) 11372–11387.

142. M. Ming, Y. Ren, M. Hu, Y. Zhang, T. Sun, Y. Ma, X. Li, W. Jiang, D. Gao, J. Bi, Promoted effect of alkalization on the catalytic performance of $Rh/alk-Ti_3C_2X_2$ (XO, F) for the hydrodechlorination of chlorophenols in base-free aqueous medium, Applied Catalysis B: Environmental, 210 (2017) 462–469.

143. X. Li, G. Fan, C. Zeng, Synthesis of ruthenium nanoparticles deposited on graphene-like transition metal carbide as an effective catalyst for the hydrolysis of sodium borohydride, International Journal of Hydrogen Energy, 39 (2014) 14927–14934.

144. X. Li, C. Zeng, G. Fan, Ultrafast hydrogen generation from the hydrolysis of ammonia borane catalyzed by highly efficient bimetallic RuNi nanoparticles stabilized on $Ti_3C_2X_2$ (X= OH and/or F), International Journal of Hydrogen Energy, 40 (2015) 3883–3891.

145. Z. Guo, T. Liu, Q. Wang, G. Gao, Construction of cost-effective bimetallic nanoparticles on titanium carbides as a superb catalyst for promoting hydrolysis of ammonia borane, RSC Advances, 8 (2018) 843–847.

146. W. Zheng, P. Zhang, W. Tian, Y. Wang, Y. Zhang, J. Chen, Z. Sun, Microwave-assisted synthesis of $SnO_2-Ti_3C_2$ nanocomposite for enhanced supercapacitive performance, Materials Letters, 209 (2017) 122–125.

147. Y. Gao, L. Wang, Z. Li, A. Zhou, Q. Hu, X. Cao, Preparation of MXene-Cu_2O nanocomposite and effect on thermal decomposition of ammonium perchlorate, Solid State Sciences, 35 (2014) 62–65.

148. Q. Xue, Z. Pei, Y. Huang, M. Zhu, Z. Tang, H. Li, Y. Huang, N. Li, H. Zhang, C. Zhi, Mn_3O_4 nanoparticles on layer-structured Ti_3C_2 MXene towards the oxygen reduction reaction and zinc–air batteries, Journal of Materials Chemistry A, 5 (2017) 20818–20823.

149. R.B. Rakhi, B. Ahmed, D. Anjum, H.N. Alshareef, Direct chemical synthesis of MnO_2 nanowhiskers on transition-metal carbide surfaces for supercapacitor applications, ACS Applied Materials & Interfaces, 8 (2016) 18806–18814.

150. X. Guo, X. Xie, S. Choi, Y. Zhao, H. Liu, C. Wang, S. Chang, G. Wang, Sb_2O_3/MXene ($Ti_3C_2T_x$) hybrid anode materials with enhanced performance for sodium-ion batteries, Journal of Materials Chemistry A, 5 (2017) 12445–12452.

151. Q.X. Xia, J. Fu, J.M. Yun, R.S. Mane, K.H. Kim, High volumetric energy density annealed-MXene-nickel oxide/MXene asymmetric supercapacitor, RSC Advances, 7 (2017) 11000–11011.

152. A. Rozmysłowska-Wojciechowska, E. Karwowska, S. Poźniak, T. Wojciechowski, L. Chlubny, A. Olszyna, W. Ziemkowska, A. Jastrzębska, Influence of modification of Ti_3C_2 MXene with ceramic oxide and noble metal nanoparticles on its antimicrobial properties and ecotoxicity towards selected algae and higher plants, RSC Advances, 9 (2019) 4092–4105.

153. S. Cao, B. Shen, T. Tong, J. Fu, J. Yu, 2D/2D heterojunction of ultrathin MXene/ Bi_2WO_6 nanosheets for improved photocatalytic CO_2 reduction, Advanced Functional Materials, 28 (2018) 1800136.

154. M. Shao, Y. Shao, J. Chai, Y. Qu, M. Yang, Z. Wang, M. Yang, W.F. Ip, C.T. Kwok, X. Shi, Synergistic effect of 2D Ti_2C and gC_3N_4 for efficient photocatalytic hydrogen production, Journal of Materials Chemistry A, 5 (2017) 16748–16756.

155. P.A. Rasheed, R.P. Pandey, K.A. Jabbar, K.A. Mahmoud, $Nb_4C_3T_x$ (MXene)/Au/DNA aptasensor for the ultraselective electrochemical detection of lead in water samples, Electroanalysis, (2022).

156. Z. Liu, H. Lin, M. Zhao, C. Dai, S. Zhang, W. Peng, Y. Chen, 2D superparamagnetic tantalum carbide composite MXenes for efficient breast-cancer theranostics, Theranostics, 8 (2018) 1648–1664.

157. J. Zou, J. Wu, Y. Wang, F. Deng, J. Jiang, Y. Zhang, S. Liu, N. Li, H. Zhang, J. Yu, Additive-mediated intercalation and surface modification of MXenes, Chemical Society Reviews, (2022).

158. M. Khazaei, A. Ranjbar, M. Arai, T. Sasaki, S. Yunoki, Electronic properties and applications of MXenes: a theoretical review, Journal of Materials Chemistry C, 5 (2017) 2488–2503.

159. R. Jiang, M. Liu, H. Huang, L. Mao, Q. Huang, Y. Wen, Q.-y. Cao, J. Tian, X. Zhang, Y. Wei, Ultrafast construction and biological imaging applications of AIE-active sodium alginate-based fluorescent polymeric nanoparticles through a one-pot microwave-assisted Döbner reaction, Dyes and Pigments, 153 (2018) 99–105.

160. L. Huang, S. Yang, J. Chen, J. Tian, Q. Huang, H. Huang, Y. Wen, F. Deng, X. Zhang, Y. Wei, A facile surface modification strategy for fabrication of fluorescent silica nanoparticles with the aggregation-induced emission dye through surface-initiated cationic ring opening polymerization, Materials Science and Engineering: C, 94 (2019) 270–278.

161. Q. Wan, Q. Huang, M. Liu, D. Xu, H. Huang, X. Zhang, Y. Wei, Aggregation-induced emission active luminescent polymeric nanoparticles: non-covalent fabrication methodologies and biomedical applications, Applied Materials Today, 9 (2017) 145–160.

162. P. Lazar, F. Karlicky, P. Jurecka, M.s. Kocman, E. Otyepková, K.r. Šafářová, M. Otyepka, Adsorption of small organic molecules on graphene, Journal of the American Chemical Society, 135 (2013) 6372–6377.

163. A. Schlierf, P. Samorì, V. Palermo, Graphene–organic composites for electronics: optical and electronic interactions in vacuum, liquids and thin solid films, Journal of Materials Chemistry C, 2 (2014) 3129–3143.

164. J.M. Englert, C. Dotzer, G. Yang, M. Schmid, C. Papp, J.M. Gottfried, H.-P. Steinrück, E. Spiecker, F. Hauke, A. Hirsch, Covalent bulk functionalization of graphene, Nature Chemistry, 3 (2011) 279–286.

165. T. Habib, X. Zhao, S.A. Shah, Y. Chen, W. Sun, H. An, J.L. Lutkenhaus, M. Radovic, M.J. Green, Oxidation stability of $Ti_3C_2T_x$ MXene nanosheets in solvents and composite films, npj 2D Materials and Applications, 3 (2019) 1–6.

166. Y. Xie, Y. Dall'Agnese, M. Naguib, Y. Gogotsi, M.W. Barsoum, H.L. Zhuang, P.R. Kent, Prediction and characterization of MXene nanosheet anodes for non-lithium-ion batteries, ACS Nano, 8 (2014) 9606–9615.

167. H. Lin, Y. Wang, S. Gao, Y. Chen, J. Shi, Theranostic 2D tantalum carbide (MXene), Advanced Materials, 30 (2018) 1703284, DOI: 10.1002/adma.201703284.

4 Solid-Solution MXenes and Their Properties

Nasima Khatun and Somnath C. Roy
Semiconducting Oxide Materials, Nanostructures,
and Tailored Heterojunction (SOMNaTH) Lab
Department of Physics and 2D Materials and Innovation Group
Indian Institute of Technology Madras
Chennai, Tamil Nadu, India

CONTENTS

4.1 AN INTRODUCTION TO MXenes AND THEIR SOLID SOLUTION

The discovery of graphene has opened a new era for the search for ultrathin-layered structured two-dimensional (2D) materials, which include transition metal dichalcogenides (TMDs), metal-organic frameworks (MOFs), transition metal halides (TMHs), hexagonal boron nitrides, phosphorene, silicene, and many more [1–6]. The excellent physical, chemical, and electronic properties make them useful for several applications. MXenes (transition metal carbides, nitrides, or carbonitrides) are recently added 2D material in this family. Since the invention of the first MXene in 2011 [7], the number has been increasing continuously, which is motivated by unusual physical and chemical properties such as high electrical conductivity

$(Ti_3C_2T_x - 20,000$ S cm^{-1}) [8], excellent hydrophilicity [7], high mechanical stiffness $(Nb_4C_3T_x - 386 \pm 13$ GPa) [9], and so on. Due to the coexistence of such properties, MXenes are already used in many fields, such as energy storage and conversion [10], electromagnetic interference (EMI) shielding [11], sensing [12, 13], catalysis [14], optoelectronics [15], biomedical applications [16, 17], and so on, and have shown promising performances. The general formula of MXene is $M_{n+1}X_nT_x$ ($n = 1$–4), where M stands for early transition metals, X stands for carbon (C) and/or nitrogen (N), and T_x stands for functional groups (-O, -F, -Cl, -OH, Br, etc.) on the surface of the outer transition metal layers [18–21]. MXenes are synthesized by selective removal of the A layer from $M_{n+1}AX_n$ (MAX) phase materials, where A is mostly a 13–16 group element [22, 23]; however, according to theoretical predictions, it can vary from group 8–16 [24]. At present, more than 155 individual MAX phases have already been synthesized, and after the discovery of the solid solution MAX phase, a large number of such phases have been predicted [24]. Moreover, all three sites (M, A, and X) can also be occupied randomly by more than one element without forming any impurity phase, resulting in solid solution MAX phases [25]. Similar to other MXenes, solid solution MXenes can be obtained by selective removal of weakly bonded A layer from the solid solution of MAX phases (M and X sites) [26]. In most cases, in situ and ex situ, hydrofluoric acid (HF) etching protocols followed by intercalation and delamination were used to prepare monolayer MXene flakes from their corresponding MAX phase (Figure 4.1).

Before further discussing solid solution MXenes, we would like first to present the general concept of the solid solution. In the book *Materials Science and Engineering: An Introduction*, Callister and Rethwisch mention that "A solid solution forms when, as the solute atoms are added to the host material, the crystal structure is maintained and no new structures are formed" [27]. A solid solution forms following four Hume-Rothery rules [28], which include the atomic size of solute elements, crystal structure, electronegativity of the elements, and valences [27]. Depending on these four parameters, two types of solid solutions are possible: substitutional and interstitial. In a solid solution, the solute atoms are uniformly distributed in the parent matrix to form a compositionally homogeneous matter.

Sokol et al. [24] have presented a total of 46 single-site (either M or X) solid-solution MAX phases that have already been synthesized; among these 31 are M and X site solid solutions. To date, about 20 solid-solution MXenes have been synthesized (Table 4.1), and many more are theoretically predicted [29]. There are a total of 14 elements for the M site, such as Sc, Ti, V, Cr, Zr, Nb, Mo, Hf, Ta, and so on [24], and

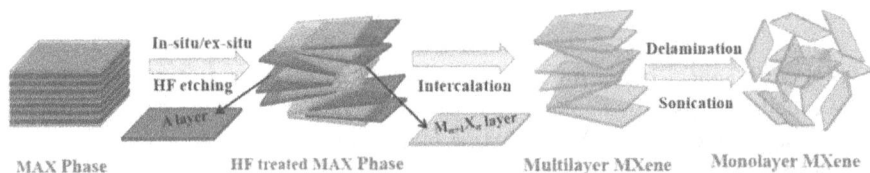

MAX Phase HF treated MAX Phase Multilayer MXene Monolayer MXene

In-situ/ex-situ HF etching A layer Intercalation $M_{n+1}X_n$ layer Delamination Sonication

FIGURE 4.1 A schematic to prepare monolayer MXene from MAX phase by in situ/ex situ HF treatment followed by intercalation and delamination.

TABLE 4.1

A List of Solid Solutions of MXenes Experimentally Synthesized to Date

$(M',M'')_{n+1}(X',X'')_n$ $(n = 1-4)$	M Site	X Site
$n = 1$ (21)	$(Ti_{0.5}Nb_{0.5})_2C$ [30], (NbTi)C [31], $(V_yTi_{1-y})_2C$ ($y = 1, 0.7, 0.5, 0.3, 0$) [32], (TiV)C [33], [34] $(Ti_{2-y}Nb_y)C$, $(Ti_{2-y}V_y)C$, and $(V_{2-y}Nb_y)C$ ($y = 0.4, 0.8, 1.2,$ and 1.6) [29], [35], [36], $(Ti_{2-y}Nb_y)C$ ($y = 0.4, 0.8, 1.2,$ and 1.6) [37], $(Ti_{1/5}V_{1/5}Zr_{1/5}Nb_{1/5}Ta_{1/5})_2C$ [38],	$(Ti_{1/3}V_{1/6}Zr_{1/6}Nb_{1/6}Ta_{1/6})_2C_yN_{1-y}$ [22]
$n = 3$ (32)	$(V_{0.5}Cr_{0.5})_3C_2$ [30], $(Ta_{1-y}Ti_y)_3C_2$ ($y = 0.4, 0.62, 0.75, 0.91,$ or 0.95) [25], $(Ti_{0.5}V_{0.5})_3C_2$ [14], [39], $(Ti_2V_{0.9}Cr_{0.1})C_2$ [40], [41],	$Ti_3(C_{0.5}N_{0.5})_2$ [30], $Ti_3C_{1.8}N_{0.2}$, and $Ti_3C_{1.6}N_{0.4}$ [42], Ti_3CN [43], [36],
$n = 3$ (43)	$(Nb_{0.8}Ti_{0.2})_4C_3$ and $(Nb_{0.8}Zr_{0.2})_4C_3$ [44], $(Mo_yV_{4-y})C_3$ ($y = 1, 1.5, 2,$ and 2.7) [45], (TiVNbMo)C_3 and (TiVCrMo)C_3 [46],	-
$n = 4$ (54)	$(Mo_4V)C_4$ [47]	-

All the solid solution MXenes presented here have Al elements for 'A' layers. T_x stands for surface functional groups but is not shown in the table for simplicity.

the X site can be occupied by a combination of C and/or N, which leads to a large number of compositions, and opens an opportunity to obtain a variety of new solid-solution MXenes.

In Table 4.1, it is observed that for all the solid solution MAX phases, 'A' layers are composed of Al atoms, and the M–Al bond is metallic, hence it is easy to selectively remove the Al layers and convert the MAX phases into MXenes.

The heterogeneity of solid-solution MXenes provides an immense scope to modulate their properties by controlling the elemental composition. The wide elemental diversity of both the M and X sites results in plenty of structures and properties, which can be useful for many applications.

4.2 STRUCTURE, COMPOSITIONS, AND PROCESSING

MXenes show hexagonal crystal structures that have space group $P6_3/mmc$ symmetry, which comes from their parent MAX phases [48]. In the basal planes of the MAX phase, M elements are arranged in a nearly closed-packed (hcp) crystal structure, with X elements in the octahedral interstitial sites between the M element planes [18]. A schematic of all types of MXenes is provided in Figure 4.2. The first row shows that MXenes after etching are terminated by functional groups. The first column displays different structured MXenes that have only one transition metal,

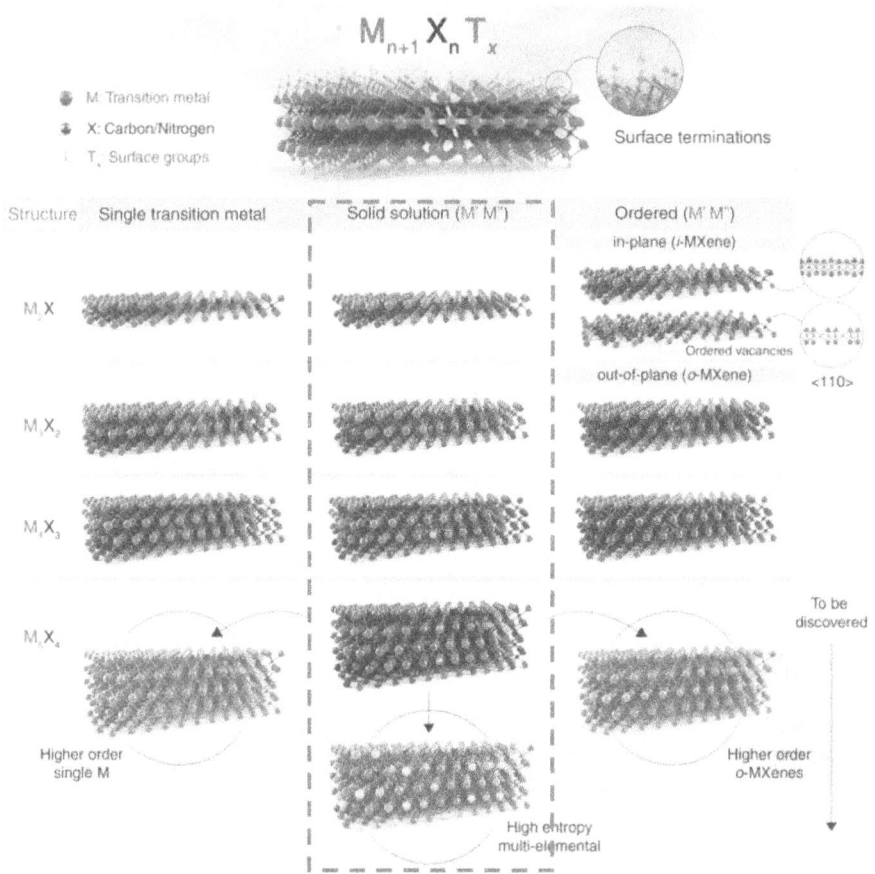

FIGURE 4.2 The top row displays MXenes after etching terminated by functional groups. The left column shows different types of MXene structures and their compositions depending on the n values from 1 to 4. The middle column shows solid-solution MXenes that have double and multiple transition metals for only M sites. The right column displays MXenes containing double transition metals (DMTs) for different n values. It shows two types of structures: in-plane (ordered vacancies) and out-of-plane DTM-ordered structures. Here, Tx is the surface functional group.

(Reproduced with permission from Ref. [18], © (2021) The American Association for the Advancement of Science.)

with a maximum n value of 4. The second column shows solid solution MXenes that have more than one transition element for M sites only. However, X sites can also be occupied by C and/or N. The third column shows MXenes that have double transition metals (DTMs). It shows two types of crystal structures: in-plane and out-of-plane ordered structures.

To date, $n = 4$ is the highest value of n for MXenes investigated by both computational and experimental approaches. Depending on the n values, four groups (21, 32, 43, and 54) of solid-solution MXenes have already been synthesized, which include both M and X sites. Further, depending on the M and X sites, the formula of solid-solution MXenes becomes $(M',M'')_{n+1}X_nT_x$, $M_{n+1}(X',X'')_nT_x$, and $(M',M'')_{n+1}(X',X'')_nT_x$ (both M'' and X'' can have more than one element).

Next, we present each group of solid-solution MXenes depending on the n value, which has been experimentally synthesized (for simplicity, we did not use T_x for all the MXenes in the rest of our discussions).

4.2.1 21 GROUP SOLID-SOLUTION MXenes

The 21 group MXenes are the thinnest, where one X layer is sandwiched between the two consecutive M layers, and the layered arrangements appear as an ABABAB sequence. The first solid solution MXene (($Ti_{0.5}Nb_{0.5})_2C$) of this group was synthesized by Naguib et al. [30] in 2012. $(Ti_{0.5}Nb_{0.5})_2AlC$ MAX phase was dispersed in 50% aqueous HF solution for 28 h at room temperature (RT) to obtain $(Ti_{0.5}Nb_{0.5})_2C$ solid-solution MXene. Here it is important to note that the etching condition was different than that of Ti_2AlC (10% HF, 10 h, and RT) and Nb_2AlC (50% HF, 90 h, and RT); the single M MAX phases [30, 49]. This is because the bond strength between Ti–Al and Nb–Al is different. Wang et al. [32] have shown that the addition of Ti in V_2AlC drastically reduces the etching time (V_2AlC – 48 h, ($V_{0.7}Ti_{0.3})_2AlC$ – 36 h, ($V_{0.5}Ti_{0.5})_2AlC$ – 24 h, ($V_{0.3}Ti_{0.7})_2AlC$ – 5 h, and Ti_2AlC – 1 h) to prepare $(V_yTi_{1-y})_2C$ solid solution MXenes. This happened because the bond strength of Ti–Al is much lower compared to that of V–Al [50]. In subsequent work, Yazdanparast et al. [33] reported the synthesis of the same solid solution MXene (TiV)C by in situ (1.9 M Lif/12HCl) and ex situ (50% HF) HF etching processes at RT and have shown the effect of synthesis condition on the quality of MXene. Han et al. [35] have also shown a systematic study of etching conditions (solution types, their concentration, temperature, and time) to prepare $(Ti_{2-y}Nb_y)C$, $(Ti_{2-y}V_y)C$, and $(V_{2-y}Nb_y)C$ solid solution MXenes for $y = 0.4$, 0.8, 1.2, and 1.6.

However, in solid-solution MXene, the M site can be occupied by multiple elements from groups 13–16. Till the year 2020, only DMT solid-solution MXenes were synthesized [26, 51]. Recently (2021), two 21 groups of solid-solution MXenes $(Ti_{1/5}V_{1/5}Zr_{1/5}Nb_{1/5}Ta_{1/5})_2C$ and $(Ti_{1/3}V_{1/6}Zr_{1/6}Nb_{1/6}Ta_{1/6})_2C_yN_{1-y}$ were synthesized for the first time by Du et al. [22, 38], where M sites are occupied by five elements, such as Ti, V, Zr, Nb, and Ta, and X sites are occupied by C and CN. These are called high entropy (HE) solid-solution MXenes.

In general, a high entropy material (HEM) can be defined in two ways – based on composition and on entropy. When five or more principal elements in a single phase persist with an equimolar or near to equimolar ratio having a concentration of each element between 5 to 35 at.%, the resultant material is called HEM [52, 53]. On the other hand, according to entropy theory, when all elements in a single phase are present in an equimolar ratio, then the maximum molar configurational entropy, $S = R \, lnN$, where R is the gas constant and N is the number of elements [54]. For five equimolar elements, $S = 1.61R$, hence the material is called HEM when $S \geq 1.61R$ [53]. The HEMs are required not only to be entropy stabilized but also to be enthalpy stabilized as well [54]. However, the HEMs are not always restricted by these rules and the approach depends on the targeted application [53, 55, 56].

Initially, a DFT calculation was performed before preparing $(Ti_{1/5}V_{1/5}Zr_{1/5}Nb_{1/5}Ta_{1/5})_2AlC$ HE solid-solution MAX phase, to see how formation enthalpies change with changes in the number of transition metals [38]. It was observed that with an increasing number of transition metals, the corresponding enthalpy per unit cell decreases (Ti_2AlC: -0.55 eV, $TiNbAlC$: -0.61 eV, $(Ti_{1/4}Zr_{1/4}Nb_{1/4}Ta_{1/4})_2AlC$: -0.80 eV, and $(Ti_{1/5}V_{1/5}Zr_{1/5}Nb_{1/5}Ta_{1/5})_2AlC$: -1.12 eV) and shows the lowest value for five elements. Hence, the HE $(Ti_{1/5}V_{1/5}Zr_{1/5}Nb_{1/5}Ta_{1/5})_2AlC$ MAX phase is considered enthalpy established as well. Similar to other MXenes, the HE $(Ti_{1/5}V_{1/5}Zr_{1/5}Nb_{1/5}Ta_{1/5})_2AlC$ MXene was synthesized by selective removal of the Al layer by an in situ HF process (LiF/HCl). The amount of Ti, V, Zr, Nb, and Ta concentration calculated on the above MXenes were ~20 at %, which satisfied the compositional definition of HEMs (5–35 at.%).

Du et al. [22] reported another 21 groups of solid solution MXene $(Ti_{1/3}V_{1/6}Zr_{1/6}Nb_{1/6}Ta_{1/6})_2C_yN_{1-y}$ containing the same transition elements but with slightly different molar ratio, and X site occupied by both C and N. They have performed DFT calculations on four MAX phase materials and shown that HE MAX phase had the lowest Gibbs free energy compared to their medium and lower entropy counterparts. To prepare HE $(Ti_{1/3}V_{1/6}Zr_{1/6}Nb_{1/6}Ta_{1/6})_2AlC_yN_{1-y}$ MAX phase, they used one medium entropy MAX phase $(Zr_{1/3}Nb_{1/3}Ta_{1/3})_2AlC$ and two mono-M MAX phase materials (Ti_4AlN_3 and V_2AlC) to solve the phase segregation issue at the time of formation. The concentration of each transition element in the MAX phase (Ti -34.1, V -16.3, Zr -16.2, Nb -16.7, and Ta -16.7 at.%) confirmed the HE feature. The $(Ti_{1/3}V_{1/6}Zr_{1/6}Nb_{1/6}Ta_{1/6})_2C_yN_{1-y}$ MXene was synthesized by using a similar in situ HF etching protocol as used for only M site solid solution [38] and the concentration of transition elements was Ti -28.3, V -19.6, Zr -8.7, Nb -19.6, and Ta -23.8 at.%, which showed HE feature. However, the concentration of Zr in MXene considerably decreased correspondingly to that of the MAX phase due to the easy dissolution of Zr in acid solution. The same research group also reported solid-solution MXenes having six elements (Ti, V, Zr, Nb, Hf, and Ta) on the M site [38].

It is proposed that HE solid-solution MXenes manifest promising physicochemical features due to their high mechanical strength, extraordinary electrochemical behavior, and so on, which are useful for energy storage and conversion applications [57, 58]. But the synthesis of HE solid-solution MXenes is still limited due to the formation of separated phases at the time of MAX phase preparation [22].

There are other solid-solution MAX phases such as $(Ti_{0.9}Mo_{0.1})AlC$ [59], $(Cr_{1-y}V_y)_2AlC$ ($y=0.5, 1, 1.5$) [60], $(Nb_{1-y}Zr_y)_2AlC$ ($0 < y < 1$) [61, 62], and $(Nb_{2/3}Sc_{1/3})_2AlC$, which have already been synthesized. As Zr, Cr, Mo, and Sc are easily dissolved in acid solution, suitable synthesis routes are required to prepare their derivative MXenes. However, it's expected that the synthesis challenges will be overcome and a large number of solid-solution MXenes can be synthesized with tailored physical and chemical properties, opening a way for their widespread applications.

4.2.2   32 GROUP SOLID-SOLUTION MXenes

The MXene family started from the 32 group (Ti_3C_2), which is mono-transition metal MXene. As shown in Figure 4.2, there are three M layers, and two X layers consecutively sandwiched between the M layers. The first synthesized double transition metal (DTM) solid solution MXene of 32 groups is $(V_{0.5}Cr_{0.5})_3C_2$, where the M site is occupied by V and Cr in the equimolar ratio [30]. However, complete removal of Al layers from its MAX phase $((V_{0.5}Cr_{0.5})_3AlC_2)$ remains a challenge, and partial removal of Al layers is achieved by treating in 50% aqueous HF solution for 69 h at RT. Hence, there is a wide scope for the investigation to find a better etching protocol in terms of suitable reagents, concentration, and temperature, which not only will lead to the complete removal of the Al layers but will also be efficient on the time scale. Subsequently, synthesis of the solid-solution MXene $Ti_3(C_{0.5}N_{0.5})_2$, where X site is occupied by equimolar C and N from its MAX phase $Ti_3Al(C_{0.5}N_{0.5})_2$, using 30% aqueous HF for 18 h at RT has been reported [30]. Recently, solid-solution MXene containing a different molar ratio of C and N, such as $Ti_3C_{1.8}N_{0.2}$, and $Ti_3C_{1.6}N_{0.4}$ [42], have been synthesized. As far as the DTM solid-solution MXenes are concerned, Shen et al. [14] successfully synthesized $(Ti_{0.5}V_{0.5})_3C_2$ from its MAX phase $(Ti_{0.5}V_{0.5})_3C_2$ by using 48% aqueous HF solution for 72 h at RT. Recently, Rigby-Bell et al. [25] have synthesized another DTM solid solution $(Ta_{0.38}Ti_{0.62})_3C_2$ with an almost negligible impurity phase. To date, only a few 32 DTM solid solutions are synthesized, and solid-solution MXene containing a higher number of transition metals and high entropy MXene remains to be achieved.

4.2.3 43 GROUP SOLID-SOLUTION MXenes

The 43 group family of MXenes began with the discovery of mono-transition metal MXene (Ta_4C_3) by Naguib et al. [30] in 2012. Subsequently, in 2016, two DTM solid-solution MXenes $((Nb_{0.8}Ti_{0.2})_4C_3$ and $(Nb_{0.8}Zr_{0.2})_4C_3)$ of 43 groups were synthesized by Yang et al. [44] from the parent MAX phases $(Nb_{0.8}Ti_{0.2})_4AlC_3$ and $(Nb_{0.8}Zr_{0.2})_4AlC_3$. In 2020, another bimetallic solid solution $((Mo_yV_{4-y})C_3)$ was synthesized with $y = 1, 1.5, 2$, and 2.7 from their MAX phase $(Mo_yV_{4-y})AlC_3$. Recently, in 2021, two solid-solution MXenes, $(TiVNbMo)C_3$ and $(TiVCrMo)C_3$, were synthesized from their MAX phase $(TiVNbMo)AlC_3$ and $(TiVCrMo)AlC_3$, where the M site is occupied by four transition metals (Figure 4.3). These are the first HE MXenes.

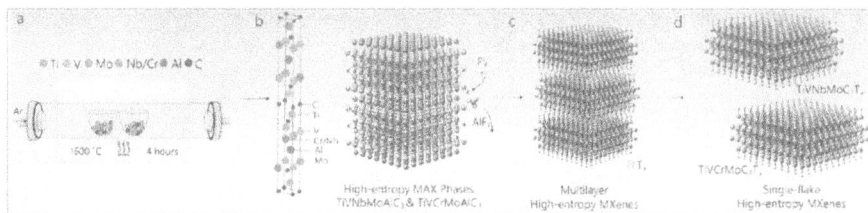

FIGURE 4.3 (a) Schematic of synthesis process of HE $(TiVNbMo)AlC_3$ and $(TiVCrMo)$ AlC_3 MAX phases, (b) distribution of atoms on unit cell for the corresponding HE-MAX phases, (c) multilayer HE MXenes with surface functional groups, and (d) single-flake $(TiVNbMo)C_3$ and $(TiVCrMo)C_3$ MXenes after delamination by organic molecules.

However, the synthesis needed one of the harshest etching conditions (48% HF for 96 h at 55°C) that has been implemented to date [46]. They also have shown that the concentration of Cr and Mo is slightly less compared to the other two transition metals, as they tend to dissolve in acid solution. Also, Cr and Mo prefer to occupy the outer transition metal atomic layers as observed in the ordered DTM MAX phases and MXenes [63–66]. Hence, HF etching directly affects these metals, with a higher removal rate and thereby creating vacancies in the outer M layers.

4.2.4 54 GROUP SOLID-SOLUTION MXenes

The first and only 54-family solid solution MXene was discovered in 2019 ((Mo_4V) C_4) [47]. In these types of MXenes, four C and/or N layers are consecutively present in between five transition metal atomic layers. Deysher et al. [47], using DFT calculations, have shown that the solid solution $(Mo_4V)AlC_4$ MAX phase shows the lowest formation energy per unit cell compared to their ordered structure (Figure 4.4). $(Mo_4V)C_4$ is a special type of MXene, where the center Mo/V layer is a twin plane and forms a herringbone-type structure. This structure is unique and uncommon. However, several other DTM and HE solid-solution MXenes from this group are expected soon.

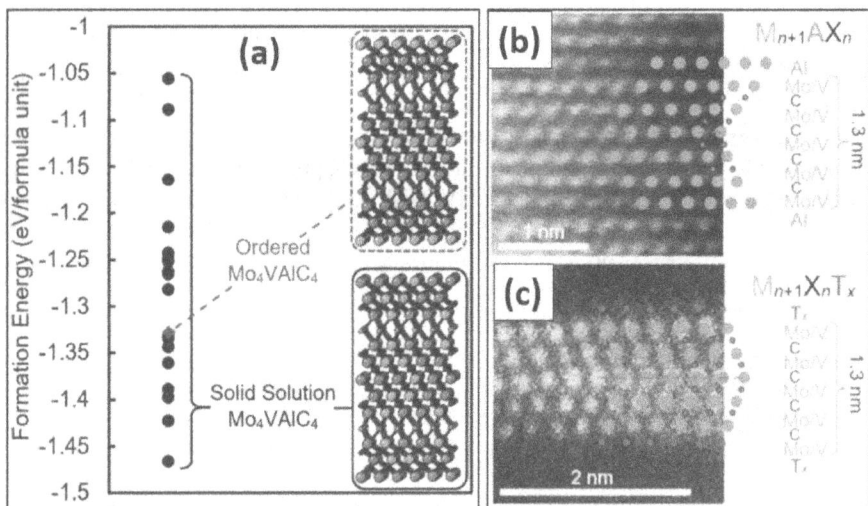

FIGURE 4.4   (a) Shows the formation energy of both solid solution (black dots) and ordered (grey diamond) $(Mo_4V)AlC_4$ MAX phases respectively. (b) STEM image of $(Mo_4V)AlC_4$ MAX phase shows the layered structure and the Mo/V, Al, and C atoms are represented by solid circles on the right side. (c) STEM image of $(Mo_4V)C_4$ MXene and a herringbone-type layered structure is observed. The atomic distribution by solid circles shows a mirror plane in the structure.

(Reproduced with permission from Ref. [47], © (2020) American Chemical Society.)

4.3 SURFACE CHEMISTRY

When 'A' layers are selectively removed from the MAX phase to prepare MXene, the outer transition metal layers are fully terminated by different types of surface functional groups, such as O, OH, F, Cl, and so on [33, 67]. This is because MXenes show large negative formation energy when attached to functional groups on their surfaces (Figure 4.5) [68]. This also signifies that functional groups are bonded strongly with the transition metals and stabilize the MXenes. The stabilization of functional groups on the surface of MXenes is further confirmed by positive phonon frequencies [69]. The types of functional groups are dependent on the synthesis process, which includes types of M atoms, etching reagents, washing protocols, post-annealing treatments, and its storing condition. The surface functional groups have three feasible sites to be attached: (i) on top of M; (ii) hollow site ① with adjacent C, and/or N atoms below M (metals); and (iii) hollow site ② on top of C, and/or N [48, 70]. The site ① is the most stable due to the low level of retardation between M and the functional group [71]. The site ② becomes energetically more favored when M cannot transfer sufficient charges to both C and/or N and surface functional groups. The O functional group needs two electrons to stabilize locally; however, OH and F functional groups require one electron from surface M. In the case of MXenes with low valence M, O functional groups can occupy both the positions ② or mixed ① and ② [48].

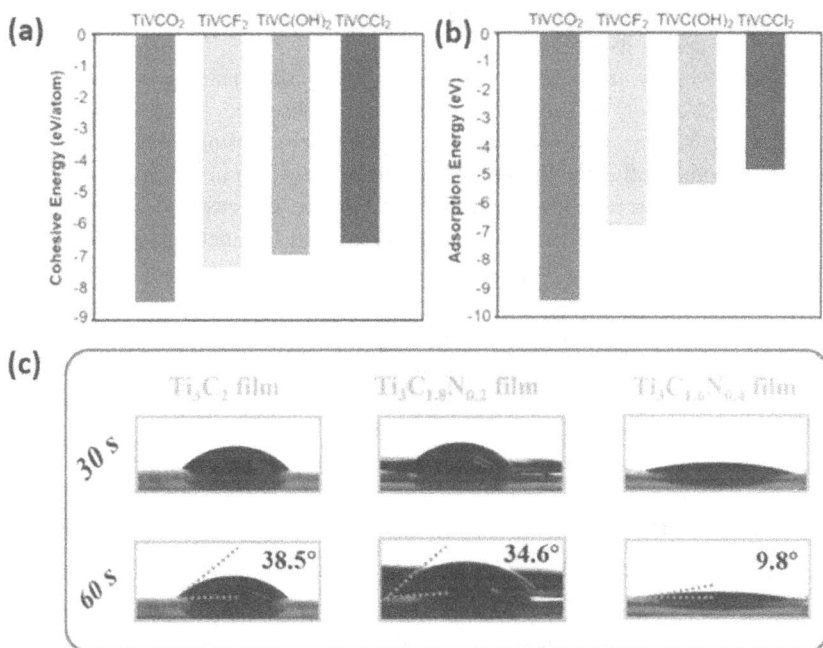

FIGURE 4.5 (a) and (b) DFT calculated cohesive energies and adsorption energies for TiVC MXene functionalized by O, F, OH, and Cl. (Reproduced with permission from Ref. [33], © (2020) American Chemical Society.) (c) shows the contact angle of 1 M KOH electrolyte with Ti_3C_2, $Ti_3C_{1.8}N_{0.2}$, and $Ti_3C_{1.6}N_{0.4}$ MXenes films and reveals that film becomes hydrophilic when N concentration increases [42].

Moreover, the MXenes are not only functionalized with only one but also with many functional groups together [72]. Similar to MXenes, solid-solution MXenes are also terminated by many functional groups, such as O, OH, F, Cl, and so on [30, 33, 43]. The functional groups are characterized by X-ray photoelectron spectroscopy (XPS), scanning transmission electron microscopy (STEM), Fourier transforms infrared spectroscopy (FTIR), neutron scattering, nuclear magnetic resonance spectroscopy (NMR), electron energy loss spectroscopy (EELS), Raman spectroscopy, and so on [35, 37, 38, 43]. Yazdanparast et al. [33] performed DFT calculations to show that for (TiV)C solid-solution MXene, O functional group is more favorable when compared with F, OH, Cl, and their combination (Figure 4.5 (a and b)). XPS studies have shown that O concentration is higher (~34%) in (TiV)C when synthesized by LiF/HCl compared to only HF-synthesized MXene (~23%), which is also true for Ti_3C_2 and V_2C [73, 74]. It has been observed that transition metals also have the ability to control the comparative ratio of functional groups attached to the MXene surface. For example, F concentration on Mo-based MXenes (Mo_2C, Mo_4VC) is lower compared to that of on the Ti-based MXene [21, 47, 75]. In spite of the availability of several characterization techniques, surface functional groups of solid solution MXenes are poorly understood. This is because the solid-solution MXene surfaces are highly complex and sensitive to the nature of transition metals. Even in the case of HE solid solution, five or more transition metals are there, where each element contains different valence electrons and acts differently to stabilize the functional groups on its surfaces.

More importantly, these functional groups highly dominate the MXene properties. In the case of bare MXenes, metallic nature is observed, but functional groups transform the MXene into semi-metallic and semiconducting. Further, the functional groups also influence the nature of optical band gap and surface wettability (hydrophilic or hydrophobic) (Figure 4.5(c)) [30]. In the case of electrochemical energy-storage applications, it has been reported that O functional group helps to enhance the energy-storage capacity, whereas F and OH groups prevent the ion transfer and thereby decrease the capacity [33].

It is therefore understood that controlling over MXenes' functional groups is absolutely essential because most of the properties are affected by these functional groups, such as optoelectronic, mechanical, magnetic, and so on, and a detailed dissection is provided later.

4.4 PROPERTIES

Applications of any material depend on its physical and chemical properties, which applies to solid-solution MXenes as well. Properties of MXenes depend on their crystal structure, composition, interlayer spacing, and surface chemistry. The unusual and unique physical and chemical properties are showing promising performances in many applications, which we have previously mentioned. In the case of solid-solution MXenes, the heterogeneity in the compositions and a large number of M elements in a single $M_{n+1}X_n$ slab open up a new avenue for their expected widespread applications. Here we have discussed the properties of solid-solution MXenes.

4.4.1 ELECTRONIC TRANSPORT PROPERTIES

Like MXenes, their solid solutions without surface functional groups (bare) exhibit metallic conductivity driven by the free electrons from transition metals [18]. It is predicted that some MXenes, with functional groups such as Ti_2C, Zr_2C, Cr_2C, Cr_2TiC, Mo_2ZrC_2, Hf_2VC_2, and so on behave like semiconductors [68, 69, 76–78]. This is attributed to the transfer of an electron from surface M to electronegative functional groups, which causes a significant shift in the density of state (DOS) at the Fermi level of the surface M [68, 79, 80], which, in turn, leads to the band-gap opening and initiates semiconducting behavior. Different functional groups affect the electronic structure differently, because the number of electrons received from surface transition metals is different; for example, -OH and -F receive one electron, whereas O receives two electrons [68]. It is theoretically established that the band gap can be tuned from 2.45 to 1.15 eV in $Hf_{2-2y}Ti_{2y}CO_2$ solid-solution MXene depending on the Ti concentration [81]. However, experimental verification is yet to be obtained. It is predicted that mechanical strain can change the electronic band structure of MXenes, which needs experimental confirmation.

Due to the high conductivity of solid-solution MXenes, these are already used in a variety of fields such as supercapacitors, EMI shielding, electrode materials for Li-ion batteries, and so on (Figure 4.6(b and d)) [29, 32, 36, 38, 44]. The conductivity of solid-solution MXenes varies from <1 S/cm to >1000 S/cm (Figure 4.6(c)) [34, 41, 42, 45]. Such a large variation in the conductivity is affected by the synthesis conditions, intercalation/delamination, and so on. Han et al. [35] have calculated the electrical conductivity of $(Ti_{2-y}Nb_y)C$, $(Ti_{2-y}V_y)C$, and $(V_{2-y}Nb_y)C$ solid-solution MXenes by both DFT technique and experimental measurements (on vacuum-filtered films). However, the relationship between experimentally obtained conductivity is yet to match the trend predicted by DFT calculations. In the case of $(Ti_{2-y}Nb_y)C$ and $(V_{2-y}Nb_y)C$, the electrical conductivity decreases with the increasing concentration of Nb. However $(Ti_{2-y}Nb_y)C$ and $(Ti_{2-y}V_y)C$ MXenes show higher conductivity at lower y values. Moreover, theirmMono-M MXenes have relatively high electrical conductivity (>1000 S/cm)) as observed for Ti_2C and V_2C, whereas lower values are obtained for Nb_2C.

In another report, He et al. have shown that the electrical conductivity of TiVC discs was 1112 S/cm, whereas the theoretical conductivity of functionalized TiVC is higher than the functionalized V_2C and Ti_2C [34]. A significant difference between the experimental and theoretical calculated conductivity trend of solid-solution MXenes suggests a more complex phenomenon responsible for it. In the case of DFT calculations, only bare MXenes (or single functional group terminated) are considered, but, in practice, there are no MXenes without surface functional groups (like -O, -OH, -F, etc.), which drastically affects the electrical conductivity. Also, inter-flake transport, flake size, and defects (mostly vacancies in the M and X cite) created during MAX phase synthesis or during etching protocol can change the conductivity of the MXene film. The electronic structure calculations suggest that the nitride MXene (TiCN) possesses higher conductivity than that of the carbides [43]. ME and HE MXenes also show higher electrical conductivity, and these are useful in energy storage and conversion applications [22, 38, 41]. Ma et al. [41] reported ultra-high

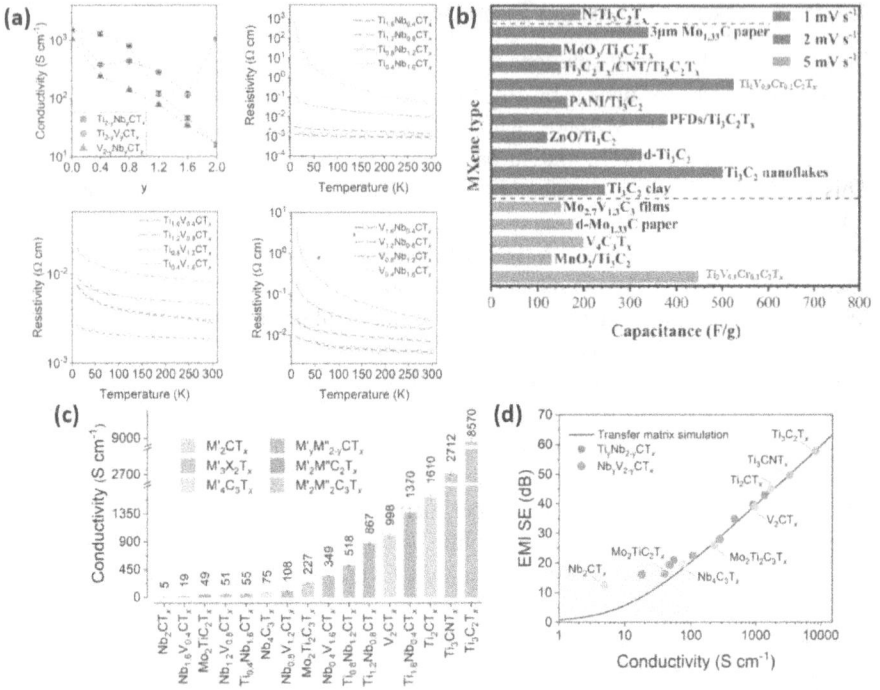

FIGURE 4.6 (a) The electrical conductivity (Ti$_{2-y}$Nby)C, (Ti$_{2-y}$Vy)C, and (V$_{2-y}$Nby)C MXenes with varying y concentration and resistivity vs. temperature plot from 0–300 K for the corresponding MXenes and shows a negative slope, which signifies the semiconducting behavior of the MXenes. (b) A comparison of the capacitance of Ti$_2$V$_{0.9}$Cr$_{0.1}$C$_2$ MXene with other MXenes and composite materials at different scan rates. (c) A comparison of electrical conductivity of solid solution MXenes including other MXenes on vacuum-filtration assisted films. (d) A plot between EMI shielding effectiveness and electrical conductivity of different MXenes on 5 μm thick films at 10 GHz and their comparison with transfer matrix simulation.

(Reproduced with permission from Ref. [35, 41], and [36], © (2020) American Chemical Society, © (2022) Elsevier, and © (2020), American Chemical Society, respectively.)

gravimetric capacitance ~553.27 F g^{-1} at 2 mV s^{-1} for ME Ti$_2$V$_{0.9}$Cr$_{0.1}$C$_2$ MXenes due to their high electrical conductivity and low resistance (8.12 Ω), which is even higher than the well-studied Ti$_3$C$_2$ MXenes (Figure 4.6(b)).

In most cases, a four-probe method has been used, but a more accurate value of conductivity can be obtained by the van der Pauw configuration. Compressed multilayer MXene discs, filtration-assisted free-standing films, and supported films are used for conductivity measurement, where conductivity is governed by intra- and interflake resistance. In the case of monolayer MXene flakes, conductivity solely depends on intraflake resistance and contact resistance. However in-depth studies on conductivity variation are yet to be reported.

The transport property of MXenes has also been quantified by resistivity measurements. The resistivity values are affected by the measurement and material configuration; for example, the resistivity of bulk sintered Ti$_3$AlC$_2$ is ~0.39 μΩ m, whereas its cold-pressed discs (by using a pressure of 1 GP) show a value of ~1200 $\mu\Omega$ m, implying an increase of ~3000 times [82]. In general, the resistivity of metals decreases with temperature $\left(\frac{d\rho}{dT} > 0\right)$ and increases for semiconductor $\left(\frac{d\rho}{dT} < 0\right)$. Despite the metallic nature, the resistivity measurement of some MXenes shows semiconductor-like temperature dependence (Figure 4.6 (a)). For example, $(Ti_{2-y}Nb_y)$ C, $(Ti_{2-y}V_y)C$, $(V_{2-y}Nb_y)C$, and Mo_4VC_4 solid-solution MXenes display negative $\frac{d\rho}{dT}$ [35, 47]. This may be attributed to interflake hopping of electrons and or charge localization at the disordered M-site of the MXenes flakes. From the same report, it was also observed that resistivity increased with increasing Nb concentration for both $(Ti_{2-y}Nb_y)C$ and $(V_{2-y}Nb_y)C$, while $(Ti_{2-y}V_y)C$ showed a slight increase in resistivity followed by a decrease and increase at higher V concentration.

The energy difference between the Fermi level and the vacuum level is called the work function. It is an important parameter that controls the direction of charge transfer and extraction of charges at the material-electrode junction. Depending on the work function value of MXenes, they can be used as transparent conductive electrodes in organic light-emitting and photonic diodes, thin-film transistors, logic circuits, and so on. Surface functional groups and compositions induce a large variation in the work function (2–8 eV) of MXenes [18]. To date, investigation on the work function of solid-solution MXenes has been limited, and only one solid-solution TiVC (work function -4.22 eV) MXene has been experimentally investigated [34]. However, the effects of variation in transition metals, their compositions, C, and/or N ratio, and surface functional groups on the work function are yet to be investigated.

It is therefore evident that an in-depth investigation is imperative to understand the role of structure, types, compositions, and the number of M and X elements and functional groups to determine the properties of solid-solution MXenes.

4.4.2 Optical Properties

Many applications are dependent on the optical properties of the materials. Similar to other MXenes, solid solutions also exhibit longitudinal and transverse surface plasmon modes ranging from the visible (vis) to the near-infrared (NIR) region [18, 23, 83]. The transverse modes are not dependent on the flake's size. Due to interband transitions, MXenes exhibit strong absorption in the ultraviolet (UV) range. Structure, types of transition metals and their compositions, carbon and/or nitrogen ratio, and functional groups play crucial roles in determining the optical properties of MXenes.

The color of any material qualitatively signifies the optical properties of that material. In the case of $(Ti_{2-y}Nb_y)C$, the color of colloidal solutions changes from bronze to gray, whereas for $(Ti_{2-y}V_y)C$, it is from purple to gray with a change in chemical composition (Figure 4.7) [35]. However, for $(V_{2-y}Nb_y)C$ there is a small change in color with compositional changes. For $(Ti_{2-y}Nb_y)C$ solid solution UV–vis spectroscopy shows one absorption peak in the visible and another in the NIR, which originate from two separate surface plasmon resonance. With increasing Ti concentration UV–vis peak becomes prominent and redshifted (at 391 nm for $Ti_{0.8}Nb_{1.2}C$ and

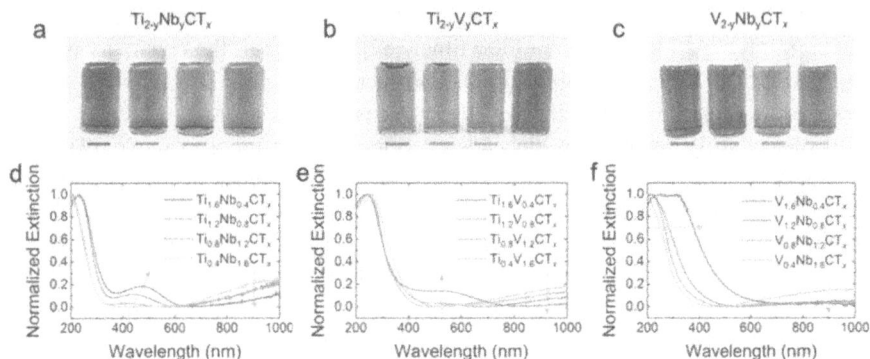

FIGURE 4.7 (a, b, and c) Digital photograph of $(Ti_{2-y}Nby)C$, $(Ti_{2-y}Vy)C$, and $(V_{2-y}Nby)C$ MXenes colloidal solution in deionized water shows a change in color with a change in compositions. (d, e, and f) show the UV–vis spectrum of the corresponding MXenes normalized to a maximum excitation of 1. It reveals multiple types of features with a nonlinear change from UV to NIR with a change in composition.

(Reproduced with permission from Ref. [35], © (2020) American Chemical Society.)

472 nm for $Ti_{1.6}Nb_{0.4}C$) and the NIR peak at ~915 nm vanishes. Similar phenomena are visible for $(Ti_{2-y}V_y)C$ and $(V_{2-y}Nb_y)C$, respectively.

Initial investigations on optical properties were limited to $n = 2$ [47]. In 2019, it was observed that solid-solution MXene $(Mo_4V)C_4$ with $n = 4$ displayed a featureless absorption spectrum in the vis-NIR region, indicating fundamental differences in light absorption with changes in the n value [47]. However, more studies are necessary to understand the relation between MXene atomic layer thickness and light-matter interaction.

4.4.3 PLASMONIC PROPERTIES

Due to the surface plasmon feature of MXenes, they were used as a substrate for surface-enhanced Raman scattering (SERS) [84–88]. In general, plasmons are collective electron oscillations resulting from light-matter interaction. Surface plasmons (SP) originate from the collective oscillations of free electrons that exist at a metal–semiconductor interface. Most of the SERS experiments were performed on mono-M MXenes using the SP effect. The literature review suggests only one report related to SERS on solid-solution MXene (TiVC) [34]. When a 532 nm laser light was irradiated on solid-solution TiVC MXene coated with rhodamine 6G dye (R6G) (1 μM), a femtomolar level of detection was observed with SERS enhancement factor of order $\sim 10^{12}$ (Figure 4.8) [34]. The abundant DOS near the Fermi level and the strong interaction between the TiVC–R6G is responsible for promoting the intermolecular charge transfer resonance in the TiVC–R6G complex, resulting in significant SERS enhancement. This opens up the possibility of using solid-solution MXenes for detecting even a single

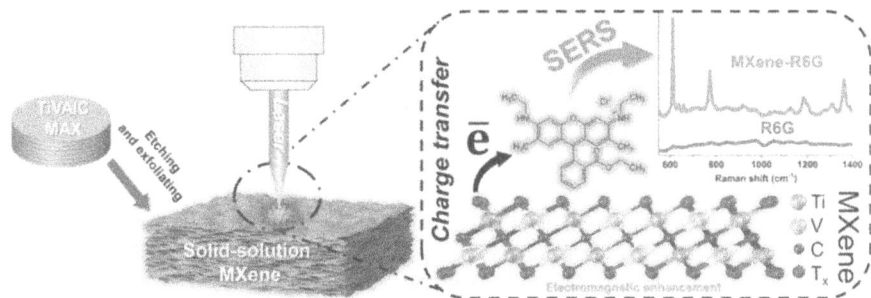

FIGURE 4.8 A schematic of producing TiVC MXenes from their parent MAX phase and then using these MXenes as a substrate for R6G SERS measurement using a leaser of 532 nm wavelength. Raman spectra show a huge enhancement in the intensity when MXene was used as a substrate for R6G.

(Reproduced with permission from Ref. [34], © (2022) American Chemical Society.)

molecule through SERS. Naturally, the SERS effect in various kinds of MXenes has a wide scope of investigation.

4.4.4 PHOTOTHERMAL PROPERTIES

Photothermal property (conversion of light to heat) of MXenes is a relatively recent development. When a colloidal solution of MXenes is irradiated with a laser, it absorbs light and converts it into heat energy. The light-to-heat conversion efficiency reaches ~100% for Ti_3C_2 colloidal solution [89]. Photothermal properties of MXenes have recently been used for some biomedical applications (photothermal therapy), such as for tumor treatment in mice, targeted treatment of cancer cells, and so on [17, 90, 91]. Solid-solution MXenes also exhibit a high absorption coefficient for a wide range of wavelength from UV–vis to NIR (Figure 4.7), which can be used for such applications. However, these aspects remain to be investigated.

4.4.5 MECHANICAL PROPERTIES

Recently, the mechanical properties of single-flake 2D materials have gained tremendous interest, because of the miniaturization of devices. The mechanical properties such as Young's modulus and fracture strength are measured by the nanoindentation technique (Figure 4.9). Individual MXene flakes are deposited on Si/SiO_2 substrate over open holes, and an atomic force microscope is used for the measurement [92]. The investigation of mechanical properties of MXenes started with $Ti_3C_2T_x$, and the maximum obtained Young's modulus for single flake $Ti_3C_2T_x$ MXene to date is ~0.33 TPa [92]. The value is still lower compared to that of theoretically predicted values (Ti_3C_2 – 0.50 TPa) [93] and could be attributed to the presence of functional groups and defects created on it [94]. Among all mono transition metal MXenes, Nb_4C_3 (single flake) shows the highest Young's modulus ~0.39 TPa [9], which is the

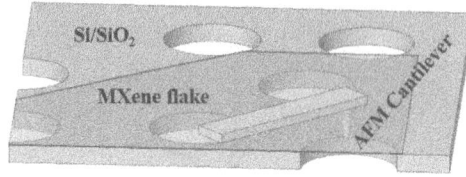

FIGURE 4.9 A schematic of nanoindentation technique using AFM cantilever tip to calculate Young's modulus and fracture strength. A monolayer of MXene flake is deposited on a pre-micro holes Si/SiO_2 substrate and a force-indention experiment is carried out for mechanical properties analysis.

best value among all the solution-processed materials. However, investigations of the mechanical properties of solid-solution MXenes are in nascent stages, and there has been no experimental report so far.

Computational results have shown that in HE MXenes (($Ti_{1/3}V_{1/6}Zr_{1/6}Nb_{1/6}Ta_{1/6})_2C_yN_{1-y}$ and ($Ti_{1/5}V_{1/5}Zr_{1/5}Nb_{1/5}Ta_{1/5})_2C$), five or more transition metals in a single M_2X slab create a distinct lattice distortion, which leads to a high mechanical strain in the atomic layers [22, 38]. The compressive and tensile strain in these HE MXenes arises due to the smaller size of Ti and V, and the bigger size of Zr, Nb, and Ta, respectively. It is expected that the experimental verification of these HE, along with other solid-solution MXenes, will be reported soon.

4.4.6 MAGNETIC PROPERTIES

Due to their unique structures and chemical compositions, many MXenes possess an intrinsic magnetism [95]. Some of them are already proven experimentally, and the magnetic properties of several other MXenes are theoretically predicted [1, 68, 77, 95–99]. The unpaired d-orbital electrons of transition metals are mainly responsible for ground-state intrinsic magnetism [95]. In most cases, ferromagnetic and antiferromagnetic behavior has been observed for MXenes [68, 77, 96, 98]. For example, in the case of nitride MXenes, it contains an extra electron compared to that of the carbide counterpart, which signifies more DOS at the Fermi level and supposes to show strong ferromagnetism. Similar to other properties, surface functional groups (O, OH, and F) have the potential to modify and change magnetic properties [97, 100]. Strain or impurities (doping and defects) also induces magnetism in non-magnetic MXenes and sometimes change their phase [101–104]. In solid-solution MXenes, a wide variety of transition metals, a combination of C, N, or CN together and the functional groups allow for the tailoring of the magnetic properties with a wide scope of parameters, which can be used for spintronics, quantum computation, ans so on.

Depending on all the properties, solid-solution MXenes are already used as an anode material in supercapacitors, EMI shielding, and catalysis (Figure 4.10). However, the solid-solution journey has just started, and we are expecting that the coexistence of their unusual properties has great potential for many applications.

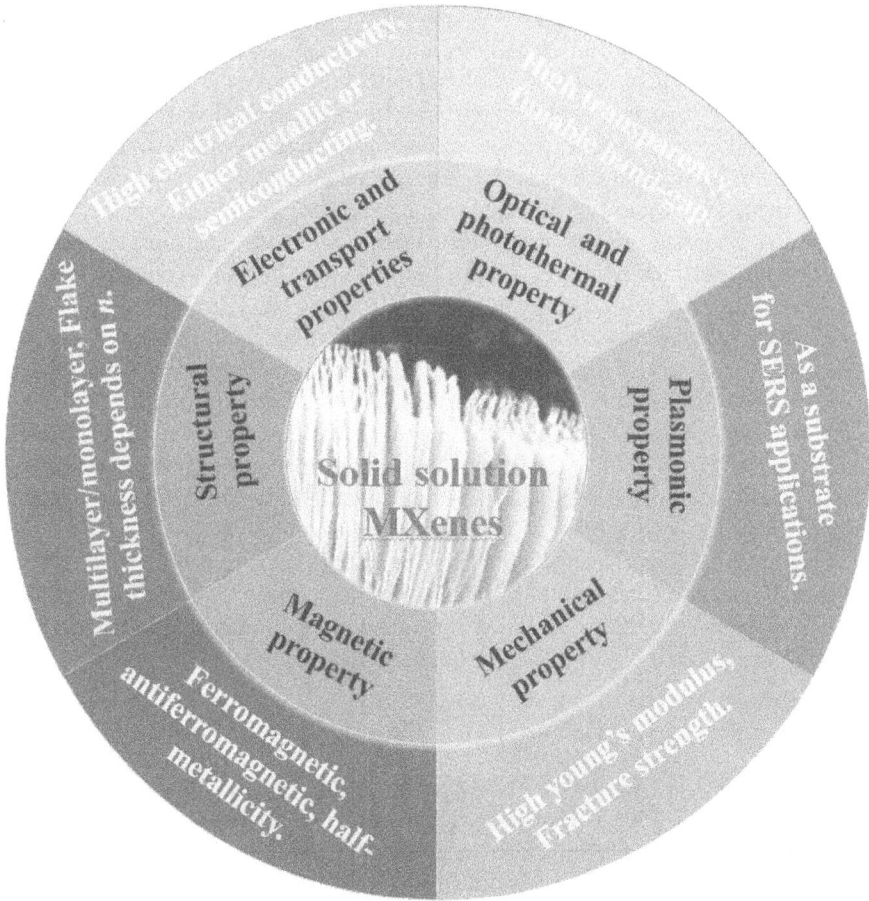

FIGURE 4.10 Coexistence of many unique and uncommon properties in solid-solution MXenes.

4.5 SUMMARY AND OUTLOOK

In this chapter, we have discussed the concept of solid-solution MXenes and how they are different from the other MXenes, including single M, DTM in-plane, and out-of-plane ordered structures. The occupancy of more than one element for both M and X sites in a single $M_{n+1}X_n$ slab provides heterogeneity in their compositions as well as bond strength between M, A, and X. This leads to a change in their synthesis conditions, which include etching agents, concentration, time, and temperature. Here, we have discussed the structure, compositions, and processing parameters of each solid-solution MXenes, depending on the n values, from 1 to 4. Functional groups terminated on surfaces of MXenes are inevitable for the synthesized MXenes, and they influence the properties. Also, the number of functional groups and their positions play an important role in forming stable MXenes, which is detailed in the surface chemistry section of this chapter. Finally, electronic and transport, optical, plasmonic, photothermal, mechanical, and magnetic properties are elaborated for both experimentally synthesized and theoretically predicted solid-solution MXenes. The coexistence of these unusual properties makes the solid solutions promising for many applications such as energy conversion and storage, EMI shielding, catalysis, antennas, and so on. However, further in-depth research on solid-solution MXenes is required to understand the structure–property relationship, phase stability, and so on., for successful fabrication and application to any specific field.

The 2D materials caught the world's attention due to their high surface-to-volume ratio, outstanding physical and chemical properties, and potential for low-power devices. However, high processing cost and low yield during synthesis limit commercialization prospects. On the other hand, MXenes can be fabricated cost-effectively along with a relatively higher yield. Their high electrical conductivity, wide variation in the work function, surface functional group governed hydrophilicity, and outstanding mechanical and photo-thermal properties are already used in a wide variety of applications, including supercapacitors, Li-ion/Na-ion batteries, room temperature gas sensors, biomedical investigation, water desalination, and environment purification. The wide variety of M elements, C, and/or N ratio, and their compositions, and especially the discovery of HE MXenes pave the way to finding many new solid solution MXenes and their potential applications. It is expected that the contents of this chapter will guide and serve as a reference for researchers planning to explore the field of solid-solution MXenes.

ACKNOWLEDGMENTS

Nasima Khatun acknowledges the Department of Physics, Indian Institute of Technology Madras for financial support under the Institute Postdoctoral Fellowship (IPDF) scheme. Nasima Khatun also thanks Sk. Md. Shaheed Hossain for helping in the literature survey and schematic drawing.

REFERENCES

1. Sethulakshmi, N., et al., *Magnetism in two-dimensional materials beyond graphene.* Mater. Today, 2019. **27**: p. 107–122.
2. Liu, X., et al., *Two-dimensional nanostructured materials for gas sensing.* Adv. Funct. Mater., 2017. **27**(37): p. 1702168.
3. Khan, K., et al., *Recent developments in emerging two-dimensional materials and their applications.* J. Mater. Chem. C, 2020. **8**(2): p. 387–440.
4. Di, J., et al., *Ultrathin two-dimensional materials for photo- and electrocatalytic hydrogen evolution.* Mater. Today, 2018. **21**(7): p. 749–770.
5. Khatun, N., et al., *Enhanced H_2 evolution through water splitting using TiO_2/ultrathin g-C_3N_4: A type II heterojunction photocatalyst fabricated by in situ thermal exfoliation.* Appl. Phy. Lett., 2021. **119**(9): p. 093901.
6. Nasima Khatun, S.D. and S.C. Roy, *Improved photoelectrochemical performance of ultra thin g-C_3N_4 nanosheet: A comparative study from bulk to nanoscale.* AIP Conf. Proc., 2021.
7. Naguib, M., et al., *Two-dimensional nanocrystals produced by exfoliation of Ti_3AlC_2.* Adv. Mater., 2011. **23**(37): p. 4248–4253.

8. Mathis, T.S., et al., *Modified MAX phase synthesis for environmentally stable and highly conductive Ti_3C_2 MXene.* ACS Nano, 2021. **15**(4): p. 6420–6429.

9. Lipatov, A., et al., *Electrical and elastic properties of individual single-layer $Nb_4C_3T_x$ MXene flakes.* Adv. Electron. Mater., 2020. **6**(4): p. 1901382.

10. Nan, J., et al., *Nanoengineering of 2D MXene-based materials for energy storage applications.* Small, 2019: p. 1902085.

11. Shahzad, F., et al., *Electromagnetic interference shielding with 2D transition metal carbides (MXenes).* Science, 2016. **353**(6304): p. 1137–1140.

12. Lee, S.H., et al., *Room-temperature, highly durable $Ti_3C_2T_x$ MXene/Graphene hybrid fibers for NH_3 gas sensing.* ACS Appl. Mater. Interfaces, 2020. **12**(9): p. 10434–10442.

13. Khatun, N., S. Rani, G. C. Behera and S. C. Roy, *Gas sensing of partially oxidized $Ti_3C_2T_x$ MXene in an argon atmosphere.* J. Mater. Nanosci., 2022. **9**(1): p. 74–78.

14. Shen, Z., et al., *A novel solid-solution MXene $(Ti_{0.5}V_{0.5})_3C_2$ with high catalytic activity for hydrogen storage in MgH_2.* Materialia, 2018. **1**: p. 114–120.

15. Cheng, J., et al., *Recent advances in optoelectronic devices based on 2D materials and their heterostructures.* Adv. Opt. Mater., 2019. **7**(1): p. 1800441.

16. Kurapati, R., et al., *Biomedical uses for 2D materials beyond graphene: current advances and challenges ahead.* Adv. Mater., 2016. **28**(29): p. 6052–6074.

17. Lin, H., et al., *Two-dimensional ultrathin MXene ceramic nanosheets for photothermal conversion.* Nano Lett., 2017. **17**(1): p. 384–391.

18. VahidMohammadi, A., J. Rosen and Y. Gogotsi, *The world of two-dimensional carbides and nitrides (MXenes).* Science, 2021. **372**(6547).

19. Kamysbayev, V., et al., *Covalent surface modifications and superconductivity of two-dimensional metal carbide MXenes.* Science, 2020. **369**(6506): p. 979–983.

20. Li, Y., et al., *A general Lewis acidic etching route for preparing MXenes with enhanced electrochemical performance in non-aqueous electrolyte.* Nat. Mater., 2020. **19**(8): p. 894–899.

21. Joseph Halim, K.M.C., M. Naguib, P. Eklund, Y. Gogotsi, J. Rosen and M. W. Barsoum, *X-ray photoelectron spectroscopy of select multi-layered transition metal carbides (MXenes).* Appl. Surface Sci., 2016. **362**: p. 406–417.

22. Du, Z., et al., *High-entropy carbonitride MAX phases and their derivative MXenes.* Adv. Energy Mater., 2021. **12**(6): p. 2103228.

23. Khatun, N., et al., *$Ti_3C_2T_x$ MXene functionalization induced enhancement of photoelectrochemical performance of TiO_2 nanotube arrays.* Mater. Chem. Phys., 2022. **278**: p. 125651.

24. Sokol, M., et al., *On the chemical diversity of the MAX phases.* Trends Chem., 2019. **1**(2): p. 210–223.

25. Rigby-Bell, M.T.P., et al., *Synthesis of new M-layer solid-solution 312 MAX phases $(Ta_{1-x}Ti_x)_3AlC_2$ (x = 0.4, 0.62, 0.75, 0.91 or 0.95), and their corresponding MXenes.* RSC Adv., 2021. **11**(5): p. 3110–3114.

26. Naguib, M., M.W. Barsoum and Y. Gogotsi, *Ten years of progress in the synthesis and development of MXenes.* Adv. Mater., 2021. **33**(39): p. e2103393.

27. Callister, W.D. and D.G. Rethwisch, Materials science and engineering: An introduction. 9th ed. 2013: Wiley New York.

28. Hume-Rothery, W. and B.R. Coles, Atomic theory for students of metallurgy. The Institute of Metals, 1 Carlton House Terrace, London SW 1 Y 5 DB, UK, 1988. **428**(1988).

29. Wang, L., et al., *Adjustable electrochemical properties of solid-solution MXenes.* Nano Energy, 2021. **88**: p. 106308.

30. Naguib, M., et al., *Two-dimensional transition metal carbides.* ACS Nano, 2012. **6**(2): p. 1322–1331.

31. Wang, Z., et al., *In situ formed ultrafine NbTi nanocrystals from a NbTiC solid-solution MXene for hydrogen storage in MgH₂*. J. Mater. Chem. A, 2019. **7**(23): p. 14244–14252.

32. Wang, Y., et al., *Preparation of (Vₓ, Ti₁₋ₓ)₂C MXenes and their performance as anode materials for LIBs*. J. Mater. Sci., 2019. **54**(18): p. 11991–11999.

33. Yazdanparast, S., et al., *Synthesis and surface chemistry of 2D TiVC solid-solution MXenes*. ACS Appl. Mater. Interfaces, 2020. **12**(17): p. 20129–20137.

34. He, Z., et al., *Two-dimensional TiVC solid-solution MXene as surface-enhanced Raman scattering substrate*. ACS Nano, 2022. **16**(3): p. 4072–4083.

35. Han, M., et al., *Tailoring electronic and optical properties of MXenes through forming solid solutions*. J. Am. Chem. Soc., 2020. **142**(45): p. 19110–19118.

36. Han, M., et al., *Beyond Ti₃C₂Tₓ: MXenes for electromagnetic interference shielding*. ACS Nano, 2020. **14**(4): p. 5008–5016.

37. Foucher, A.C., et al., *Shifts in valence states in bimetallic MXenes revealed by electron energy-loss spectroscopy (EELS)*. 2D Mater., 2022. **9**(2): p. 025004.

38. Du, Z., et al., *High-entropy atomic layers of transition-metal carbides (MXenes)*. Adv Mater, 2021. **33**(39): p. e2101473.

39. Zhang, X., et al., *A novel complex oxide TiVO₃.₅ as a highly active catalytic precursor for improving the hydrogen storage properties of MgH₂*. Int. J. Hydrog. Energy, 2018. **43**(52): p. 23327–23335.

40. Babak Anasori, W.H. and S.K. Nemani, *High entropy MXenes and methods of making thereof*. US Patent, 2022. **US 2022/0115660 A1**.

41. Ma, W., et al., *A new Ti₂V₀.₉Cr₀.₁C₂Tₓ MXene with ultrahigh gravimetric capacitance*. Nano Energy, 2022. **96**: p. 107129.

42. Tang, Y., et al., *The effect of in situ nitrogen doping on the oxygen evolution reaction of MXenes*. Nanoscale Adv., 2020. **2**(3): p. 1187–1194.

43. Sun, W., et al., *Multiscale and Multimodal characterization of 2D titanium carbonitride MXene*. Adv. Mater. Interfaces, 2020. **7**(11): p. 1902207.

44. Yang, J., et al., *Two-dimensional Nb-based M₄C₃ solid solutions (MXenes)*. J. Am. Ceram. Soc., 2016. **99**(2): p. 660–666.

45. Pinto, D., et al., *Synthesis and electrochemical properties of 2D molybdenum vanadium carbides – solid solution MXenes*. J. Mater. Chem. A, 2020. **8**(18): p. 8957–8968.

46. Nemani, S.K., et al., *High-entropy 2D carbide MXenes: TiVNbMoC3 and TiVCrMoC3*. ACS Nano, 2021. p. 12815–12825.

47. Deysher, G., et al., *Synthesis of Mo₄VAlC₄ MAX phase and two-dimensional Mo₄VC₄ MXene with five atomic layers of transition metals*. ACS Nano, 2020. **14**(1): p. 204–217.

48. Hantanasirisakul, K. and Y. Gogotsi, *Electronic and optical properties of 2D transition metal carbides and nitrides (MXenes)*. Adv. Mater., 2018. **30**(52): p. e1804779.

49. Naguib, M., et al., *New two-dimensional niobium and vanadium carbides as promising materials for Li-ion batteries*. J. Am. Chem. Soc., 2013. **135**(43): p. 15966–15969.

50. Zhou, J., et al., *Ti-enhanced exfoliation of V₂AlC into V₂C MXene for lithium-ion battery anodes*. Ceram. Inter., 2017. **43**(14): p. 11450–11454.

51. Hong, W., et al., *Double transition-metal MXenes: Atomistic design of two-dimensional carbides and nitrides*. MRS Bulletin, 2020. **45**(10): p. 850–861.

52. Yeh, J.W., et al., *Nanostructured high-entropy alloys with multiple principal elements: novel alloy design concepts and outcomes*. Adv. Eng. Mater., 2004. **6**(5): p. 299–303.

53. Miracle, D.B. and O.N. Senkov, *A critical review of high entropy alloys and related concepts*. Acta Materialia, 2017. **122**: p. 448–511.

54. Harrington, T.J., et al., *Phase stability and mechanical properties of novel high entropy transition metal carbides*. Acta Materialia, 2019. **166**: p. 271–280.

55. Yeh, J.-W., *Recent progress in high-entropy alloys*. Annales de Chimie Science des Matériaux, 2006. **31**(6): p. 633–648.

56. Senkov, O.N., et al., *Mechanical properties of $Nb_{25}Mo_{25}Ta_{25}W_{25}$ and $V_{20}Nb_{20}Mo_{20}Ta_{20}W_{20}$ refractory high entropy alloys*. Intermetallics, 2011. **19**(5): p. 698–706.

57. Lal, M.S. and R. Sundara, *High entropy oxides-A cost-effective catalyst for the growth of high yield carbon nanotubes and their energy applications*. ACS Appl. Mater. Interfaces, 2019. **11**(34): p. 30846–30857.

58. Lal, M.S. and R. Sundara, *Multifunctional high entropy oxides incorporated functionalized biowaste derived activated carbon for electrochemical energy storage and desalination*. Electrochimica. Acta, 2022. **405**: p. 139828.

59. Pan, R., J. Zhu and Y. Liu, *Synthesis, microstructure and properties of $(Ti_{1-x}, Mo_x)_2AlC$ phases*. Mater. Sci. Technol., 2018. **34**(9): p. 1064–1069.

60. Caspi, E.a.N., et al., *Ordering of (Cr,V) Layers in Nanolamellar $(Cr_{0.5}V_{0.5})_{n+1}AlC_n$ Compounds*. Mater. Res. Lett., 2014. **3**(2): p. 100–106.

61. Lapauw, T., et al., *The double solid solution $(Zr, Nb)_2(Al, Sn)C$ MAX phase: a steric stability approach*. Sci. Rep., 2018. **8**(1): p. 12801.

62. Naguib, M., et al., *New solid solution MAX phases: $(Ti_{0.5}, V_{0.5})_3AlC_2$, $(Nb_{0.5}, V_{0.5})_2AlC$, $(Nb_{0.5}, V_{0.5})_4AlC_3$ and $(Nb_{0.8}, Zr_{0.2})_2AlC$*. Mater. Res. Lett., 2014. **2**(4): p. 233–240.

63. Anasori, B., et al., *Two-Dimensional, ordered, double transition metals carbides (MXenes)*. ACS Nano, 2015. **9**(10): p. 9507–16.

64. Anasori, B., et al., *Experimental and theoretical characterization of ordered MAX phases Mo_2TiAlC_2 and $Mo_2Ti_2AlC_3$*. J. Appl. Phys., 2015. **118**(9): p. 094304.

65. Liu, Z., et al., *$(Cr_{2/3}Ti_{1/3})_3AlC_2$ and $(Cr_{5/8}Ti_{3/8})_4AlC_3$: New MAX-phase compounds in Ti–Cr–Al–C system*. J. Am. Ceram. Soc., 2014. **97**(1): p. 67–69.

66. Dahlqvist, M. and J. Rosen, *Order and disorder in quaternary atomic laminates from first-principles calculations*. Phys. Chem. Chem. Phys., 2015. **17**(47): p. 31810–31821.

67. Khatun, N. and S.C. Roy, *Optimization of etching and sonication time to prepare monolayer $Ti_3C_2T_x$ MXene flakes: A structural, vibrational, and optical spectroscopy study*. Micro and Nanostructures, 2022. **167**: p. 207256.

68. Khazaei, M., et al., *Novel electronic and magnetic properties of two-dimensional transition metal carbides and nitrides*. Adv. Funct. Mater., 2013. **23**(17): p. 2185–2192.

69. Khazaei, M., et al., *Topological insulators in the ordered double transition metals $M_2'M''C_2$ MXenes (M'=Mo, W; M''=Ti, Zr, Hf)*. Phys. Rev. B, 2016. **94**(12).

70. Enyashin, A.N. and A.L. Ivanovskii, *Two-dimensional titanium carbonitrides and their hydroxylated derivatives: Structural, electronic properties and stability of MXenes $Ti_3C_{2-x}N_x(OH)_2$ from DFTB calculations*. J. Solid State Chem., 2013. **207**: p. 42–48.

71. Khazaei, M., et al., *Electronic properties and applications of MXenes: A theoretical review*. J. Mater. Chem. C, 2017. **5**(10): p. 2488–2503.

72. Näslund, L.-Å., et al., *Chemical bonding of termination species in 2D carbides investigated through valence band UPS/XPS of $Ti_3C_2T_x$ MXene*. 2D Mater., 2021. **8**(4): p. 045026.

73. Hope, M.A., et al., *NMR reveals the surface functionalisation of Ti_3C_2 MXene*. Phys. Chem. Chem. Phys., 2016. **18**(7): p. 5099–5102.

74. Liu, F., et al., *Preparation of high-purity V_2C MXene and electrochemical properties as Li-Ion batteries*. J. Electrochem. Soc., 2017. **164**(4): p. A709–A713.

75. Seh, Z.W., et al., *Two-Dimensional Molybdenum Carbide (MXene) as an Efficient Electrocatalyst for Hydrogen Evolution*. ACS Energy Lett., 2016. **1**(3): p. 589–594.

76. Si, C., J. Zhou and Z. Sun, *Half-metallic ferromagnetism and surface functionalization-induced metal-insulator transition in Graphene-like two-dimensional Cr_2C crystals*. ACS Appl. Mater. Interfaces, 2015. **7**(31): p. 17510–17515.

77. Yang, J., et al., *Tunable electronic and magnetic properties of $Cr_2M'C_2T_2$ (M' = Ti or V; T = O, OH or F)*. Appl. Phys. Lett., 2016. **109**(20): p. 203109.

78. Dong, L., et al., *Rational design of two-dimensional metallic and semiconducting spin-tronic materials based on ordered double-transition-metal MXenes*. J. Phys. Chem. Lett., 2017. **8**(2): p. 422–428.

79. Xie, Y. and P.R.C. Kent, *Hybrid density functional study of structural and electronic properties of functionalized $Ti_{n+1}X_n(X=C, N)$ monolayers*. Phys. Rev. B, 2013. **87**(23).

80. Ivanovskii, A.L. and A.N. Enyashin, *Graphene-like transition-metal nanocarbides and nanonitrides*. Russ. Chem. Rev., 2013. **82**(8): p. 735–746.

81. Zhang, Y., et al., *Prediction of MXene based 2D tunable band gap semiconductors: GW quasiparticle calculations*. Nanoscale, 2019. **11**(9): p. 3993–4000.

82. Barsoum, M., *Physical properties of the MAX phases*. Encyclopedia Mater. Sci. Technol., 2006.

83. El-Demellawi, J.K., et al., *Tunable multipolar surface plasmons in 2D $Ti_3C_2T_x$ MXene flakes*. ACS Nano, 2018. **12**(8): p. 8485–8493.

84. Xie, H., et al., *Electrostatic self-assembly of $Ti_3C_2T_x$ MXene and gold nanorods as an efficient surface-enhanced Raman scattering platform for reliable and high-sensitivity determination of organic pollutants*. ACS Sens., 2019. **4**(9): p. 2303–2310.

85. Peng, Y., et al., *Charge-transfer resonance and electromagnetic enhancement synergistically enabling MXenes with excellent SERS sensitivity for SARS-CoV-2 S protein detection*. Nanomicro Lett., 2021. **13**: p. 52.

86. Soundiraraju, B. and B.K. George, *Two-dimensional titanium nitride (Ti_2N) MXene: synthesis, characterization, and potential application as surface-enhanced Raman scattering substrate*. ACS Nano, 2017. **11**(9): p. 8892–8900.

87. Li, G., et al., *Surface-modified two-dimensional titanium carbide sheets for intrinsic vibrational signal-retained surface-enhanced Raman scattering with ultrahigh uniformity*. ACS Appl. Mater. Interfaces, 2020. **12**(20): p. 23523–23531.

88. Yang, K., et al., *$Ti_3C_2T_x$ MXene-loaded 3D substrate toward on-chip multi-gas sensing with surface-enhanced Raman spectroscopy (SERS) barcode readout*. ACS Nano, 2021.

89. Li, R., et al., *MXene Ti_3C_2: An effective 2D light-to-heat conversion material*. ACS Nano, 2017. **11**(4): p. 3752–3759.

90. Lin, H., et al., *A two-dimensional biodegradable niobium carbide (MXene) for photo-thermal tumor eradication in NIR-I and NIR-II biowindows*. J. Am. Chem. Soc., 2017. **139**(45): p. 16235–16247.

91. Lin, H., et al., *Theranostic 2D tantalum carbide (MXene)*. Adv. Mater., 2018. **30**(4).

92. Lipatov, A., et al., *Elastic properties of 2D $Ti_3C_2T_x$ MXene monolayers and bilayers*. Sci. Adv., 2018. **4**(6): p. eaat0491.

93. Borysiuk, V.N., V.N. Mochalin and Y. Gogotsi, *Molecular dynamic study of the mechanical properties of two-dimensional titanium carbides $Ti_{n+1}C_n$ (MXenes)*. Nanotechnology, 2015. **26**(26): p. 265705.

94. Plummer, G., et al., *Nanoindentation of monolayer $Ti_{1+n}C_nT_x$ MXenes via atomistic simulations: The role of composition and defects on strength*. Comput. Mater. Sci., 2019. **157**: p. 168–174.

95. Ansori, B. and Y. Gogotsi, 2D metal carbides and nitrides (MXenes): Structure, properties and applications. 2019: Springer Nature Switzerland AG.

96. Hantanasirisakul, K., et al., *Evidence of a magnetic transition in atomically thin $Cr_2TiC_2T_x$ MXene*. Nanoscale Horiz., 2020. **5**(12): p. 1557–1565.

97. Allen-Perry, K., et al., *Tuning the magnetic properties of two-dimensional MXenes by chemical etching*. Materials (Basel), 2021. **14**(3): p. 694.

98. Jiang, X., et al., *Recent progress on 2D magnets: Fundamental mechanism, structural design and modification*. Appl. Phys. Rev., 2021. **8**(3): p. 031305.

99. Kumar, H., et al., *Tunable magnetism and transport properties in nitride MXenes*. ACS Nano, 2017. **11**(8): p. 7648–7655.

100. Hu, J., et al., *Investigations on V_2C and V_2CX_2 (X = F, OH) monolayer as a promising anode material for Li Ion batteries from first-principles calculations.* J. Phys. Chem. C, 2014. **118**(42): p. 24274–24281.

101. Zhao, S., W. Kang and J. Xue, *Manipulation of electronic and magnetic properties of M_2C (M = Hf, Nb, Sc, Ta, Ti, V, Zr) monolayer by applying mechanical strains.* Appl. Phys. Lett., 2014. **104**(13): p. 133106.

102. Gao, G., et al., *Monolayer MXenes: promising half-metals and spin gapless semiconductors.* Nanoscale, 2016. **8**(16): p. 8986–8994.

103. Iqbal, M., et al., *Co-existence of magnetic phases in two-dimensional MXene.* Mater. Today Chem., 2020. **16**: p. 100271.

5 Composites of MXenes

Sayan Ganguly
Bar-Ilan Institute for Nanotechnology and
Advanced Materials, Department of Chemistry
Bar-Ilan University
Ramat-Gan, Israel

CONTENTS

5.1 INTRODUCTION

Two-dimensional materials, often known as 2D materials, have garnered a lot of attention since since the discovery of graphene in 2004 [2, 3]. 2D materials have shown many intriguing properties that are not found in their bulk counterparts. These materials hold tremendous promise for a wide variety of applications, including electronic and opto-electronic devices, along with electrochemical catalysis. This is due to the reduction in dimension and size that 2D materials have undergone. In current history, significant strides have been made in the development of synthetic processes, which has led to the successful production of an increasing number of 2D materials beyond graphene [4].

An expanding family of 2D transition metal carbides, carbonitrides, and nitride compounds known as MXenes has gotten a lot of attention recently for their intriguing physical and chemical characteristics, which are tightly linked to the wide range of elements and surface terminations that make up the compounds. Further, the ease with which MXenes may form composites with other materials, such as polymers, oxides, and carbon nanotubes, makes it possible to effectively modify the characteristics of MXenes for use in a variety of applications [5]. Because of their high conductivity, reducibility, and biocompatibility, MXenes and MXene-based composites have not only gained popularity as electrode materials in the field of energy storage, as is common knowledge, but they have also shown significant potential in applications connected to the environment, such as electro/photocatalytic water splitting, photocatalytic reduction of carbon dioxide, water purification, and sensors. These applications include electro/photocatalytic water splitting. MXenes are the term given to a recently found vast family of 2D early transition metal carbides and/

DOI: 10.1201/9781003281511-5

or nitrides, and they are quickly becoming a star among these compounds [6, 7]. MXenes are primarily produced by removing the A layers from MAX phases, which are ternary carbides or nitrides and have the general chemical formula $M_{n+1}AX_n$. In this formula, M represents an early transition metal, A represents an element from group IIIA or IVA, X can be either carbon or nitrogen, and n can be any of the numbers 1, 2, or 3 [8, 9]. The hexagonal structures of MAX phases are layered, and the $M_{n+1}X_n$ units and the A layers are piled atop one another in an alternating fashion [10]. As M–X bonds are so much stronger than M–A bonds, the A layers may be selectively etched chemically without damaging the M–X ties, resulting in weakly bonded $M_{n+1}X_n$ layers that can be easily separated by sonication. This is because the M–X bonds are significantly stronger than the M–A bonds [11]. The newly generated 2D materials are referred to as MXenes to emphasize the loss of A layers from the parent MAX phase and to highlight the similarities between their 2D nature and that of graphene. In the etching process, the surfaces of the M_n+1X_n units are invariably coated with functional groups such as oxygen ([double bond, length as oxygen (=O), hydroxyl (–OH), and/or fluorine (–F)]. This fact should be emphasized [12]. In order to describe the chemical formula of MXenes, we may write it as $M_{n+1}X_nT_x$, where T_x is meant to represent the surface functional groups. In the laboratory, the relative abundances of the various functional groups on the MXene surfaces are hard to predict and vary depending on the etching conditions. MXenes are endowed with a plethora of fascinating mechanical, electrical, magnetic, and electrochemical capabilities as a result of the diverse chemistry that underpins them. In particular, the high flexibility of MXenes, in conjunction with their 2D morphology and layered structures, makes it simple for MXenes to combine with other materials to form composites. This opens the door to the possibility of combining the exceptional qualities of a variety of materials in a manner that is mutually beneficial. Results, not only of MXenes but also of composites based on MXenes, have gained a significant amount of scientific attention, as they offer a great deal of potential for a wide variety of applications. MXenes and MXene-based composites initially found applications in energy storage as high-performance electrode materials for lithium–sulfur batteries, sodium-ion batteries, and supercapacitors. These applications are possible because of the high conductivity and excellent electrochemical activity of MXenes and MXene-based composites [13].

This review of the different polymer composites of MXene extensively covers the present applications while analyzing their remarkable outcomes and comparing them to various other 2D materials that are utilized in the same area of application. Although other reviews contain the information supplied about MXene functionalized composites, this review of the various polymer composites of MXene is one of the few that has this information. This offers a peek at their capabilities as well as a broad insight into their potential future uses of these adaptable and cost-effective materials.

5.2 BACKGROUND OF MXene-BASED COMPOSITES

Even though the uses of the precursor are fairly restricted, when exfoliated into 2D MXene, these carbon nitrides can reach several extraordinary characteristics that are of significant scientific interest. Researchers are particularly interested in the exceptional electrical, magnetic, optical, electrochemical, and water barrier capabilities of MXene, which are among the many applications that are now making use of this material [14]. Consequently, scientists are currently struggling with the process of synthesizing MXene since an ideal MXene crystal structure would be devoid of both M and X vacancies. However, this has not yet been realized [15]. Coats and the fabrication done with Ti_3C_2 are vulnerable to oxidation and chemical deterioration, which ultimately results in the formation of TiO_2. Touseef Habib conducted research on the impact of several environmental factors on the deterioration of MXene and demonstrated that the process moves more quickly when exposed to UV light [16]. MXene, for example, is being considered as a potential method for bringing large-scale practical applications to lithium-ions batteries, through aided delamination; non-lithium batteries (anodes for electrochemical processes); and for the next generation of energy storage technologies [17, 18]. MXene is gaining speed in other sectors of research where, historically, other 2D materials have ruled supreme. This is outside the purview of the electrochemical applications that have been developed for it. Polymers have been around long enough to qualify as a respectable contender for the inclusion of 2D materials that were developed before MXene [19]. This is because polymers have the ability to self-assemble into three-dimensional structures. These lengthy chains of carbon molecules are integrated into a variety of other compounds in order to change the chemical and mechanical characteristics of those molecules. The sheer variety of polymers is one of the reasons why the study of them continues to be a rapidly expanding topic in research [20]. These composites are distinguished by their resistance to abrasion, resistance to corrosion, resistance to fatigue, resistance to impact, great strength, resistance to fracture, and great stiffness. The popularity of these composites can be attributed to the fact that they are easily manufactured and do not require a high initial investment [21]. Due to the interfacial tension of polymer matrix composites (PMC) and the orientation of dispersed phase inclusions such particles, flakes, laminates, and fibers, polymer matrix composites (PMC) have enabled exceptional achievements in the field of engineering. MXene could surely have been implanted on polymers, which made them a good choice for the composite role. Because of the widespread interest that MXene has garnered in the field of energy storage and transmission, conducting polymers such as polypyrrole have been the subject of much research. MXenes are being placed on a wide variety of different polymers for uses that are both distinctive and exciting. This is possible because of their 2D structure. Despite the limited scope of these investigations, they have proven to be beneficial and have made a significant contribution in opening up previously unexplored fields of science and technology to future research and development. The various groups of researchers have synthesized and investigated polymer composites, which have then been used to a broad variety of appositeness, such as photothermal conversion, wastewater treatment, increase in mechanical capabilities, and so on [22–24]. The multitude of manufacturing procedures and grafting techniques that were utilized to generate these composites awarded them their one-of-a-kind features, which were then employed to overcome the obstacles that limited the technologies that were available at the time.

5.3   FABRICATIONS OF MXene–POLYMER COMPOSITES

As a result of the rise in popularity of portable and wearable electronic devices, new critical research directions for EMI shielding materials include making them light-weight, having a low density, being flexible, and having great mechanical stability. MXenes have the potential to drastically cut down on consumption in order to keep up with the latest trends in the creation of electronic components. MXenes have, up to this point, gained a significant amount of attention and applications in the realm of EMI, and they have become the newest and most popular EMI shielding materials, following in the footsteps of carbon nanotubes and graphene [25]. The manufacture of MXenes/polymer EMI shielding composites with low MXenes loading, light weight, remarkable, good EMI shielding performances, and beautiful mechanical characteristics may be realized by combining MXenes with polymer matrix to produce the effective 3D electrically conductive networks. Methods such as physical mixing, freeze-drying, pre-support molding, vacuum-assisted filtration, and alternating vacuum-assisted filtration are frequently used in the preparation process.

MXenes and polymer matrix may be mechanically blended together by the process of physical blending (Figure 5.1), which is a method that can be used to manufacture MXenes/polymer EMI shielding composites. Between MXenes and the polymer matrix, there is no chemical crosslinking that takes place. Using a process known as solvent-assisted mixed compression, Rajavel et al. were able to successfully create $Ti_3C_2T_x$/polyvinylidene fluoride (PVDF) nanocomposites [26]. The EMI SE of the composites reached a value of 48.47 db when the quantity of $Ti_3C_2T_x$ was 22.55 vol% and the thickness was 2 mm. Electrostatic assembly and press molding were the two methods that Sun et al. used to manufacture $Ti_3C_2T_x$/polystyrene (PS) composites with an isolated structure [27]. When the quantity of Ti3C2Tx was 1.9 vol% and the thickness was 2 mm, the $Ti_3C_2T_x$/PS nanocomposites had σ of 1081 S/m and an EMI SE of 54 dB. This was possible due to the synergistic impact of its efficient electrically conductive network. Epoxy resins are utilized extensively in

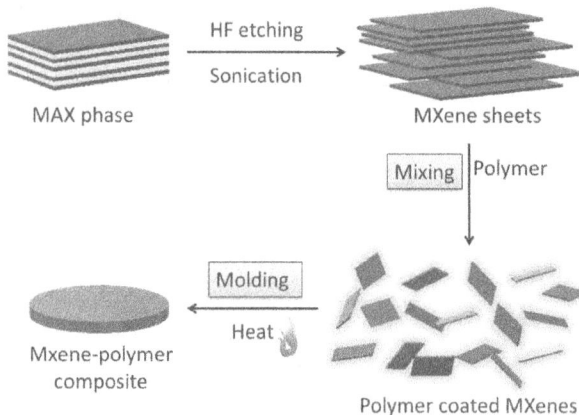

FIGURE 5.1   Schematic illustration of MXene-based physical mixing method for composite preparation.

electronic equipment, aircraft, and a variety of other sectors because of their exceptional chemical stability, great mechanical qualities, and competitive pricing [28].

When it comes to preparing MXenes/polymer EMI shielding composite foams, freeze-drying is one of the ways that is utilized most frequently. The fundamental idea behind freeze-drying is to first freeze the suspension after it has completed self-assembly in order to encourage the formation of solid organic molecules, and then to sublimate the ice or solid organic molecules in order to allow them to escape through vacuum drying in order to obtain the porous MXenes/polymer EMI shielding 3D foam materials. This process is known as freeze-drying. Freeze-drying was used by Xu et al. in order to generate porous MXene/PVA EMI shielding composite foams. These foams were formed of a few layers of Ti_2CT_x MXenes and PVA [29] (Figure 5.2). The researchers then rolled the foams in order to obtain the matching MXene/PVA EMI shielding composite films. The EMI specific shielding effectiveness (SSE) of MXene/PVA EMI shielding composite foams was 5136 dB cm^{-2} g^{-1} when the quantity of MXenes was 0.15 vol% and the thickness was 2 mm. The EMI shielding efficiency (SE) was 28 dB when the amount of MXenes was 0.15 vol%. To manufacture silver nanowire/MXene/MF (AgNW/MXene/MF) EMI shielding sponges, Weng et al. employed melamine formaldehyde sponge (MF) as a coating

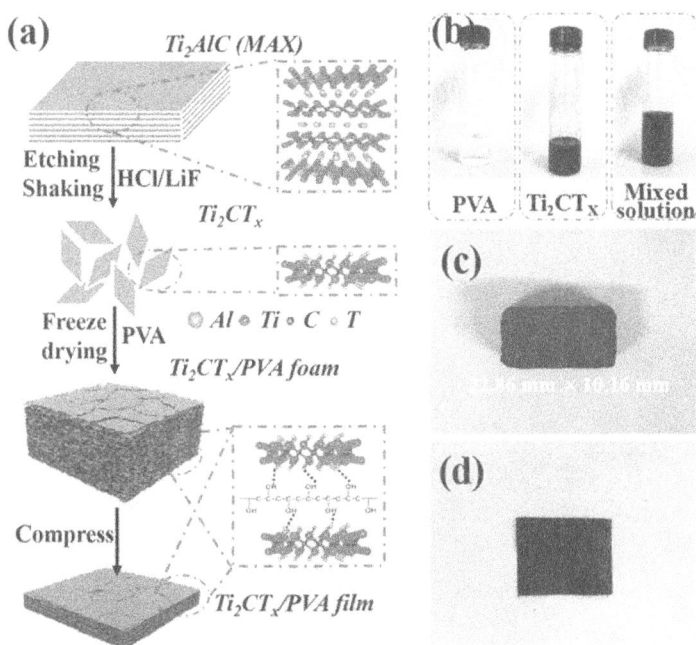

FIGURE 5.2   (a) Illustration of the preparation process of f-Ti2CTx/PVA composite foam and film. (b) Bottle of PVA solution that is nearly transparent, bottle of f-Ti2CTx solution with pure black color, and bottle of f-Ti2CTx/PVA solution. (c) Photograph of f-Ti2CTx/PVA foam. (d) Typical photograph of f-Ti2CTx/PVA film.

template. This was done using a combination of dip coating and directed freeze-drying [30]. At 49.5 mg cm^{-3} and 2 mm thickness, the AgNW/MXene/MF shielding composite sponges demonstrated good EMI shielding performance of 52.6 dB.

Freeze-drying is able to ensure the porous condition of three-dimensional materials to the maximum extent possible without harming the internal structure of the materials because ice crystals have the ability to sublimate immediately. The MXenes/polymer EMI shielding composite foams that are made using this technology have the features of high porosity, tunable pore size distribution, and stable three-dimensional multistage pore architectures. At the same time, due to the construction of entire electrically conductive networks, it is possible to acquire the ideal and EMI SE at a lower MXenes quantity. This is possible because of the fact that the ideal EMI SE may be obtained. Poor air stability of MXenes/polymer EMI shielding composite foam materials is a typical issue. This is due to the fact that MXenes are susceptible to oxidize and the foam structures have a large specific surface area.

Backfilling polymer matrix into MXenes that have created 3D electrically conductive networks is the process that takes place during the molding technique known as pre-support molding, which is used to generate MXenes/polymer EMI shielding composites. Fabrication of 3D porous $Ti_3C_2T_x$ structures was accomplished by Zhao et al. by the use of GO-assisted hydrothermal assembly and directed freeze-drying (Figure 5.3), followed by backfilling with epoxy resins to enhance the material's mechanical characteristics [31].

The authors Wang et al. used the sol-gel process to make $Ti_3C_2T_x$/C hybrid foam (MCF), which was then backfilled with epoxy resins to make MCF/epoxy composites [32]. MCF/epoxy electromagnetic interference (EMI) shielding composites had an optimum of 184 S/m and an EMI SE of 46 dB when the amount of MCF was 4.25 wt% and the thickness was 3 mm. Additionally, the rGO-MXenes

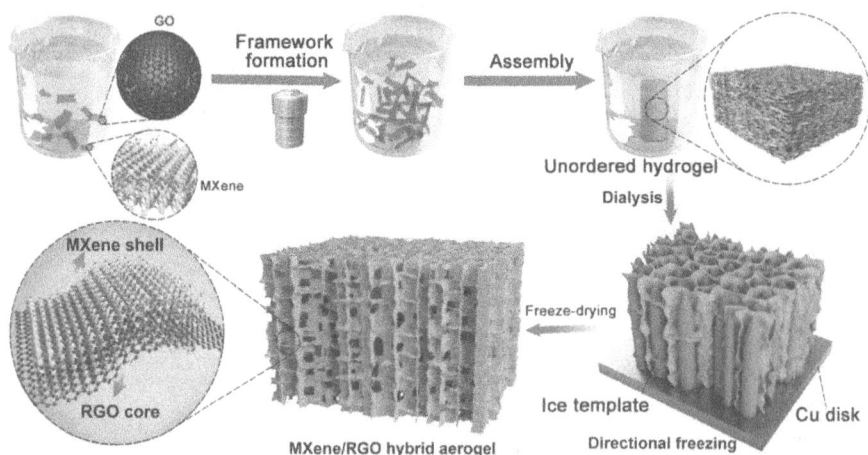

FIGURE 5.3 Schematic showing $Ti_3C_2T_x$ MXene/RGO hybrid aerogel fabrication via GO-assisted hydrothermal assembly, directional freezing, and freeze-drying.

(Reproduced with the permission from Ref. [31], © 2018 Americal Chemical Society.)

honeycomb structure (rGMH) was initially generated via template induction and electrostatic self-assembly. Subsequently, the rGMH/epoxy EMI shielding composites were obtained by backfilling with epoxy resins [33]. The rGMH/epoxy composites exhibited the highest (387.1 S/m) EMI SE values when the quantity of rGO and MXenes was 1.2 wt% and 3.3 wt%, respectively, and the thickness was 3 mm (55 dB). MXenes formed 3D electrically conductive skeletons in the pre-support molding, and the backfill of the polymer matrix can give it better mechanical properties without destroying the original electrically conductive networks. This allows for high and EMI SE to be achieved with low MXenes loading, which is desirable. At the same time, the polymer matrix has the ability to completely cover the MXenes and shut off contact between the MXenes and the air, which prevents the MXenes from being oxidized. However, the corresponding mechanical characteristics of the MXenes/polymer EMI shielding composites are subpar. This is due to the absence of a chemical link between the polymer matrix and the MXenes.

The MXenes/polymer EMI shielding composite films can be prepared using a technique called vacuum-assisted filtration. This technique speeds up the process of removing liquid from a matrix consisting of uniformly mixed MXenes and polymer by utilizing the negative pressure that is created when suction is applied. By using a self-assembly process that was induced by vacuum-assisted filtration, Cao et al. were able to generate d-$Ti_3C_2T_x$/CNF (Figure 5.4) EMI shielding composite sheets that were very thin and flexible [34].

Via the use of vacuum-assisted filtration, Xie et al. were able to generate d-$Ti_3C_2T_x$/ aramid nanofiber (ANF) composite sheets that possessed exceptional mechanical characteristics, and EMI shielding composite sheets made with d-$Ti_3C_2T_x$ and ANF had a shielding effectiveness (SE) of 28 dB when the amount of d-$Ti_3C_2T_x$ was 60 wt% and the thickness was 17 μm [35]. By using vacuum-assisted filtering, Zhan et al. were able to manufacture ultra-thin MXene/TOCNF EMI shielding composite

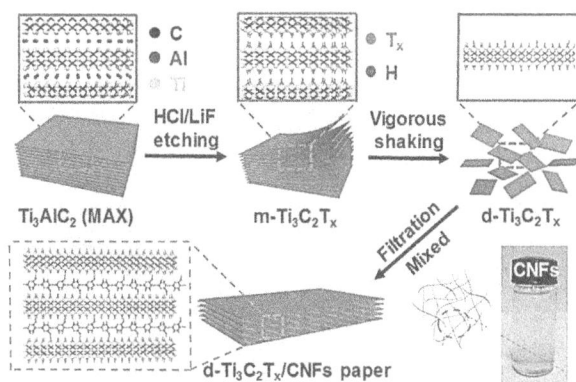

FIGURE 5.4   HCl and LiF etch Ti_3AlC_2 (MAX) to produce m-$Ti3C_2T_x$. Under vigorous shaking, m-Ti3C2Tx sediment is disseminated in deionized water to produce d-$Ti_3C_2T_x$ nanosheets. Then CNFs are added to d-$Ti_3C_2T_x$. The d-$Ti_3C_2T_x$/CNF composite paper is made from a 24-h-stirred suspension.

(Reproduced with the permission from Ref. [34], © 2018 American Chemical Society.)

films using 2, 2, 6, 6-tetramethyl-1-Piperidinyloxy (TEMPO) oxidized cellulose nanofiber (TOCNF) [36]. The EMI SE of the Ti3C2Tx/TOCNF EMI shielding composite sheets was 39.6 dB when the thickness was 38 m and the quantity of Ti3C2Tx was 50 wt%.

5.4   SYNTHETIC THERMOPLASTICS-BASED MXene COMPOSITES

The structure of polyvinyl butyral, often known as PVB, is seen in Figure 5.4. PVB is an acetal and results from the reaction that takes place when aldehyde and alcohol are combined. As a result of their transparency, PVB have found widespread application as a resin glue in the glass sector, where they are used to manufacture laminated glass sheets, which are now a regular sight in vehicles and high-rise structures. Once PVB is sprayed in the space between two sheets of glass, it not only provides an elegant solution for anti-shock and anti-break-in capabilities in high-rise structures, but it also protects against broken glass, which may be hazardous in the aftermath of automobile accidents. PVB possesses a wide variety of applications in the glass composites industry, which range from frosted glass panels to ornamental glass. After another four hours of ball milling, the mixture was given PVB to function as a binder, and polyethylene glycol/diethyl phthalate to operate as a plasticizer. After the combination had been ball milled, the resulting slurry was used to tape cast a thin sheet with a thickness of only 200 μm, and repeated tape casting led to the production of the multilayer polymer composite [37]. In terms of its morphology, Co_2Z had a structure similar to a plate, and its particle size ranged from 200 μm to 250 μm, whereas MXene had a structure that was similar to graphene in that it had a layered structure. This composite material has the potential to be exploited in radar absorption technologies, which are currently prevalent in low-observable aircrafts.

The category known as ultra-high molecular weight polyethylene, or UHMWPE for short, is a subset of thermoplastic polyethylene that is distinguished by its characteristically long chains and high molecular weight, which is measured in the millions. The combination of heat and pressure can be used to reform thermoplastics into a new shape. These polymers can have an amorphous or semi-crystalline structure, and their molecules can either be branched or linear. Their intermolecular interactions are weak, but their intramolecular bonds are strong. UHMWPE has numerous desired features, including resistance to corrosive chemicals, minimal moisture absorption, and excellent abrasion resistance, which makes them an intriguing possibility for an MXene polymer composite and for a number of applications ranging from armor to fishing nets [38]. In a study that was carried out by Heng Zhang et al., a composite material made of Ti_3C_2/UHMWPE was developed in order to examine its mechanical and anti-friction capabilities [39]. After the mixture had been homogeneously combined with UHMWPE with the assistance of a high-speed mixer, the composite was manufactured by pressing the mixture into the desired shape using a press vulcanizer. The morphology of Ti_3C_2 reveals that the material is composed of microparticles that provide several nanosheets with a lamellar structure comparable to that of graphite or clay. During a study of the morphology of the surfaces of the nanocomposites using a SEM, spherulites with diameters ranging from 150 to 200 μm and Ti_3C_2 at the center were detected. These spherulites had Ti_3C_2 at the core. Because the development of spherulites was more noticeable at greater Ti_3C_2 concentrations, this indicated that MXene particles served as the nucleation nuclei for polymer chains that eventually formed many spherulites. With 2.0 wt% Ti_3C_2, the composite has the highest level of recoverable compliance and creep resistance. This indicates that the impact becomes more severe at increasing concentrations. As the amount of Ti_3C_2 in the composite was raised, the anti-frictional capabilities of the MXene/UHMWPE material showed a considerable decline as well. Each subsequent addition of Ti_3C_2 brings about an increasing degree of crystallinity in the composite, which in turn brings about a reduction in the frictional coefficient. The resemblances in morphology between graphene and Ti_3C_2 translate to weak van der Waals bonding in multilayer structures, which, when placed under stress shear, exfoliates into finer layers, thereby performing a lubricating action. Graphene is a two-dimensional material, while Ti_3C_2 is a three-dimensional material. Because of the high concentration of MXene that was included in the composite, the mechanical characteristics of a flexible polymer such as UHMWPE were significantly improved. When all of the enhanced qualities of the MXene/polymer composite are taken into consideration, it becomes clear that the Ti3C2/UHMWPE composite offers a significant advancement above the pure polymer matrix. Additionally, the use of this composite might be observed in systems like conveyor belts, equipment used in maritime environments due to its low capacity for absorbing moisture, and even in the food processing industry.

A high-temperature engineering thermoplastic and an amorphous polymer, polyether sulfone (PES) is processed using conventional plastics handling equipment. Amorphous and minimal mold shrinkage allows it to tolerate high temperatures in water and air for long periods of time. PES has excellent thermal, electrical, mechanical, chemical, and flame resistance, sterilizing resistance, and good hydrolytic properties, all of which are supported by excellent optical clarity [21]. Due to its permselectivity, heat resilience, and ultrafiltration (UF) qualities, PES membranes have been widely employed in desalination and wastewater treatment. In an attempt to segregate Congo red dye and inorganic ions, Han et al. constructed an MXene/PES composite membrane with high flux and extraordinary hydrophilicity. This membrane has the potential to achieve their goal [40]. Etching and ultrasonicating Ti_3AlC_2 in this research led to the production of $Ti_3C_2T_x$. After that, it was put to use in the process of preparing the MXene composite membrane by employing a straightforward filtering technique that was carried out at 0.2 MPa on a polyethersulfone (PES) ultrafiltration membrane. Both outstanding flow (115 L m^{-2} h^{-1}) and acceptable rejection of Congo red dye are demonstrated by the MXene composite membrane (92.3% at 0.1 MPa). At a pressure of 0.1 MPa, the membrane showed a rejection of inorganic salts of less than 23% while allowing a flow of more than 432 L m^{-2} h^{-1}. The composite membrane, on account of the loose lamellar structure it possesses, is in a position to exhibit effective permselectivity in the process of separating salts and dyes. In addition, the lamellar hydrophilic MXene contributes to the composite membrane's remarkable hydrophilicity and flow, both of which are exceptional.

Poly(vinylidene fluoride) (PVDF) is a fluoroplastic material that is not based on TFE and is made up almost entirely of $-CH_2CF_2CH_2CF_2-$ units. It has a fusion temperature of around 373 K in and a high dielectric value. PVDF is used in every-day life as an electrical insulator and as a decorative coating for steel and aluminum in residential and commercial structures so that these materials can better withstand the effects of weather. It is produced by the same nonaqueous and aqueous dispersion polymerization processes as other fluoropolymers, with the difference that greater temperatures and pressures are applied throughout the manufacturing process. Kashif Rasool et al. researched the antibacterial characteristics of MXene/PVDF membrane, taking into consideration the ultrafiltration qualities of the polymer that are suitable for use in biotechnical applications [41]. The production of TiO_2/C on the surface of an old membrane is the source of the surface oxidation that is responsible for the growth inhibition seen in that kind of membrane. Because of their architecture, the stacked nanosheets produced narrow and consistent 2D channels. This not only reduces the number of macro- and meso-holes on the surface of the membrane, but it also makes the channels more uniform. The ease of operation of photothermal conversion has led to increased research interest in a variety of applications, including steam production, desalination, cancer therapy, and others. Renyuan Li et al. were able to create a layered MXene layer over hydrophilic PVDF by using a straightforward process using vacuum-assisted filtering. In addition to having a single crystallinity, the exfoliated MXene layers possessed a hexagonal symmetry in their structure [42]. The hydrophilicity of PVDF led the membrane to sink when it was placed in water; as a result, poly(dimethylsiloxane), also known as PDMS, was implanted into the membrane surface in order to reduce the surface energy while maintaining the ability to be wetted by water. The filtration process was assisted along by the membrane's pore size, which measured 22 μm, while the flexibility of the PVDF membrane contributed to the structural stability of the system. According to the findings, MXene had extremely high light-to-heat conversion rates of nearly 100%, while the self-floating memberanes with stacked MXene layers displayed an evaporation efficiency of 84%. This indicates that MXene has the potential to become a state-of-the-art photothermal material. The production of steam, the purification of water, and the treatment of wastewater might all stand to benefit from these recent discoveries.

Shaobo Tu et al. published their findings in 2018 under the title "Enhanced Dielectric Permittivity of PVDF Composites Using MXene as Nanofillers." This was accomplished by demonstrating that the build-up of charges at the interface results in the formation of tiny dipoles [43] (Figure 5.5). They revealed that percolative composites based on poly(vinylidene fluoride) (PVDF) and employing 2D MXene nanosheets as fillers display dramatically increased dielectric permittivity. Near the percolation limit of approximately 15.0 wt% of MXene loading, the poly(vinylidene fluoride-trifluoro-ethylene-chlorofluoroehylene) (P[VDF-TrFE-CFE]) polymer that has 2D $Ti_3C_2T_x$ nanosheets embedded within it achieves a dielectric permittivity as high as 10^5. This is measured by the dielectric constant. This has the potential to out-perform all other composites built of carbon-based fillers in the same polymer that have been described earlier. When loaded with MXene, it has been established that the action that increases the dielectric constant is present in other types of polymers as well. We demonstrate that the increase in the dielectric constant is mostly driven by the buildup of charge that is brought about by the development of tiny dipoles at

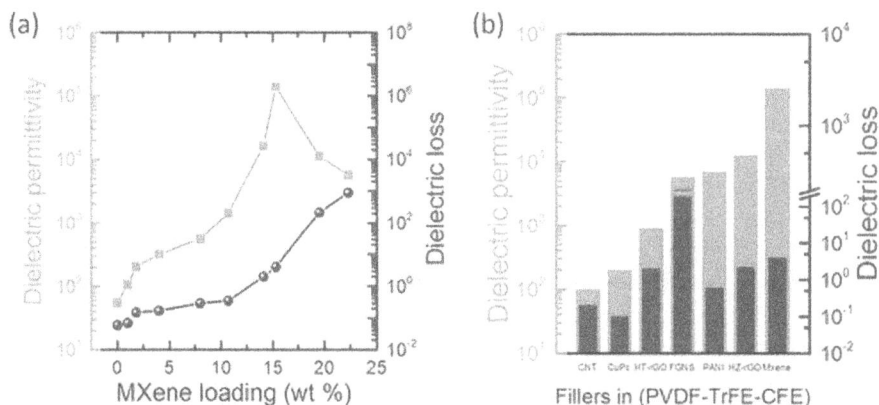

FIGURE 5.5 (a) Permittivity and dielectric loss of MXene/P(VDF-TrFE-CFE) at room temperature and 1 kHz; (b) bar charts comparing maximum dielectric permittivity and dielectric loss for P(VDF-TrFE-CFE) with different conductive fillers.

(Reproduced with the permission from Ref. [43], © 2018 American Chemical Society.)

the interfaces between the MXene sheets and the polymer matrix when an externally applied electric field is present.

5.5 CONDUCTING POLYMER-BASED MXene COMPOSITES

Polypyrrole (PPy), a heterocyclic polymer, has been the subject of research interest ever since its earliest reported synthesis in 1968. This is due to the distinctive electric and electronic properties that the material possesses, despite the fact that the structure of the material causes problems related to the material's physical and material properties. In the instance of the commercially available PPy with tosylate counter-ion, the electrical and mechanical characteristics of the polymer are highly conductive at 15 S/cm under ambient circumstances. The electrical properties of the polymer are dependent on the counter-ion that is utilized. An organ-like Ti_3C_2/PPy nanocomposite was produced by Wenling Wu et al. for the goal of serving as electrode materials for supercapacitors. This was accomplished by the in situ polymerization of PPy monomers. [44]. The morphology of the polymer revealed that it was composed of spherical agglomerated particles. This type of particle plays an important function in the electrolyte ion diffusion process because it generates a broader pathway inside the interlamellar space. The increased specific capacity is due to a synergy that exists between the electric double-layer capacitor, the organ-like Ti_3C_2 nanosheets, and the PPy nanoparticles PPy conduct as well as adaptability. In an experiment that was quite similar to this one, which was carried out by Minshun Zhu et al., the composite material was produced by intercalating PPy into layered MXene [45]. It was stated that the capacitance increased to 203 mF cm^{-2} and that it retained 100% of its original value even after 20,000 cycles of charging and discharging. It was clear that the l-Ti_3C_2 particles stabilized the PPy backbones after studying the morphology of the film after charging and discharging cycles. This was demonstrated by the fact that there was no discernible change in the film's morphology. This stability might be attributed to the limiting of intercalated polymer backbones imposed during the cycling process by the bonds formed with MXene sheets. After going through 20,000 charging and discharging cycles, the pure PPy saw a dramatic shift in its shape. The ripples were no longer there, and aggregation could be seen. In addition, the PPy/ MXene film was utilized in the production of an all-solid-state supercapacitor. This all-solid-state supercapacitor demonstrated an excellent capacity of over 35 mF cm^{-2} and stable bending states in excess of 10,000 charging/discharging cycles. Because of the positive electrochemical properties of PPy, it is frequently employed in the production of flexible supercapacitors. The advantages seen following the integration of MXene have the potential to enhance research interest and product development to an exponentially higher level. By reducing the amount of time needed to charge, rapid charging and discharging cycles can be of significant assistance to wearable technologies such as smart watches, wireless earphones, and many other technologies that are still in the development stage.

Researchers have been interested in polyanilines as a result of their high conductivity; despite the fact that they were discovered more than 150 years ago, they belong to the family of semi-flexible rod polymers. Polyanilines are a type of polymer. PANIs are increasingly finding applications in the conductive coatings of yarns, which allows for the production of intelligent and multifunctional fibers. PANI can be easily synthesized by adjusting the pH value, however the process of its polymerization and the nature of its oxidation are complex. In the research that Huawei and his colleagues conducted, they made a composite material out of $Ti_3C_2T_x$ MXene and polyaniline (PANI) and gave it a sandwich intercalated structure. This was done so that they could analyze the microwave absorption capabilities of the material [46]. In the realm of energy storage, there is a lot of interest in the concept of a composite system that is harmonized with three different components and still manages to keep the qualities of each individual component. A graphene-encapsulated MXene Ti_2CT_x@polyaniline composite (GMP) material was proposed by Fu et al. as a material for supercapacitor electrodes. This material was achieved in a systematically stable configuration using several ternary nanomaterials [47]. The graphene-encapsulated MXene Ti_2CT_x@polyaniline composite (GMP) is a material that is made up of an accordion-like MXene Ti_2CT_x, an ionic-interactive PANI that is attached tightly onto the MXene, and a sheet of chemically converted graphene (CCG) for complete encapsulation (Figure 5.6).

By combining a suspension of MXene with distillate aniline monomers, several MXene/PANI composites with varying concentrations of PANI were produced. These composites were then introduced into the product. After the APS had been added for oxidative polymerization, the mixture was stirred for a prolonged period of time while submerged in a bath of freezing water. After that, the precipitate was washed with HCl and distilled water in order to create composites with varying water-to-solids ratios. After the polymerization process, it was discovered that minute particles were scattered throughout the surface. By manipulating the quantity of aniline present, it was possible to alter both the morphology of the sample and the size of its constituent particles. It has been hypothesized that the dielectric characteristics of the composite can be directly influenced by the amount of aniline that is

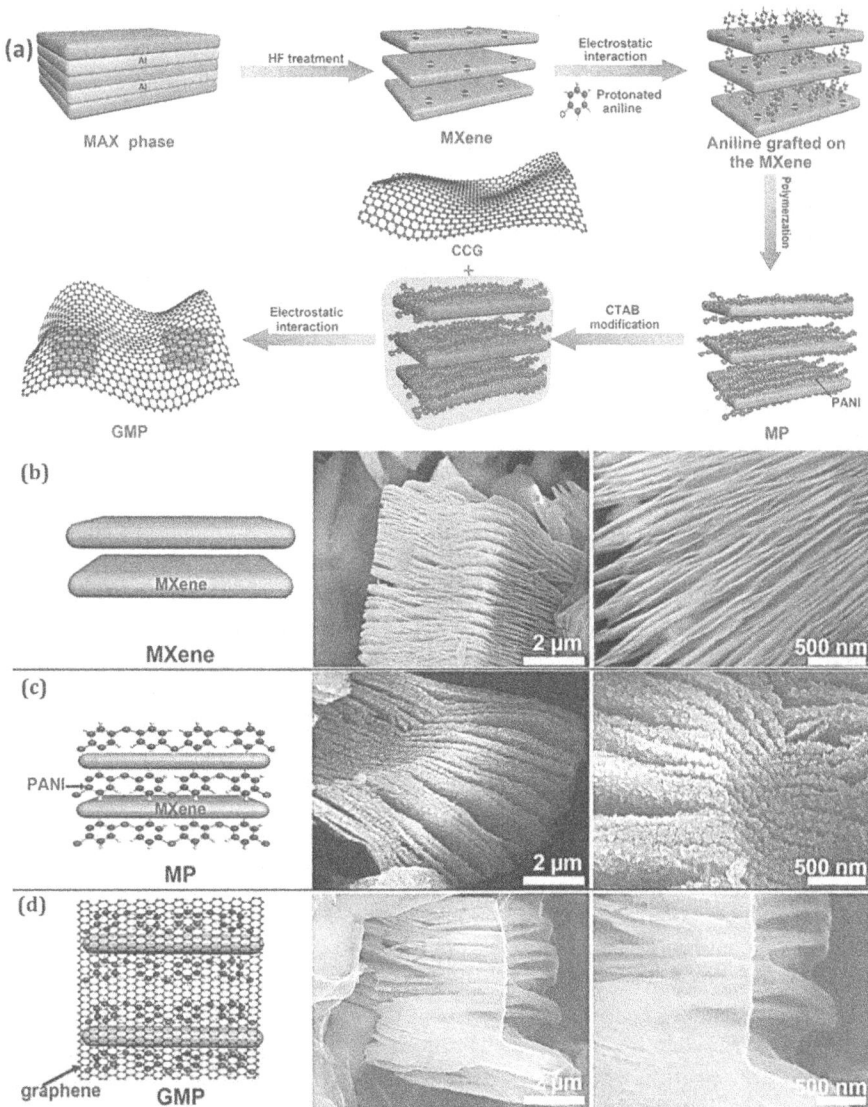

FIGURE 5.6 (a) The overall process of producing GMP as a diagram. Schematic illustrations and SEM images of (b) MXene Ti2CTx, (c) MP, and (d) GMP.

(Reproduced with the permission from Ref. [47], © 2021 American Chemical Society.)

present on its surface. Due to the unequal charge distribution and electromagnetic attenuation that occurred at each reflection site, the MXene/PANI exhibited a comparable model when compared to a model of a resistor-capacitor circuit. As a result of an increase in MXene's conductive routes, the sample with a mass ratio of 1:3 exhibited strong microwave absorption, which resulted in an Rl value of 56.3 dB at 13.80 GHz and an absorption efficiency of 99.9999% [46]. Because of its high rate, the efficiency of the absorption might be considered a possible state of the art in the future.

Thin $Ti_3C_2T_x$/PEDOT:PSS hybrid film, which provides good volumetric capacitance for asymmetric supercapacitors, may be made by filtration of $Ti_3C_2T_x$ inks and sulfuric acid treatment to remove the insulating PSS component. The self-restacking propensity of the flakes of MXene under the influence of intermolecular force boundaries the interaction of these carbon-nitride-based electrodes electrochemically [48]. As a result, the construction of this hybrid film offered the conductive PEDOT not only as a support but also plays a vital function as a conductive pathway to fast-track the electrochemical reaction. The hybrid film had achieved an interlayer spacing of 15.1 angstroms, which, when measured against the interlayer spacing in a pure MXene layer, is much bigger. This finding lends credence to the assertion that $Ti_3C_2T_x$ NSs are capable of obstructing the restacking process. The rapid diffusion and movement of electrolyte ions was considerably aided by the alternating layers of MXene and polymers, as well as the increase in interlayer spacing that was utilized. This, in turn, contributed to the enhancement of the capacity for rapid charging storage. $Ti_3C_2T_x$/poly(3,4-ethylenedioxythiophene):polystyrene sulfonate (PEDOT:PSS) conductive fibers were manufactured by the team under the supervision of Jizhen Zhang et al., and they showed exceptional results in conductivity, volumetric capacitance, and rate performance [49]. After the formation of the composite fiber, the morphology of the new fiber takes on the appearance of a corrugated surface. This is in contrast to the pure polymer fibers, which have a surface that is smooth and free of debris. In light of the fact that it became more difficult to gather fibers as a result of breakage with MXene loading larger than 70%, the team evaluated the qualities of fibers with a loading of less than 70%. Those made from hybrid fibers.

5.6 BIOPOLYMER-BASED MXene COMPOSITES

Natural polymer composites are of ecological interest owing to their biodegradability and bio-compatibility, which leads researchers into one of nature's amino polysaccharides widely known as Chitosan (CS). Among the different efforts to introduce MXene into polymers, natural polymer composites are of ecological value because of their biodegradability and bio-compatibility. CS is obtained by the deacetylation of chitin derived from a wide variety of insects and crustaceans. It possesses antioxidant activity, antimicrobial and hypoglycemic properties, as well as the ability to trap cholesterol and triglycerides [50]. Because of its molecular weight and the degree to which it is acetylated, the natural polymer's properties, such as its solubility, materials-forming capacity, and biodegradability, are highly desired in the fields of agricultural engineering, food engineering, and environmental engineering around the world [51].

Cellulose is one of the natural polymers that may be found in nature in the greatest abundance. Cellulose nanofiber, often known as CNF, is a relatively new type of material that is gaining popularity due to the structural roles it plays in trees. CNFs are the primary units that bear weight and are the most durable components in plants; as a result, CNFs find their place in today's world as bio-based construction materials. As the use of electronic communication tools continues to grow in prevalence, several preventative methods against electromagnetic shielding have also been developed and implemented. For increased EMI shielding, Cao et al. synthesized an MXene/CNF paper with a structure that was inspired by nacre [34]. This approach is only one of many that have been presented thus far. A vacuum-filtration-induced self-assembly procedure was used to create an ultrathin and flexible high-tensile and fracture-strain-resistant composite material. Other desirable properties of this composite are its ultrathin profile and its flexibility. The ultimate tensile strength was attained at 135.4 MPa and the strain at fracture was observed to be 16.7% by keeping the weight ratio between these two components. $Ti_3C_2T_x$ was synthesized from delaminated Ti_3AlC_2, and CNFs were isolated from garlic husk. For the purpose of demonstrating the composite paper's folding endurance, the paper was folded up to 14,260 times. This was possible because the layered structure, which resembled nacre, provided a strong interaction between the MXene and the CNF and resembled a "brick-mortar" reinforcing mechanism. Electrically, the composite paper with a thickness of just 47 μm demonstrated a conductivity of 739.4 S m^{-1}, and its efficacy in shielding electromagnetic interference was measured to be 25.8 dB at 12.4 GHz when it included 80% d-$Ti_3C_2T_x$. Regardless of the amount of d-Ti3C2Tx present, composite paper exhibited a high level of microwave absorption and a relatively low level of microwave reflection, which can be interpreted as a predominant electromagnetic interference shielding mechanism due to absorption. As a result, the use of this composite may be found to predominate in the devices that make use of microwaves, such as microwave ovens, cellular technologies, and remote sensing systems.

5.7 POSSIBILITIES FOR THE FUTURE

Researchers from all around the world have generated findings that show an increasing tendency that points to an increase in both the existing technologies and the mechanical qualities. This pattern is indicative of an improvement. MXene, despite the fact that it is still in the testing phase, has been used in a wide variety of scientific applications. These applications range from those that save lives in the medical and biological fields to those that treat wastewater. When combined with polymers, the developing 2D material has shown itself to be more than capable of supplying high power densities. However, when these power storage technologies are mixed with polymers, they are included with flexibility, biocompatibility, and many other advantages. As a result, MXene has proven to be the technology of the next generation for both the storage of energy and wearable technology.

5.8 SUMMARY

In this chapter, we focused on the polymer composites of the newly discovered 2D carbon nitrides known as MXene. These composites have demonstrated their value in applications such as energy storage, conductive coatings, and advances in mechanical qualities. MXene has produced outstanding results or, at the very least, a noticeable improvement over the technologies that are currently being used in a variety of fields, such as photothermal conversion, flexible supercapacitors, modified yarn energy storage, ultra-high flux and anti-fouling wastewater treatment membranes, solvent-resistant nanofiltration, electromagnetic interference (EMI) shielding, flame retardancy, and so on. These results mark the beginning of a new era. MXene has been acclaimed for its electrical characteristics; nevertheless, its high dispersion capabilities in solvents such as DMF have been shown to be successful in many industries. One of these fields is flame retardancy, in which the same substance was employed to reduce harmful flames. MXene, as a result of its multilayer structure, has demonstrated outstanding EMI and microwave absorption. Additionally, the fact that it is two-dimensional gives it an advantage over the conventional thick shielding options. When it comes to flexible super-capacitors, the same thing applies: the power density, the capacitance retention, the conductivity, and the flexibility have all seen significant improvements over their forebears, and when compared to the various 2D materials, MXene is either ahead of or is catching up to these titans, who have already made significant strides in these areas. Furthermore, MXene has ensured the beginning of a new era of power storage and distribution systems for the future generations in our power-hungry world, and they are on track to open up a universe of possibilities that are limited only by human imagination.

REFERENCES

1. Ganguly S, Bhawal P, Ravindren R, Das NC. Polymer nanocomposites for electromagnetic interference shielding: a review. Journal of Nanoscience and Nanotechnology 2018;18:7641–69.
2. Shao Y, Wang J, Wu H, Liu J, Aksay IA, Lin Y. Graphene based electrochemical sensors and biosensors: a review. Electroanalysis: An International Journal Devoted to Fundamental and Practical Aspects of Electroanalysis 2010;22:1027–36.
3. Mattevi C, Kim H, Chhowalla M. A review of chemical vapour deposition of graphene on copper. Journal of Materials Chemistry 2011;21:3324–34.
4. Bhimanapati GR, Lin Z, Meunier V, Jung Y, Cha J, Das S, et al. Recent advances in two-dimensional materials beyond graphene. ACS Nano 2015;9:11509–39.
5. Ganguly S, Kanovsky N, Das P, Gedanken A, Margel S. Photopolymerized thin coating of polypyrrole/graphene nanofiber/iron oxide onto nonpolar plastic for flexible electromagnetic radiation shielding, strain sensing, and non-contact heating applications. Advanced Materials Interfaces 2021;8:2101255.
6. Tang X, Guo X, Wu W, Wang G. 2D metal carbides and nitrides (MXenes) as high-performance electrode materials for Lithium-based batteries. Advanced Energy Materials 2018;8:1801897.
7. Anasori B, Lukatskaya M, Gogotsi Y. 2D metal carbides and nitrides (MXenes) for energy storage. Nature Reviews Materials 2017;2:16098.
8. Barsoum MW, El-Raghy T. The MAX phases: Unique new carbide and nitride materials: ternary ceramics turn out to be surprisingly soft and machinable, yet also heat-tolerant, strong and lightweight. American Scientist 2001;89:334–43.
9. Barsoum MW. The MN+ 1AXN phases: a new class of solids: thermodynamically stable nanolaminates. Progress in Solid State Chemistry 2000;28:201–81.
10. Zhang X, Xu J, Wang H, Zhang J, Yan H, Pan B, et al. Ultrathin nanosheets of MAX phases with enhanced thermal and mechanical properties in polymeric compositions: $Ti_3Si_{0.75}Al_{0.25}C_2$. Angewandte Chemie 2013;125:4457–61.
11. Naguib M, Kurtoglu M, Presser V, Lu J, Niu J, Heon M, et al. Two-dimensional nanocrystals produced by exfoliation of Ti_3AlC_2. Advanced Materials 2011;23:4248–53.

12. Berdiyorov GR, Mahmoud KA. Effect of surface termination on ion intercalation selectivity of bilayer $Ti_3C_2T_2$ (T= F, O and OH) MXene. Applied Surface Science 2017;416:725–30.

13. Yan J, Ren CE, Maleski K, Hatter CB, Anasori B, Urbankowski P, et al. Flexible MXene/ graphene films for ultrafast supercapacitors with outstanding volumetric capacitance. Advanced Functional Materials 2017;27:1701264.

14. Carey MS, Sokol M, Palmese GR, Barsoum MW. Water transport and thermomechanical properties of $Ti_3C_2T_x$ MXene epoxy nanocomposites. ACS Applied Materials & Interfaces 2019;11:39143–9.

15. Khazaei M, Mishra A, Venkataramanan NS, Singh AK, Yunoki S. Recent advances in MXenes: From fundamentals to applications. Current Opinion in Solid State and Materials Science 2019;23:164–78.

16. Habib T, Zhao X, Shah SA, Chen Y, Sun W, An H, et al. Oxidation stability of $Ti_3C_2T_x$ MXene nanosheets in solvents and composite films. npj 2D Materials and Applications 2019;3:1–6.

17. Mashtalir O, Lukatskaya MR, Zhao MQ, Barsoum MW, Gogotsi Y. Amine-assisted delamination of Nb2C MXene for Li-ion energy storage devices. Advanced Materials 2015;27:3501–6.

18. Xie Y, Dall'Agnese Y, Naguib M, Gogotsi Y, Barsoum MW, Zhuang HL, et al. Prediction and characterization of MXene nanosheet anodes for non-lithium-ion batteries. ACS Nano 2014;8:9606–15.

19. Das P, Ganguly S, Saha A, Noked M, Margel S, Gedanken A. Carbon-dots-initiated photopolymerization: an in situ synthetic approach for MXene/poly (norepinephrine)/ copper hybrid and its application for mitigating water pollution. ACS Applied Materials & Interfaces 2021;13:31038–50.

20. Ganguly S, Das P, Saha A, Noked M, Gedanken A, Margel S. Mussel-inspired polynorepinephrine/MXene-based magnetic nanohybrid for electromagnetic interference shielding in X-band and strain-sensing performance. Langmuir 2022;38:3936–50.

21. Jimmy J, Kandasubramanian B. Mxene functionalized polymer composites: synthesis and applications. European Polymer Journal 2020;122:109367.

22. Gore PM, Kandasubramanian B. Functionalized aramid fibers and composites for protective applications: a review. Industrial & Engineering Chemistry Research 2018;57:16537–63.

23. Gore PM, Zachariah S, Gupta P, Balasubramanian K. Multifunctional nano-engineered and bio-mimicking smart superhydrophobic reticulated ABS/fumed silica composite thin films with heat-sinking applications. RSC Advances 2016;6:105180–91.

24. Rastogi P, Kandasubramanian B. Breakthrough in the printing tactics for stimuli-responsive materials: 4D printing. Chemical Engineering Journal 2019;366:264–304.

25. Ganguly S, Ghosh S, Das P, Das TK, Ghosh SK, Das NC. Poly (N-vinylpyrrolidone)-stabilized colloidal graphene-reinforced poly (ethylene-co-methyl acrylate) to mitigate electromagnetic radiation pollution. Polymer Bulletin 2020;77:2923–43.

26. Rajavel K, Luo S, Wan Y, Yu X, Hu Y, Zhu P, et al. 2D $Ti_3C_2T_x$ MXene/polyvinylidene fluoride (PVDF) nanocomposites for attenuation of electromagnetic radiation with excellent heat dissipation. Composites Part A: Applied Science and Manufacturing 2020;129:105693.

27. Sun R, Zhang HB, Liu J, Xie X, Yang R, Li Y, et al. Highly conductive transition metal carbide/carbonitride (MXene)@ polystyrene nanocomposites fabricated by electrostatic assembly for highly efficient electromagnetic interference shielding. Advanced Functional Materials 2017;27:1702807.

28. Gu J-W, Zhang Q-Y, Li H-C, Tang Y-S, Kong J, Dang J. Study on preparation of SiO2/ epoxy resin hybrid materials by means of sol-gel. Polymer-Plastics Technology and Engineering 2007;46:1129–34.

29. Xu H, Yin X, Li X, Li M, Liang S, Zhang L, et al. Lightweight Ti_2CT_x MXene/poly (vinyl alcohol) composite foams for electromagnetic wave shielding with absorption-dominated feature. ACS Applied Materials & Interfaces 2019;11:10198–207.

30. Weng C, Wang G, Dai Z, Pei Y, Liu L, Zhang Z. Buckled AgNW/MXene hybrid hierarchical sponges for high-performance electromagnetic interference shielding. Nanoscale 2019;11:22804–12.

31. Zhao S, Zhang H-B, Luo J-Q, Wang Q-W, Xu B, Hong S, et al. Highly electrically conductive three-dimensional $Ti_3C_2T_x$ MXene/reduced graphene oxide hybrid aerogels with excellent electromagnetic interference shielding performances. ACS Nano 2018;12:11193–202.

32. Wang L, Qiu H, Song P, Zhang Y, Lu Y, Liang C, et al. 3D $Ti_3C_2T_x$ MXene/C hybrid foam/epoxy nanocomposites with superior electromagnetic interference shielding performances and robust mechanical properties. Composites Part A: Applied Science and Manufacturing 2019;123:293–300.

33. Song P, Qiu H, Wang L, Liu X, Zhang Y, Zhang J, et al. Honeycomb structural rGO-MXene/epoxy nanocomposites for superior electromagnetic interference shielding performance. Sustainable Materials and Technologies 2020;24:e00153.

34. Cao W-T, Chen F-F, Zhu Y-J, Zhang Y-G, Jiang Y-Y, Ma M-G, et al. Binary strengthening and toughening of MXene/cellulose nanofiber composite paper with nacre-inspired structure and superior electromagnetic interference shielding properties. ACS Nano 2018;12:4583–93.

35. Xie F, Jia F, Zhuo L, Lu Z, Si L, Huang J, et al. Ultrathin MXene/aramid nanofiber composite paper with excellent mechanical properties for efficient electromagnetic interference shielding. Nanoscale 2019;11:23382–91.

36. Zhan Z, Song Q, Zhou Z, Lu C. Ultrastrong and conductive MXene/cellulose nanofiber films enhanced by hierarchical nano-architecture and interfacial interaction for flexible electromagnetic interference shielding. Journal of Materials Chemistry C 2019;7:9820–9.

37. Yang H, Dai J, Liu X, Lin Y, Wang J, Wang L, et al. Layered PVB/Ba3Co2Fe24O41/Ti3C2 Mxene composite: enhanced electromagnetic wave absorption properties with high impedance match in a wide frequency range. Materials Chemistry and Physics 2017;200:179–86.

38. Park S-J, Seo M-K. Interface Science and Composites: Academic Press; 2011.

39. Zhang H, Wang L, Chen Q, Li P, Zhou A, Cao X, et al. Preparation, mechanical and anti-friction performance of MXene/polymer composites. Materials & Design 2016;92:682–9.

40. Han R, Ma X, Xie Y, Teng D, Zhang S. Preparation of a new 2D MXene/PES composite membrane with excellent hydrophilicity and high flux. RSC Advances 2017;7:56204–10.

41. Rasool K, Mahmoud KA, Johnson DJ, Helal M, Berdiyorov GR, Gogotsi Y. Efficient antibacterial membrane based on two-dimensional $Ti_3C_2T_x$ (MXene) nanosheets. Scientific Reports 2017;7:1–11.

42. Li R, Zhang L, Shi L, Wang P. MXene Ti_3C_2: an effective 2D light-to-heat conversion material. ACS Nano 2017;11:3752–9.

43. Tu S, Jiang Q, Zhang X, Alshareef HN. Large dielectric constant enhancement in MXene percolative polymer composites. ACS Nano 2018;12:3369–77.

44. Wu W, Wei D, Zhu J, Niu D, Wang F, Wang L, et al. Enhanced electrochemical performances of organ-like Ti_3C_2 MXenes/polypyrrole composites as supercapacitors electrode materials. Ceramics International 2019;45:7328–37.

45. Zhu M, Huang Y, Deng Q, Zhou J, Pei Z, Xue Q, et al. Highly flexible, freestanding supercapacitor electrode with enhanced performance obtained by hybridizing polypyrrole chains with MXene. Advanced Energy Materials 2016;6:1600969.

46. Wei H, Dong J, Fang X, Zheng W, Sun Y, Qian Y, et al. Ti3C2Tx MXene/polyaniline (PANI) sandwich intercalation structure composites constructed for microwave absorption. Composites Science and Technology 2019;169:52–9.

47. Fu J, Yun J, Wu S, Li L, Yu L, Kim KH. Architecturally robust graphene-encapsulated MXene Ti_2CT_x@ Polyaniline composite for high-performance pouch-type asymmetric supercapacitor. ACS Applied Materials & Interfaces 2018;10:34212–21.

48. Li L, Zhang N, Zhang M, Zhang X, Zhang Z. Flexible $Ti_3C_2T_x$/PEDOT: PSS films with outstanding volumetric capacitance for asymmetric supercapacitors. Dalton Transactions 2019;48:1747–56.

49. Zhang J, Seyedin S, Qin S, Wang Z, Moradi S, Yang F, et al. Highly conductive $Ti_3C_2T_x$ MXene hybrid fibers for flexible and elastic fiber-shaped supercapacitors. Small 2019;15:1804732.

50. Sarkar G, Orasugh JT, Saha NR, Roy I, Bhattacharyya A, Chattopadhyay AK, et al. Cellulose nanofibrils/chitosan based transdermal drug delivery vehicle for controlled release of ketorolac tromethamine. New Journal of Chemistry 2017;41:15312–9.

51. Lizardi-Mendoza J, Monal WMA, Valencia FMG. Chemical characteristics and functional properties of chitosan. Chitosan in the Preservation of Agricultural Commodities: Elsevier; 2016. p. 3–31.

6 Electrical Conductivity of MXenes-Based Polymer Composites

Sayani Biswas and Prashant S. Alegaonkar
Department of Physics
Central University of Punjab
Ghudda, India

CONTENTS

6.1 INTRODUCTION

The onset of extensive study of two-dimensional (2D) materials brought forward by graphene was a remarkable event [1]. The inter-dimensional scientific study is now turning out to be an important genre of material research as the market or industrial trend currently revolves around the miniaturization concept. Scaling down the dimension not only has physical or visible changes but also has great effects on the material's properties, which can be assessed by observing bond formations and atomic properties at nanoscale and then utilising the knowledge at an industrial level – the very essence of nanotechnology. According to Novoselov, Geim et al. graphene is an excellent material choice for metallic transistor because of its exciting electronic properties [1, 2]. This widely studied 2D material posed as an example that other 2D materials can also bring important changes in material science. In 2011, MXene was discovered and was subsequently declared as the new addition to

DOI: 10.1201/9781003281511-6

the two-dimensional family [3]. Since then, apart from the synthesis of MXene, the main topic of discussion has been its potential application in energy storage owing to its electronic properties. Till now the most experiments and the number of analysis made for utilising MXenes has been in the energy storage sector only. MXene is obtained from their precursor material: 3D layered solid MAX phase labelled after their composition $M_{n+1}AX_{n}$, where M is an early transition metal, A is a group 13 or 14 element and X is carbon or nitrogen with n = 1, 2, 3 generally [4]. The selective etching of the A layers from the MAX phases using a combination of etchants produces MXenes with chemical formula $M_{n+1}X_{n}T_{x}$ where T_{x} is a surface termination (-OH, -F, -O) obtained from the etchant used in the reaction [5]. The direct contrast to graphene is the unique combination of high electrical conductivity and strong hydrophilicity. The bare surface of graphene is hydrophobic, while the surface terminations in MXenes make the exposed lamella highly negative and hydrophilic in nature, which is crucial for any kind of composite formations. The fact that the surface terminations can be modified or completely altered is another point to be noted and appreciated about MXenes. Their 2D morphology and sheet-like structure and large surface area facilitates outstanding polymer composite formation. The atomic thick layers should theoretically aid in fabrication of nanocomposites (NCs) with enhanced mechanical and electrical properties. This combination has not been found in other 2D matrices with functionalised surfaces like graphene oxides, clays, and so on. A research paper explains the superior role of MXene nanosheets over graphene as hybridisation matrix producing enhanced interfacial electronic coupling [6]. Exfoliated MXene ($Ti_{3}C_{2}$) nanosheets and reduced graphene oxide (rGO) were utilised as hybridisation matrix for constructing MnO_{2}-$Ti_{3}C_{2}$ and MnO_{2}-rGO nanohybrids and then were subsequently tested as electrode materials for supercapacitors. For graphene, the serious obstacles were hydrophobicity and π-π interaction leading to strong self-stacking affinity of rGO sheets (Figure 6.1) [7]. MXenes are popularly being tested for energy storage purposes like Li-ion batteries [8], supercapacitors [9] and other applications in various sectors like environmental [10], catalysis [11], EMI shielding [12] and biomedical purposes [13], emphasising its diversified scope of usage.

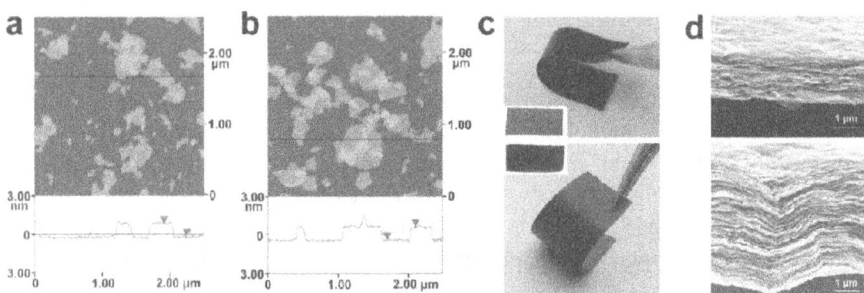

FIGURE 6.1   Characterisation images of GO and rGO nanosheets and paper. AFM images of (a) GO and (b) rGO layers; (c) images of flexible and freestanding GO and rGO paper (top and bottom respectively); (d) SEM measurements of thickness of GO and rGO.

(Reprinted with permission from [14].)

6.1.1   A BRIEF ACCOUNT OF SYNTHESIS OF MXenes

Being in the relatively early stage of development of synthesis, the concept of manu-facturing MXene is still a far shot. MXenes are derived from their precursor mate-rial MAX phases via a wet chemical etching method, as shown in Figure 6.2. MAX phases are ternary layered solids with 'A' layers which is typically a group 13 or 14 elements sandwiched between transition metal layer and carbide/nitride layer. The bond strengths vary in these solids, which makes destroying the A layer selectively possible without any significant change in the other bonds phases; making them the perfect candidates for top-down approaches applied to obtain 2D materials. The M–X bond nature is strong and directional covalent, and the M–A bond is weaker than the M–X bonds [4]. MAX phases reported till date have hexagonal-layered $P6_3$ symmetry and come in a wide variety in terms of combinations of M, A, and X atoms. The different bond strengths in MAX phases make them the perfect candi-dates for top-down approaches applied to obtain 2D materials. The first etchant used to etch out the 'A' layers was HF by Naguib and co-workers [3]. Stacked 2D MXenes were produced which could be sonicated or delaminated to obtain individual sheets as shown in Figure 6.2. Soon after, other milder etchants produced toxic wastes. The idea of using HF in large-scale production is dangerous and needs excessive safety protocols to be imposed. Ghidiu et al. implemented the etchant LiF/HCl mixed solu-tion which produced HF in situ and had the same etching capability at 35°C. The result was $Ti_3C_2T_x$ 'clay' with high volumetric capacitance [15]. Other fluoride salt and acid solutions were also experimented with, like KF/HCl [16], CaF_2, NaF [17] in HCl or H_2SO_4. In the same time frame, other methods were also sought out to avoid the use of acid altogether in order to minimise risk factors associated with them.

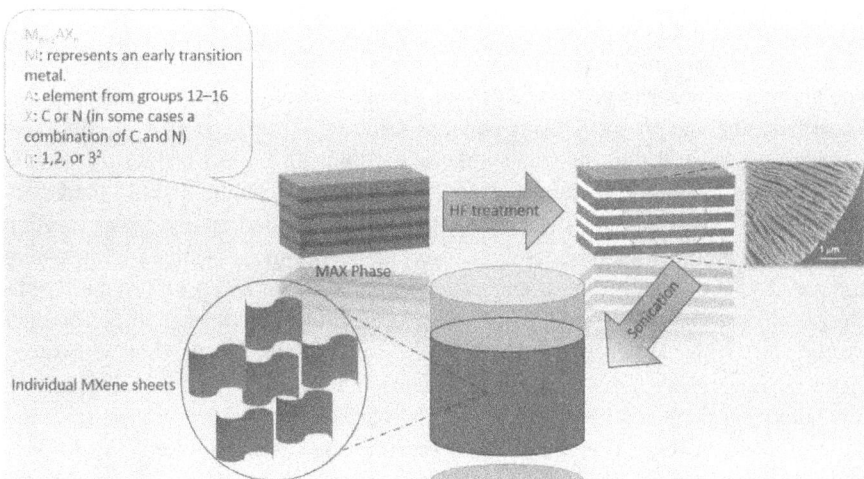

FIGURE 6.2   Schematic diagram of exfoliation of 'A' layer from MAX phase to obtain MXene using the HF treatment.

(Reprinted with permission from [19].)

Halim et al. introduced a weakly acidic and environment friendly etchant NH_4HF_2 following the in situ HF process as an etchant [18]. This method is a single step synthesis method with simultaneous intercalation of ammonium ions between the MXene layers, which is a plus point. The reaction mechanism is as follows:

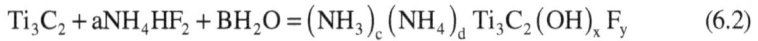

$$Ti_3AlC_2 + 3NH_4HF_2 = (NH_4)_3 AlF_6 + Ti_3C_2 + 3/2\ H_2 \qquad (6.1)$$

$$Ti_3C_2 + aNH_4HF_2 + BH_2O = (NH_3)_c (NH_4)_d Ti_3C_2 (OH)_x F_y \qquad (6.2)$$

In reaction (1) the NH_3 and NH_4^{+1} ion get intercalated between $Ti_3C_2T_x$ layers, which is great for any kind of capacitive action. This intercalation makes the delamination of MXene stacked layers easier, as it swells up the MXene. The distance between the layers increases, which can be witnessed by the c lattice parameter, which was increased by 25% compared to the films etched using HF. NH_4HF_2 is less hazardous than HF and is therefore a milder etchant. In 2017, Feng et al. further reported that the surface of the 2D flakes is negative due to the surface functional groups present. As a result, the cations (NH_4^+) are attracted to the negative surface and get attached onto the surface, enlarging the c lattice parameter of $Ti_3C_2T_x$ accordingly [20].

Contrary to the wet chemical etching method involving aqueous solutions, water-free etching is also coined as a fluorine containing etching method using organic solvents. MXenes have high negative zeta potential, which aids in forming highly concentrated colloidal solutions through molecular agglomeration intercalated between the layers simultaneously. This optimises the lengthy procedure of organic molecular intercalation using DMSO [21], Amine [22], TBAOH [23] and others. Zhao et al. used NH_4NF_2 and a series of organic solvents – dimethyl sulfoxide, N,N Dimethylformamide, propylene carbonate and N-methyl-2-pyrrolidone – to form a concoction that not only produces HF molecules for etching but also provides complexes of NH_4^+/organic solvent molecules for intercalation between MXene layers. The absence of water molecules led to the distribution of terminal groups very different from the one produced using aqueous solutions.

There are several fluorine-free synthesis methods also which primarily include hydrothermal method [24], carbon vapour deposition (CVD) method [25] and electrochemical synthesis [26]. In a detailed account provided recently, the reaction mechanisms and the diverse conditions associated with these synthesis routes were compared [27]. Although these methods provide a viable and green route to synthesise MXenes going HF free, their resultant yield and the stability of the product was considerably low when compared to the conventional HF or fluorine-assisted methods. Regardless of the synthesis route chosen, most of the work is still at laboratory stage and there is still the search for the perfect blend of preparation to create the optimised MXene. The recent trend and increased research interest in MXene polymer composites is depicted in Figure 6.3.

It is clear from the figure that the number and scale of research is very small compared to the research carried out on pure MXenes. The synthesis mechanism and the ratio of the precursor to etchant materials are very important and have significant effects in categorising MXenes for a particular application. These conditions

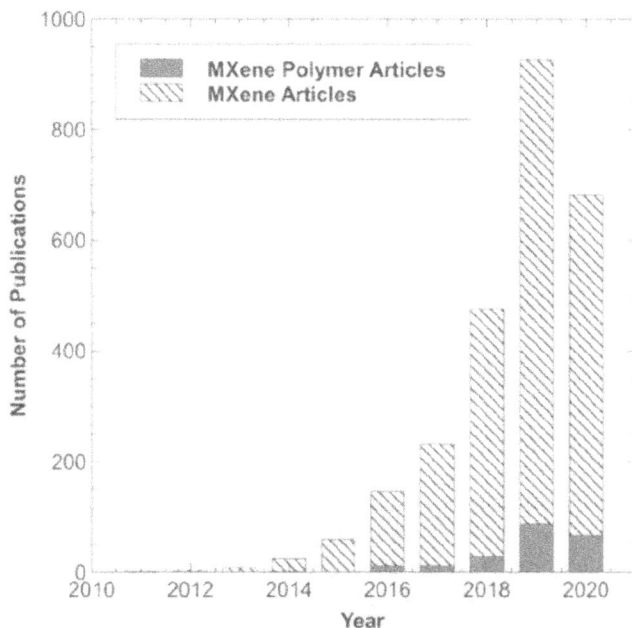

FIGURE 6.3   Number of articles published per year on MXene and MXene polymer NCs.
(Reprinted with permission from [28].)

during the synthesis dictate the distribution of surface termination groups, which is very important for further discussion on MXene modification and its potential polymerisation.

6.1.2   MXene Modification and Surface Chemistry

Tailoring the performance of MXenes and potentially expanding its application is possible for the presence of abundant terminations, sheet-like structures, and adjusting the interlayer gap [29]. Covalent and non-covalent surface modifications of pure MXene are carried out recently, resulting in functionalised MXene of choice [30]. Covalent modification is manifested by reaction between the surface functionalised groups and non-covalent modifications including hydrogen bonding, Van der Waals force and electrostatic interaction. The type and concentration of surface terminations can be controlled by the choice and concentration of etchants used and also tuning the etching time and temperature. $-O$ and $-F$ terminations can be easily obtained via conventional fluoride containing etchants or HF. Some post-etching processes like treating with hydrazine and annealing at a range of temperatures also affect the surface terminations. DFT calculations have been utilised to determine the effect of surface terminations on electronic conductivity of MXenes [31]. The calculations support the fact that the density of states at Fermi level (DOS) is affected by the nature and spread of surface terminations. While annealing at $700–775°C$ there was a significant decrease of –F concentration, which resulted in enhanced electrical conductivity [32]. In case of MXene bond formation with polymers, there are two ways of execution: *ex situ* or *in situ* polymerisation. The *ex situ* blending approach is more appealing for producing MXene/polymer composite. So hydrogen bonding and electrostatic interactions are the efficient form of interaction between MXene and polymers [33, 34]. In terms of electronic properties, pristine MXenes exhibit metallic properties, but there are MXenes that show semiconductivity and topological insulativity. MXenes with –F terminations show semiconductive properties [35] and the choice of M atoms categorise MXenes as insulators, like Mo and W-based MXenes [36, 37]. There are several reports on comparative studies of pristine and functionalised MXenes which shows that –O terminated MXenes tend to be semiconductors and –OH, –F terminated ones are metallic in nature [38, 39]. These studies are very important for understanding the future direction of MXene production and ways to improve the overall performance based on the predetermined requirements.

Polymers are being widely used in our everyday life because of their stability, excellent properties and easy to process nature. But their applications are limited as a single polymer sheet. For emerging popularity of wearable technology, polymers coated with MXene are being considered as a solution owing to their flexible nature and high mechanical/tensile strength. With proper surface modifications and desired functionalisation, MXene can be made to work well with polymers. In this chapter, the discussion has been put forward regarding MXenes as emerging nanofillers for polymer composites delivering high performance narrowing down to the electronic properties.

6.2 MXene-POLYMER COMPOSITES: SYNTHESIS AND ELECTRONIC PROPERTIES

Since the 1950s, the delaminated 2D nanosheets have made an appearance and garnered interest, but very few were tested out to be conductive [40]. After the emergence of graphene followed now by the discovery of MXenes, studies show a positive sign now about conducting sheets that can change the direction of future electronics. Composites of MXene are stirring potential ideas in technological and scientific applications worldwide. MXenes when not delaminated are multilayered structures with layers stacked one over the other, with a clear resemblance to exfoliated graphite [41]. Delaminated solo layers provide large accessible surface areas that help in reinforcing polymer composites. Among the other possible choices of composites, polymers are the more flexible option, as they are versatile, compatible and cost effective. MXenes can form stable colloidal solutions which makes them prone to dispersion in polymer-based solutions. Figure 6.4 illustrates the SEM images of MAX, multilayered MXenes and also schematics of MXene colloidal solution. Both multilayered MXene and its colloidal solution need to be studied separately in order to figure out the perfect blend for polymer nanocomposites, as they both offer different attributes to the entire process. MXenes electronic properties range from metal and semiconductor to topological insulativity. The main factor determining the property is the composition of MXenes. Other factors that have influence are tuning surface functionalisation and controllable thickness [42, 43]. The bare MXene

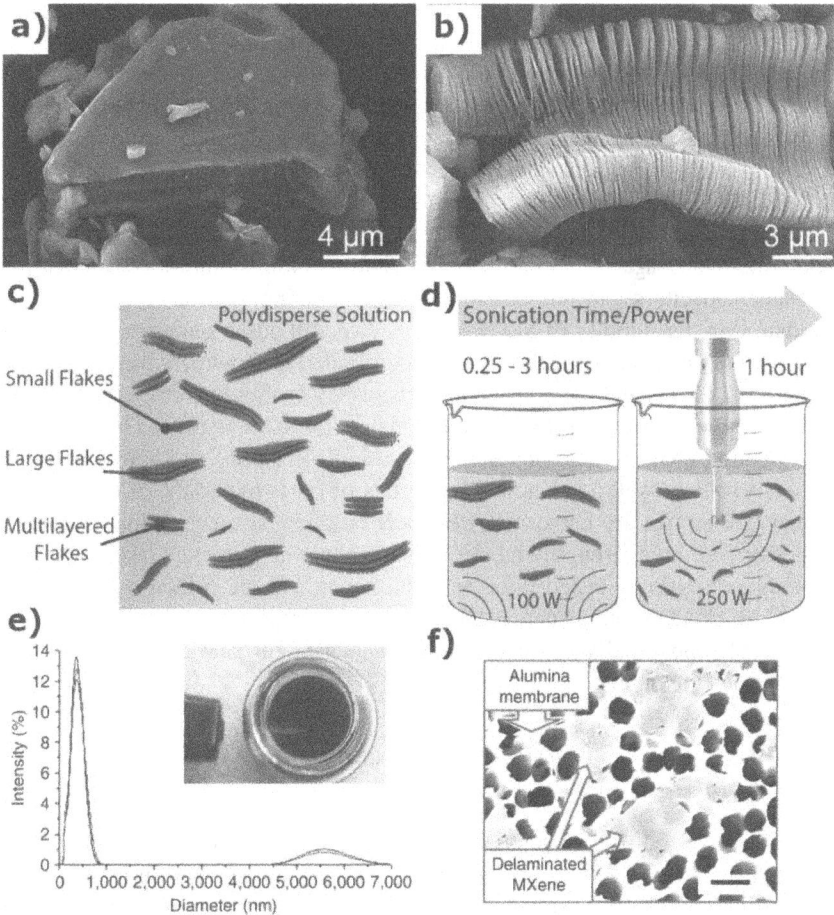

FIGURE 6.4 SEM images of (a) Ti_3AlC_2, (b) $Ti_3C_2T_x$ after HF etching, (c) schematic diagram of MXene suspension of various sized flakes, (d) schematic illustration of effect of sonication on the dimension of suspended MXene flakes over time and increased power, (e) the intensity vs. diameter graph depicting light-scattering experiments of MXene colloidal solution post-sonication with an additional image of Tyndall effect, (f) SEM micrograph of drop-cast MXene colloids on an alumina substrate showing transparent single MXene sheets.

(Reprinted with permission from: Naguib et al. (a) and (b) [44], Maleski et al. [45] (c) and (d), Mashtalir et al. [46] (e) and (f). [28].)

and most number of functionalised MXenes are found out to have metallic nature. The MXenes with –OH groups show very low work function value, less than 2.8 eV, while –O terminated ones have been shown to possess ultra-high work function, even larger than Pt, which has the highest work function among elemental metals. It has been shown that conductivity of MXenes can range from 1 Scm^{-1} to 1000 Scm^{-1} [27]. Usually for superconducting and topological insulativity states to manifest, a single type of surface functionalisation is required theoretically.

But all the MXenes produced till date have mixed terminations with percentage distribution of groups depending on synthesis mechanisms. So, MXenes are considered for making the industrial break in energy storage and conductivity because they can be moulded and altered accordingly. However MXenes have a high tendency of getting oxidised and are also prone to chemical degradation forming TiO_2. Habib et al. confirmed the speedy degradation of MXene coatings under ultraviolet radiation exposure after running a comparative study under various other conditions [47]. When the scope of application is zoomed into electrochemical and future generation energy storage purposes like Li-S batteries [48], sodium-ion batteries [49], microsupercapacitors [50] and so on, MXenes are being considered to be a viable candidate. But the stability of MXenes on its own still poses some serious issues in terms of lifetime of the devices they are incorporated in. Polymers were introduced way before MXenes and have been under scientific scrutiny long enough to be identified as the prime choice as hybridisation matrix for 2D materials. They constitute a continuously growing field in science because of versatility and ability to change the mechanical as well as chemical properties of other materials. Some merits of polymer composites are corrosion resistance, fatigue resistance, impact resistance and possessing great strength and stiffness. Conducting polymers like polypyrrole are being exhaustingly tested to be embedded with MXene, which on its own has acquired much attention in energy storage and transfer purposes. The investigations are limited but potentially of great impact, considering further development in areas of science and technology.

6.2.1 POLY(VINYL ALCOHOL) AND POLYDIALLYLDIMETHYLAMMONIUM COMPOSITE OF MXene

Ling et al. reported the fabrication of flexible conductive free-standing MXene-polymer films via VAF (vacuum-assisted filtration) that showed impressive thermal stability and electrical conductivity. The two polymers chosen were cationic polymer polydiallyldimethylammonium (PDDA) and poly(vinyl alcohol) (PVA), and MXene produced was $Ti_3C_2T_x$ from Ti_3AlC_2 MAX phase precursor. $Ti_3C_2T_x$ surface is negative, so PDDAs cationic nature will make the bond formation easier. On the other hand, PVA has an abundance of hydroxyl group and is highly soluble in water, which makes it usable in gel, composites and electrolytes [51–54]. Figure 6.5a shows the schematic diagram of the fabrication process on a surface level. The negative zeta-potential of MXene helped form a colloidal solution (35 mL; 0.34 mg.mL^{-1}) and PDDA solution was added drop wise to the solution and subsequently magnetically stirred for 24 h. The same procedure was followed for PVA where aqueous solution of PVA and MXene was mixed and sonicated in an ice bath for 15 minutes. Four different ratios of MXene to PVA mixture were prepared (90:10, 80:20, 60:40, and 40:60) and the samples were labelled respectively by the ratios [55]. From XRD analysis (Figure 6.5b), it was found out that in both the $Ti_3C_2T_x$/PVA and $Ti_3C_2T_x$/PDDA free-standing films the XRD patterns showed a downward shift to lower angles compared to pure $Ti_3C_2T_x$ films (from 6.5° to 4.7° for $Ti_3C_2T_x$/PDDA and from 6.5° to a range of 4.8° to 6° for increased PVA loading).

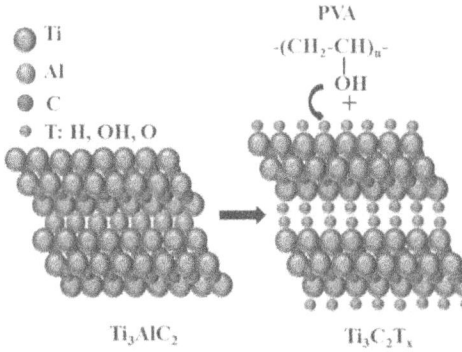

FIGURE 6.5 (a)   Schematic illustration of synthesis process of $Ti_3C_2T_x$/PVA NCs.

This was the result of intercalation of PDDA and PVA molecules between the $Ti_3C_2T_x$ flakes. After performing electrochemical testing, capacitive performances of $Ti_3C_2T_x$ based films were reported. For pure Ti_3C_2 films, the volumetric capacitances recorded earlier crossed 300 F/cm^3 which is higher than previously determined values of CDC (180 F/cm^3) [57], graphene gel films (260 F/cm^3) [58] and activated graphene (60–100F/cm^3) [59]. When $Ti_3C_2T_x$/PDDA was used as electrode material in supercapacitors, the resulting volumetric capacitance was slightly lower (296F/cm^3 at 2 mV/s) because of its slightly lower density (2.71 g/cm^3) than pristine $Ti_3C_2T_x$ (3.19 g/cm^3). Formerly PVA and KOH combination has been proved to be good for

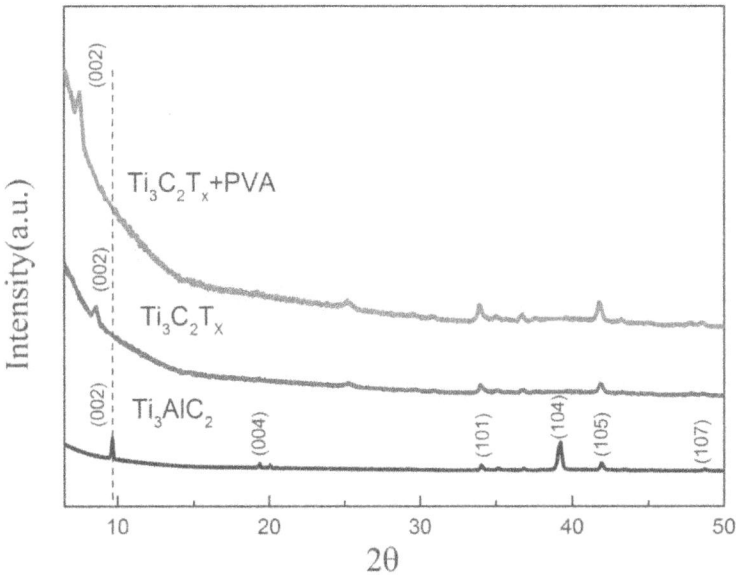

FIGURE 6.5 (b)   XRD patterns of Ti_3AlC_2, $Ti_3C_2T_x$ flakes and $Ti_3C_2T_x$/PVA NCs.

(Reprinted with permission from [56].)

electrical energy storage devices. So in case of $Ti_3C_2T_x$/PVA films, KOH was added and the X-ray spectroscopy showed uniform homogeneous dispersion of potassium over the $Ti_3C_2T_x$/PVA-KOH films. Conductivity recorded here is 11,200 S/m and the volumetric capacitance was found out to be 528 F/cm^2 at 2 mV/s and 306 F/cm^3 at 100 mV/s. After 10,000 cycles at 5 A/g, the capacitance dropped to 314 F/cm^3, which demonstrated good cyclic stability.

Another instance of MXene reinforced PVA nanofiber fabrication via electro-spinning technique was put forward in 2017. Taking into consideration the merits of PVA, $Ti_3C_2T_x$/PVA and $Ti_3C_2T_x$/CNC/PVA composite was produced for running a comparative study and assessing the ability to tap into the properties. CNC (carboxylated nanocellulose) was prepared from date palm leaves, and multilay-ered ML-$Ti_3C_2T_x$ was delaminated in 70% ethanol/water. The conductivity of pure $Ti_3C_2T_x$ ink obtained was 19.4 μS/cm. Finally, in 15 wt% solution of PVA, $Ti_3C_2T_x$ and CNC were added and the resultant solution was ultra-sonicated for 20 minutes while inside an ice bath. The nanofibers were electrospun at room temperature and at voltage 17 kV, speed 200 RPM and flow rate of 0.3 mL/h. The dc conductivity characterisation showed the characteristic behaviour of polymers. The conductivity vs. frequency curves show plateau at lower frequencies and this relates to the dc con-ductivity (Figure 6.6). When the curve rises monotonically, the conductivity shows a dependency of sth power of frequency, s ranging from 0.7 to 1.0. The electrical con-ductivity for 0.14 wt% $Ti_3C_2T_x$ in $Ti_3C_2T_x$/CNC/PVA nanocomposite was found to be 0.8 S/cm^{-1} and as the $Ti_3C_2T_x$ content increased the conductivity also increased. The samples demonstrated narrow band-gap semiconducting nature and low conductor barrier due to –F and –OH functionalisation, which is perfect for anode material in Li-ion batteries. The electrospinning process of preparation is the key factor here for obtaining the enhanced electrical conductivity [60].

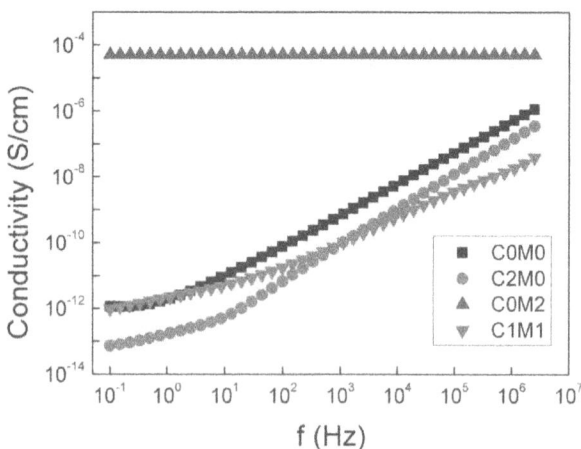

FIGURE 6.6 Graph of DC conductivity vs. frequency at 20°C for the $Ti_3C_2T_x$/CNC/PVA samples with varying ratios and compositions.

(Reprinted with permission from [60].)

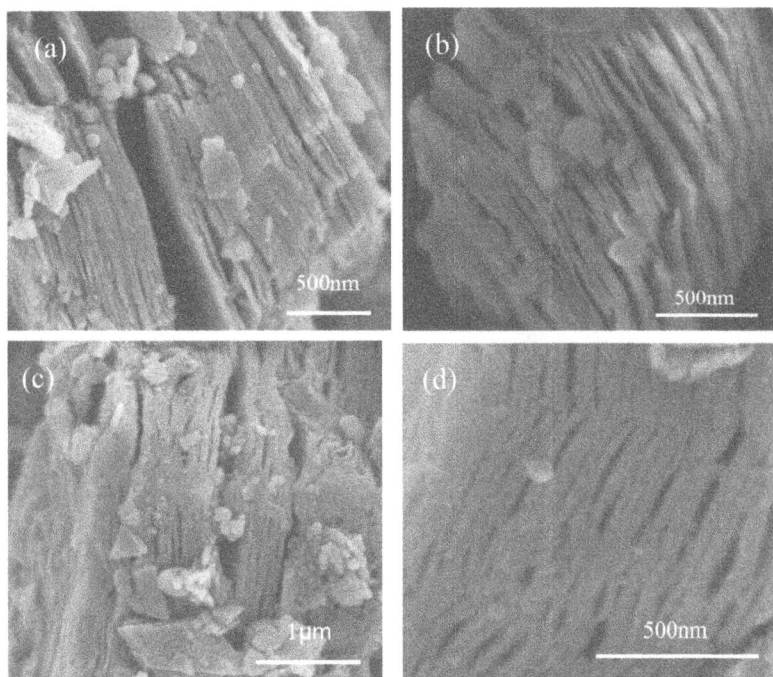

FIGURE 6.7   SEM images of (a) $Ti_3C_2T_x$ MXene, (b) zoomed in view of $Ti_3C_2T_x$, (c) $Ti_3C_2T_x$/ PVA, (d) zoomed in view of $Ti_3C_2T_x$/PVA.

(Reprinted with permission from [56].)

Liu et al. also confined polymer between MXene flakes and studied its thermal stability and thermal conductivity using polarised-laser power-dependent Raman spectroscopy and temperature-dependent Raman spectroscopy. The incorporation of PVA led to the stacking of the flakes due to reaction between hydroxyl group in PVA and –OH termination on MXene surface, resulting in Ti–O bond formation. The synthesised titanium carbide has the typical layered structure as shown in Figure 6.7. Without changing the crystal structure tremendously, this bond formation actually improved thermal stability of MXene. For $Ti_3C_2T_x$, the thermal conductivity was 55.8 W/(mK), and for $Ti_3C_2T_x$/PVA it was 47.6 W/(mK). These values are higher than most other 2D materials and pave the path for further advances in energy storage applications [56].

6.2.2   POLYANILINE COMPOSITE OF MXene

Polyaniline is another highly conductive polymer discovered over a century ago, and belongs to a family of semi-flexible rod polymers. In situ polymerisation of multilayered PANI-Ti_3C_2 composite at low temperature was performed by Ren et al. aiming at using the nanocomposite as electrode material in supercapacitors [61]. After etching Ti_3AlC_2, 100 μL aniline (ANI) was added to 0.2 g Ti_3C_2 dispersed in

30 ml 1 M HCl aqueous solution. While the solution was under magnetic stirring at 2°C for 30 minutes, 0.335g APS (ammonium persulfate) dissolved in 20 ml 1M HCl aqueous solution was added. The final mixture was stirred for 6 h at 2°C and subsequently the PANI-Ti$_3$C$_2$ precipitates were washed with deionised water. The CV curves at 10 mVs^{-1} of PANI-Ti$_3$C$_2$, PANI, Ti$_3$C$_2$ were obtained. Both Ti$_3$C$_2$ and PANI-Ti$_3$C$_2$ curves were rectangular in shape with peaks which can be associated with the redox reactions, a direct result of intercalation/deintercalation of Na$^+$ cations. The reaction mechanism of the entire process is follows:

$$PANI - Ti_3C_2 + Na^+ + e^- \longleftrightarrow PANI - Ti_3C_2Na \qquad (6.3)$$

$$Ti_3C_2 + Na^+ + e^- \longleftrightarrow NaTi_3C_2 \qquad (6.4)$$

For PANI-Ti$_3$C$_2$, the area under CV curve is larger than both pure PANI and MXene, as shown in Figure 6.8. So, specific capacitance of PANI-Ti$_3$C$_2$ electrode (126 Fg^{-1}) is higher than PANI (75 Fg^{-1}) and Ti$_3$C$_2$ (109 Fg^{-1}). GCD curves also support the higher specific capacitance of PANI-Ti$_3$C$_2$ demonstrated by CV curves.

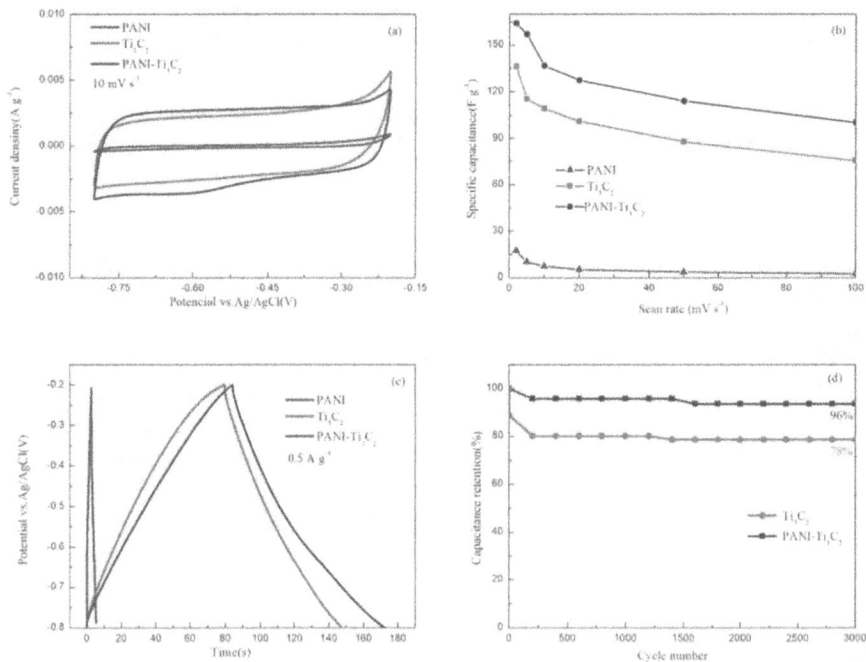

FIGURE 6.8 (a) CV curves of Ti$_3$C$_2$, PANI and Ti$_3$C$_2$T$_x$/PANI composite at scan rate of 10 mV/s; (b) specific capacitance vs. scan rate for Ti$_3$C$_2$, PANI and Ti$_3$C$_2$T$_x$/PANI composite; (c) GCD curves of Ti$_3$C$_2$, PANI and Ti$_3$C$_2$T$_x$/PANI composite at current densities of 0.5 A/g; (d) the galvanostatic cycling curves of Ti$_3$C$_2$, PANI and Ti$_3$C$_2$T$_x$/PANI electrode at 3 A/g.

At 2 mVs^{-1} specific capacitance obtained was 164 Fg^{-1} and there was 96% capacitive retention after 3000 cycles.

Another report highlighted the fabrication of hybrid Ti$_3$C$_2$T$_x$/PANI nanosheets by in situ polymerisation of aniline monomer on Ti$_3$C$_2$T$_x$ monolayers without the use of any additional oxidising chemicals. Al from Ti$_3$AlC$_2$ was etched by the minimally intensive layer delamination (MILD) method using LiF+HCl combo as etchant [62, 63]. 300 μL of aniline monomer dissolved in 20 ml aqueous solution of 1M HCl was added to MXenes on different ratios. The reason behind taking various ratios is demonstrating the improved high-rate electrochemical with decreasing the amount of polymer settled on Ti$_3$C$_2$T$_x$ flakes. Hybrid films of different thickness were fabricated by controlled deposition of PANI on Ti$_3$C$_2$T$_x$ with adjustable Ti$_3$C$_2$T$_x$ to aniline ratio during synthesis. Very high capacitances of 503 Fg^{-1} and 1682 Fcm^{-3} at 2 mVs^{-1} scan rate were obtained for 4 μm thick composite, and the reason of such enhanced value was the elevated ion transport properties of electrodes due to PANI deposition. Even at the thickness of 45 μm, the capacitance was 371 Fg^{-1} (1141 Fcm^{-3}) and 287 Fg^{-1} (884 Fcm^{-3}) at scan rates of 2 mVs^{-1} and 20 mVs^{-1} with 98% capacitance retention over 10,000 cycles, which is higher than the pristine Ti$_3$C$_2$T$_x$ electrodes [64].

In 2020, another research work showcased the synthesis of Ti$_3$C$_2$T$_x$/PANI composite through chemical oxidative polymerisation of aniline monomer on Ti$_3$C$_2$T$_x$ MXene layers under acidic conditions. Here the surface functionalisation –OH, -O is acting as nucleation sites and can aid with aniline monomer deposition. Stacked multilayered Ti$_3$C$_2$ was obtained by etching Ti$_3$AlC$_2$ powders using LiF/HCl (1 g LiF dissolved in 30 mL of 6 M HCl solution) mild combo etchant. The solution was put under magnetic stirring, and Ti$_3$AlC$_2$ powder was gradually added with parameters maintained at 40°C for 24 hours. Once the stirring commenced, the resultant sediments were washed in deionised water and vacuum dried. DMSO assisted intercalation, followed by centrifugation at 3500 rpm for 60 minutes, produced single 2D flakes. The –F termination was replaced with –OH by stirring the resultant powder in 1 M KOH solution, which causes K$^+$ ion intercalation between the layers. The final solution mixture was obtained using APS-assisted polymerisation and the black-green product of Ti$_3$C$_2$T$_x$/PANI was put under electrochemical measurements. Cyclic voltammetry (CV) and Galvanostatic Charge/Discharge tests were run and the electrolyte used was aqueous solution of 1 M H$_2$SO$_4$, which provided protons as dopants for PANI to trigger its electrical conductivity. The CV curves at the scan rate of 10 mVs^{-1} in potential range of -0.2 to 0.6 V showed that area under the MXene/PANI curve was much higher than pure PANI and pure MXene as there is a direct proportionality between specific capacitance and the integral area covered by CV curves [65]. So, it is clear that MXene/PANI electrode has higher specific capacitance. The GCD curves were all non-linear, which is indicative of pseudocapacitive behaviour [66]. The values of specific capacitance at current densities respectively are 556.2 Fg^{-1} at 0.5 Ag^{-1}, 496.5 Fg^{-1} at 1 Ag^{-1}, 467.5 Fg^{-1} at 2 Ag^{-1} and 437.5 Fg^{-1} at 5 Ag^{-1}. These values are quite high when compared to other reported values in this specific genre, like MXene CNT (150 Fg^{-1} at 2 mVs^{-1}) [67], Ti$_3$C$_2$T$_x$/PPy (416 Fg^{-1} at 5 mVs^{-1}) [68]. Further, there is a 91.6% capacitance retention capability over 5000 cycles, which is a validation of its stability [69].

6.2.3 POLYPYRROLE COMPOSITE OF MXene

Polypyrrole (PPy) is a heterocyclic polymer that has garnered interest since its appearance in 1968 mainly because of its electrochemical properties. The first instance of incorporating conducting PPy between $Ti_3C_2T_x$ layers was put forward by Boota et al. in 2016 [70]. PPy molecules get intercalated between the layers, and the PPy chains are well aligned with the layers due to hydrogen bonding. The polymerisation was strictly oxidant free and produced brittle films which are opposite to flexible $Ti_3C_2T_x$ films. CV curves of $PPY/Ti_3C_2T_x$ in the ratio 1:2 at all scan rates showed rectangular figures with no peaks and high specific capacitance suggesting pseudocapacitive nature. The highest capacitance of 416 Fg^{-1} at 5 mVs^{l} was obtained for 2:1 composite of $Ti_3C_2T_x/PPy$, which is higher than the value of pristine $Ti_3C_2T_x$. The improved and accessible MXene surfaces as well as the pseudocapacitive contribution from 8 wt% doped PPy corresponding to 88 Fg^{-1} are the reasons behind the enhanced capacitance [71, 72]. The volumetric capacitance was found to be 1000 Fcm^{-3}, the highest value of electrodes based on MXenes reported till that time. For the 1:1 composite, the increased pyrrole concentration caused a sharp decline in capacitance because of the presence of undoped PPy and the deposition of a thick polymer layer which caused hindrance in charge transport. So for the 1:2 composition, further cycle life tests showed 92% retention for 25,000 cycles at 100 mVs^{-1}, surpassing previously reported PPy-based electrodes [73–75].

Wu et al. prepared organ like Ti_3C_2/PPy composite via APS induced in situ polymerisation method with different volume percentages of pyrrole content. Here HF etching was carried out on Ti_3AlC_2 to obtain Ti_3C_2, and comparative study was run for pure Ti_3C_2, pure PPy, Ti_3C_2/PPy-2 (26μL), Ti_3C_2/PPy-1 (13 μL), Ti_3C_2/PPy-3 (52 μL), Ti_3C_2/PPy-4 (104 μL) and Ti_3C_2/PPy-5 (208 μL) (the bracketed amounts denotes the PPy content in each sample) [76]. SEM analysis showed the PPy is agglomeration of spherical particles and is deposited on Ti_3C_2 layers evenly, as shown in Figure 6.9. Increased PPy loading manifested as increased thickness of Ti_3C_2 layers.

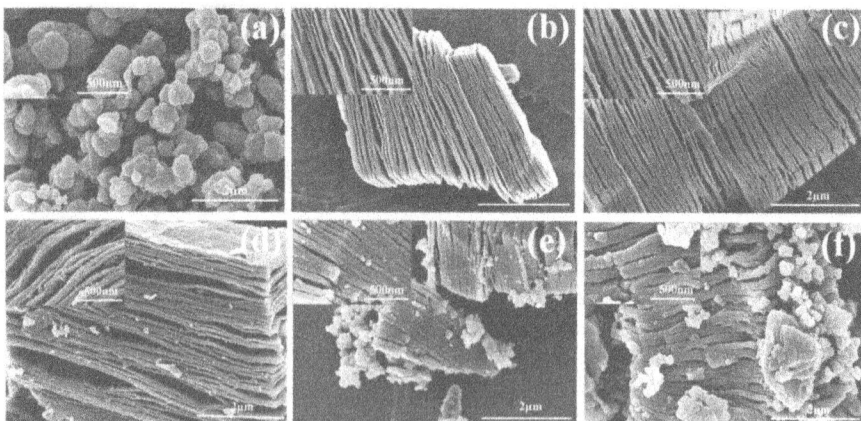

FIGURE 6.9 SEM images of (a) pure PPy, (b) pure Ti_3C_2 MXene, (c) Ti_3C_2/PPy-2, (d) Ti_3C_2/PPy-3, (e) Ti_3C_2/PPy-4, (f) Ti_3C_2/PPy-5.

(Reprinted with permission from [76].)

This particular synthesis method depends on the hydrogen bond formation and the action of electrostatic forces between the MXene surface and the long chains of PPy.

For sample Ti_3C_2/PPy-2 the loading is even without self-agglomeration of PPy at the sides of the sample. Electrochemical measurements carried out show that Ti_3C_2/ PPy-2 demonstrated the largest specific capacitance value of 184.36 Fg^{-1} at a scan rate of $2mVs^{-1}$, which is 1.37 times higher than pristine MXene electrodes. The CV curves showed no deformation in its rectangular shape at higher scan rates of 100 mVs^{-1}, suggesting far better rate performance and cycle stability. The benefits of the synergistic effect between highly conductive Ti_3C_2 and PPy with unique doping and de doping qualities really shined through in this nanocomposite. The prevention of agglomeration of PPy directly affected the charge transport and ion diffusion, leading to a higher conductivity.

6.2.4 Polypropylene Composite of MXene

Targeting the effect of EMI shielding, MXene/polypropylene nanocomposite was fabricated very recently, which turned out to be highly conductive with compact and continuous matrix MXene nanosheets. LiF/HCl solution was used for etching Ti_3AlC_2 to obtain $Ti_3C_2T_x$ MXene nanosheets. The oxygen plasma treated PP textile was first dipped in PEI solution, then washed and dried and subsequently decorated on MXene nanosheets. The dip coating process was repeated to adjust the MXene content [77]. Propylene is an insulating polymer and so its electrical conductivity measured was typically low ($2.0 * 10^{-12}$ Sm^{-1}). When the MXene content was increased, there was an observable transition from insulation to conduction with rising value of electrical conductivity. The relationship followed by MXene content and the electrical conductivity it imparted was the power law [78]:

$$\sigma = \sigma_0 \left[\left(\varphi - \varphi_c \right) \left(1 - \varphi_c \right) \right]^t$$

where σ is the electrical conductivity of the composite; σ_0 is the intrinsic conductivity constant of $Ti_3C_2T_x$ MXene; and φ, φ_c, t are the volume fraction of MXene, percolation threshold and critical exponent respectively [79, 80]. The electrical conductivity of the MXene-PP nanocomposite with 1.78 vol% of MXene in it was measured to be 192.9 Sm^{-1} and it increased to 437.5 Sm^{-1} for a MXene content of 2.12 vol%. These values, when compared to previously reported values of other MXene-polymer composites, are very impressive, despite the fact that PP is an insulating polymer. When tested for the EMI shielding effect, it was determined that the EMI shielding effectiveness (SE) elevates with an increase in MXene loading. 21 dB of EMI Se over X-band at 0.34 vol% of MXene content was obtained, which passed the nanocomposite for commercial applications. A sharp increase of EMI SE value to $\approx$ 43dB was observed at 1.27 vol% of MXene content, manifesting an excellent EMI shielding performance greater than 60 dB in the X-band. The peak value achieved was with only 2.12 vol% of MXene making the nanocomposite usable in military and civil applications. This paves a new pathway for MXene-based polymer composites to be incorporated in various other electrical and EMI shielding prospects [81].

6.2.5   Polyacrylamide Composite of MXene

Like PVA, polyacrylamide (PAM) is also a water soluble polymer and can form hydrogen-bond for which it is widely used in industrial applications [82, 83]. DMSO assisted intercalation of $Ti_3C_2T_x$ imparts considerable swelling in MXene structure, causing complete delamination of the flakes. Due to negative zeta potential of MXenes, it forms a stable colloidal solution in water, and this behaviour provides a favourable interaction environment for polymers. Naguib et al. performed the DMSO associated intercalation of $Ti_3C_2T_x$ MXene and then mixed with aqueous PAM solution to fabricate MXene/PAM nanocomposite. HF treated Ti_3AlC_2 produced $Ti_3C_2T_x$ and then 1 g of the powder was stirred with 10 mL DMSO for 18 hours at room temperature. After washing, 10 wt% aqueous solution of PAM was added in varying ratios. The estimated MXene vol% content by TGA data ranged from 1.7 to 100. At just 1.7 vol% (6 wt%) of MXene loading, the electrical conductivity was 3.3 * 10^{-2} Sm^{-1}. Further increase in conductivity was recorded, reaching a value of 3.3 Sm^{-1} at a MXene loading of 44.5 vol% (75 wt%). This work shows the potential formation of highly conductive network in the PAM matrix where the conductivity is contributed mainly by the MXene flakes. PAM in particular provided a higher rate of dispersion of MXene flakes, which shows that water-soluble polymer and hygroscopic intercalated MXene is a perfect combination for MXene-polymer composite [84].

6.2.6   Polyethylene Glycol (PEG) Composite of MXene

Over the 6000 phase change materials (PCMs) discovered till date, polyethylene glycol (PEG) has gained huge attention because of low cost and favourable thermodynamic properties like high phase transition enthalpy, moderate transition temperature and so on [85–87]. The quick phase change behaviour of PEG from solid to liquid state and its low thermal conductivity comes with the risk factor of leakage during the transition. So in order to form stable PCMs, PEG is often encapsulated to keep it in solid state during the phase transition. Lu et al. fabricated Ti_3C_2@PEG composite by direct vacuum impregnation method. In the final product maximum adsorption mass fraction of PEG was determined to be 77.5 wt%. Both pure PEG and PEG@MXene was tested for thermal conductivity and electrical conductivity. The low thermal conductivity (0.285 W/Mk) of pure PEG was improved to 7.2 times higher value (2.052 W/mK) in PEG@MXene (Figure 6.10). The MXene skeleton provided with excellent heat transfer along the interconnected networks is the reason for the enhancement in values. When the electrical conductivity was compared, PEG@MXene showed decent electrical conductivity of 10.41 S/m, which is much higher than the low value of 10^{-11} S/m for PEG [88]. So, the PEG@MXene blend provided the desired result by taking advantage of PEG and MXene individually and making a conductive NC film.

6.2.7   Poly(vinylidene fluoride) (PVDF)/Poly(vinylidene fluoride) Trifluoroethylene (P(VDF-TrFE)) Composite of MXene

Poly(vinylidene fluoride) is a type of fluoropolymer having 35–70% crystallinity (semicrystalline) and an extended crisscrossed chain. PVDF has shown good

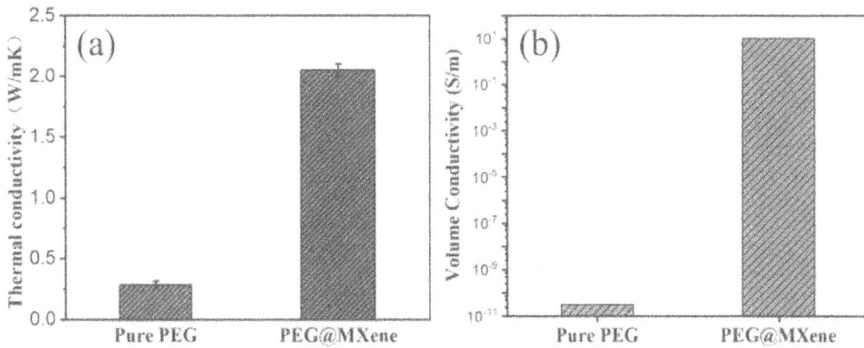

FIGURE 6.10   Bar graphs demonstrating thermal conductivity and volume conductivity of pristine PEG and PEG@MXene.

(Reprinted with permission from [88].)

compatibility with various other polymers, which makes it commercially usable as blends for acrylics. Li et al. fabricated a gradient sandwich like MXene/PVDF composite for achieving substantial improvement in dielectric constant as well as breakdown strength [89]. The $Ti_3C_2T_x$ MXene nanosheets synthesised from Ti_3AlC_2 MAX phase via MILD method were put under a series of processes to obtain the final product. Pure PVDF films were produced by spin coating technique, and prepared MXene solution was spray coated on PVDF films separately. The individual spin coated pure PVDF films and the spray coated MXene/PVDF films were detached, cut into square pieces and laid out layer by layer overlapping each other followed by hot pressing for 15 minutes at 180°C. The final product was labelled as xMXene/yPVDF where x and y denote the number of layers of MXene and PVDF respectively. To curb the dielectric loss in percolative composites offering less compatibility with polymer matrices, multilayered structure has been adapted as a strategy [90–92]. PVDF layers provided good insulating effect on the MXene layers and no visible voids were present in the SEM images, which suggests the absence of percolation phenomenon. The enhanced interfacial polarisation caused a hike in dielectric constant measured at 1 kHz from 10.5 of pure PVDF to a value of 41 for 4 MXene/5 PVDF film. Also increase in the number of layers increased the dielectric constant as well as the loss associated with it. For 4 MXene/5 PVDF film, excellent broadband dielectric behaviour was observed, which remained at a high value of 78.4% at 1 kHz and reached up to 32.2 at 1 MHz. The dielectric loss was taken down to 0.2, which is a promising result for applications.

Wu et al. also synthesised PVDF/MXene nanocomposites by dispersing exfoliated MXene in DI water followed by addition of 0.15g CTAB in 500 mL of MXene aqueous solution. PVDF/DMF solution with 10 wt% PVDF was mixed to CTAB modified MXene and the resulting nanocomposite was written as PVDF-X (X refers to wt% of MXene@CTAB) (Figure 6.11a). The dielectric properties of pristine MXene, PVDF and the PVDF-X composite were compared and the observation made was a gradual increase of dielectric permittivity value with increasing MXene@CTAB

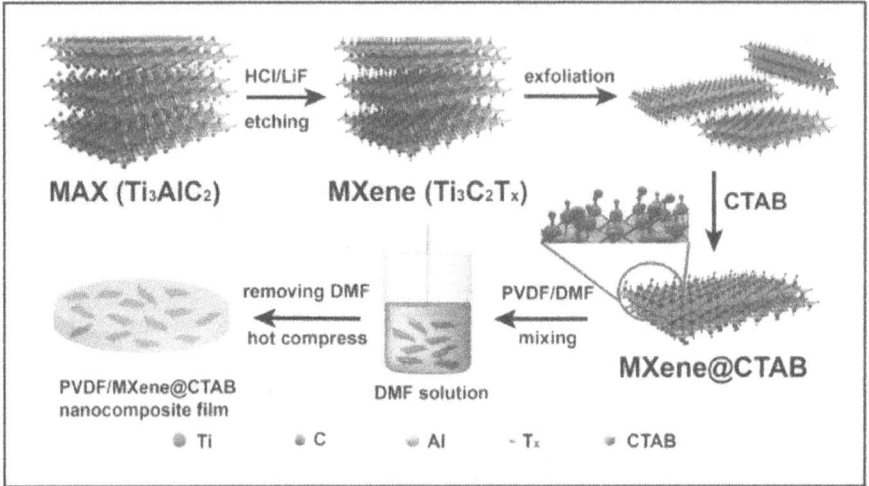

FIGURE 6.11 (a) Schematic illustration depicting preparation process of PVDF/MXene@ CTAB NC films.

concentration. The MXene@CTAB acts as miniature conductive capacitors inside PVDF matrix and also installs polarisation effects at the interface. The XRD patterns (Figure 6.11b) show the presence of a prominent shoulder due to the addition of MXene@CTAB showing the transfer formation of α-phase and β-phase of PVDF. The AC conductivity sharply increased to 3.99 10^{-6} S/m for PVDF-10 compared to 5.92 $*10^{-8}$ S/m for pure PVDF-0 at 1 kHz. 7 wt% increase in MXene content

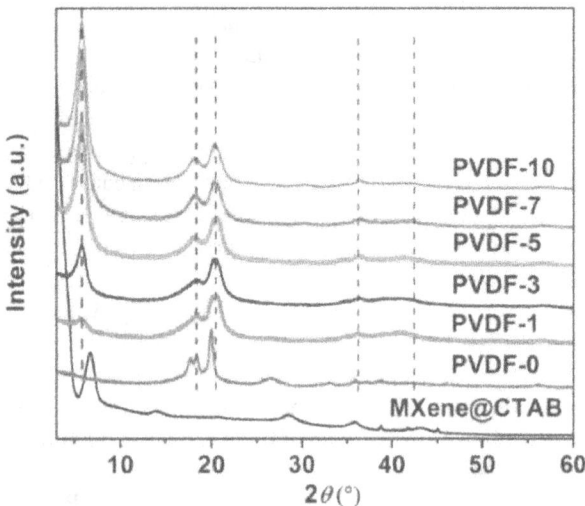

FIGURE 6.11 (b) XRD curves of different MXene@CTAB content in PVDF NCs.

(Reprinted with permission from [93].)

provided with a conductive network, and the CTAB provided insulating effect in order to suppress dielectric loss. Finally the dielectric permittivity obtained was 82.1, which is higher than pure PVDF by 8.5 times [93].

Highly sensitive pressure sensors with fast response at low energy consumption are being developed for better functioning of wearable electronic devices [94–96]. A study put forward the preparation of hydrophobic inorganic/organic polymerised films using PVDF-TrFE and multilayered $Ti_3C_2T_x$ by the spin-coating process. This material was prepared for fabricating piezoresistive pressure sensor. The PVDF-TrFE wrapped MXene resisted the spontaneous oxidation of the MXene and thus provides a stable matrix. MXene to $Ti_3C_2T_x$@P(VDF-TrFE) ratio was optimized to 1:1.6 at which the sensor showed the highest gauge factor of 817.4k/Pa from 0.072 to 0.74 kPa region. There was also 99% stability retention even after exposure to ambient air for 20 weeks. This is a great leap towards the future of wearable electronic devices engineering and conceptualisation, which is going to be quite desirable commercially and also mark great advances in technology [97].

6.3 CONCLUDING REMARKS

The combination of conducting polymer and MXene can pave pathways in diversified fields of applications without being limited in a single sector. In this review, the focus of discussion being the conductivity of MXene-polymer NCs, examples have been highlighted for futuristic application in energy storage, EMI shielding, wearable electronics and so on. Having said that, there is a real scarcity of research works in this particular field, which makes practical realisation of the NCs produced quite difficult. The literature documenting the progressive NCs still needs a lot of work, as there is a need to perfect the blend between polymer and MXene for obtaining optimised results. Also, other MXenes must be utilised in this particular cause for diversified results, as almost 99% of work till date is surrounding titanium carbide (Ti_3C_2). Nevertheless, these small breakthroughs are testament of a better future and potential mass usage of MXenes and MXene-NCs in material science.

REFERENCES

1. Novoselov, K.S.; Geim, A.K.; Morozov, S.V.; Jiang, D.; Zhang, Y.; Dubonos, S.V.; Grigorieva, I.V.; Firsov, A.A. Electric field effect in atomically thin carbon films. *Science* **2004**, *306*, 666–669, https://doi.org/10.1126/science.1102896.
2. Geim, A.; Novoselov, K. The rise of graphene, 2007. *Nat. Mater.* **2010**, *6*, 183.
3. Naguib, M.; Kurtoglu, M.; Presser, V.; Lu, J.; Niu, J.; Heon, M.; Hultman, L.; Gogotsi, Y.; Barsoum, M.W. Two-dimensional nanocrystals produced by exfoliation of Ti_3AlC_2. *Adv. Mater.* **2011**, *23*, 4248–4253, https://doi.org/10.1002/adma.201102306
4. Radovic, M.; Barsoum, M.W. MAX phases: Bridging the gap between metals and ceramics. *Am. Ceram. Soc. Bull.* **2013**, *92*, 20–27.
5. Benchakar, M.; Loupias, L.; Garnero, C.; Bilyk, T.; Morais, C.; Canaff, C.; Guignard, N.; Morisset, S.; Pazniak, H.; Hurand, S.; et al. One MAX phase, different MXenes: A guideline to understand the crucial role of etching conditions on $Ti_3C_2T_x$ surface chemistry. *Appl. Surf. Sci.* **2020**, *530*, 147209, https://doi.org/10.1016/j.apsusc.2020.147209.

6. Jin, X.; Shin, S.J.; Kim, N.; Kang, B.; Piao, H.; Choy, J.H.; Kim, H.; Hwang, S.J. Superior role of MXene nanosheet as hybridization matrix over graphene in enhancing interfacial electronic coupling and functionalities of metal oxide. *Nano Energy* **2018**, *53*, 841–848, https://doi.org/10.1016/j.nanoen.2018.09.055.

7. Jin, X.; Adpakpang, K.; Kim, I.Y.; Oh, S.M.; Lee, N. S.; Hwang, S.J. An effective way to optimize the functionality of graphene-based Nanocomposite: Use of the colloidal mixture of graphene and inorganic nanosheets, *Sci. Rep.* **2015**, *5*, 11057.

8. Liang, X.; Garsuch, A.; Nazar, L.F. Sulfur cathodes based on conductive MXene nanosheets for high-performance lithium-sulfur batteries. *Angew. Chem.* **2015**, *127*, 3979–3983, https://doi.org/10.1002/ange.201410174.

9. Lukatskaya, M.R.; Mashtalir, O.; Ren, C.E.; Dall'Agnese, Y.; Rozier, P.; Taberna, P.L.; Gogotsi, Y. Cation intercalation and high volumetric capacitance of two-dimensional titanium carbide. *Science* **2013**, *341*, 1502–1505.

10. Ren, C.E.; Hatzell, K.B.; Alhabeb, M.; Ling, Z.; Mahmoud, K.A.; Gogotsi, Y. Charge- and size-selective ion sieving through $Ti_3C_2T_x$ MXene membranes. *J. Phys. Chem. Lett.* **2015**, *6*, 4026–4031.

11. Seh, Z.W.; Fredrickson, K.D.; Anasori, B.; Kibsgaard, J.; Strickler, A.L.; Lukatskaya, M.R.; Vojvodic, A. Two-dimensional molybdenum carbide (MXene) as an efficient electrocatalyst for hydrogen evolution. *ACS Energy Lett.* **2016**, *1*, 589–594.

12. Shahzad, F.; Alhabeb, M.; Hatter, C.B.; Anasori, B.; Hong, S.M.; Koo, C.M.; Gogotsi, Y. Electromagnetic interference shielding with 2D transition metal carbides (MXenes). *Science* **2016**, *353*, 1137–1140, https://doi.org/10.1126/science.aag2421.

13. Anasori, B.; Gogotsi, Û.G. *2D Metal Carbides and Nitrides (MXenes)*; Springer: Berlin, Germany, **2019**.

14. Hu, W.; Peng, C.; Luo, W.; Lv, M.; Li, X.; Li, D.; Huang, Q.; Fan, C. Graphene-based antibacterial paper. *ACS Nano* **2010**, *4*, 4317–4323, https://doi.org/10.1021/nn101097v.

15. Ghidiu, M.; Lukatskaya, M.R.; Zhao, M.Q.; Gogotsi, Y.; Barsoum, M.W. Conductive two-dimensional titanium carbide 'clay' with high volumetric capacitance. *Nature* **2014**, *516*, 78–81.

16. Wu, M.; Wang, B.; Hu, Q.; Wang, L.; Zhou, A. The synthesis process and thermal stability of V2C MXene. *Materials* **2018**, *11*, no. 11, 2112, https://doi.org/10.3390/ma11112112.

17. Wang, X.; Garnero, C.; Rochard, G.; Magne, D.; Morisset, S.; Hurand, S.; Coutanceau, C. A new etching environment (FeF_3/HCl) for the synthesis of two-dimensional titanium carbide MXenes: A route towards selective reactivity vs. water. *J. Mater. Chem. A* **2017**, *5*, 22012–22023.

18. Halim, J.; Lukatskaya, M.R.; Cook, K.M.; Lu, J.; Smith, C.R.; Näslund, L.; May, S.J.; Hultman, L.; Gogotsi, Y.; Eklund, P.; et al. Transparent conductive two-dimensional titanium carbide epitaxial thin films. *Chem. Mater.* **2014**, *26*, 2374–2381, https://doi.org/10.1021/cm500641a.

19. Jimmy, J.; Kandasubramanian, B. MXene functionalized polymer composites: Synthesis and applications. *Eur. Polym. J.* **2019**, 109367, https://doi.org/10.1016/j.eurpolymj.2019.109367.

20. Feng, A.; Yu, Y.; Jiang, F.; Wang, Y.; Mi, L.; Yu, Y.; Song, L. Fabrication and thermal stability of NH_4HF_2-etched Ti_3C_2 MXene. *Ceram. Int.* **2017**, *43*, 6322–6328, https://doi.org/10.1016/j.ceramint.2017.02.039.

21. Mashtalir, O.; Naguib, M.; Mochalin, V.; Dall'Agnese, Y.; Heon, M.; Barsoum, M.W.; Gogotsi, Y. Intercalation and delamination of layered carbides and carbonitrides. *Nat. Commun.* **2013**, *4*, 1716, https://doi.org/10.1038/ncomms2664.

22. Mashtalir, O.; Lukatskaya, M.R.; Zhao, M.Q.; Barsoum, M.W.; Gogotsi, Y. Amine-assisted delamination of Nb2C MXene for Li-ion energy storage devices. *Adv. Mater.* **2015**, *27*, 3501–3506, https://doi.org/10.1002/adma.201500604.

23. Naguib, M.; Unocic, R.R.; Armstrong, B.L.; Nanda, J. Large-scale delamination of multi-layers transition metal carbides and carbonitrides "MXenes". *Dalton Trans.* **2015**, *44*, 9353–9358, https://doi.org/10.1039/c5dt01247c.

24. Wang, L.; Zhang, H.; Wang, B.; Shen, C.; Zhang, C.; Hu, Q.; Zhou, A.; Liu, B. Synthesis and electrochemical performance of $Ti_3C_2T_x$ with hydrothermal process. *Electron. Mater. Lett.* **2016**, *12*, 702–710, https://doi.org/10.1007/s13391-016-6088-z.

25. Xu, C.; Wang, L.; Liu, Z.; Chen, L.; Guo, J.; Kang, N.; Ma, X.L.; Cheng, H.M.; Ren, W. Large-area high-quality 2D ultrathin Mo_2C superconducting crystals. *Nat. Mater.* **2015**, *14*, 1135–1141, https://doi.org/10.1038/nmat4374.

26. Yang, S.; Zhang, P.; Wang, F.; Ricciardulli, A.G.; Lohe, M.R.; Blom, P.W.M.; Feng, X. Fluoride-free synthesis of two-dimensional titanium carbide (MXene) using a binary aqueous system. *Angew. Chem. Int. Ed.* **2018**, *57*, 15491–15495, https://doi.org/10.1002/anie.201809662.

27. Biswas, S.; Alegaonkar, P.S. MXene: Evolutions in chemical synthesis and recent advances in applications. *Surfaces* **2022**, *5*, no. 1, 1–34, https://doi.org/10.3390/surfaces5010001.

28. Carey, M.; Barsoum, M.W. MXene polymer nanocomposites: A review. *Mater. Today Adv.* **2021**, *9*, 100120, https://doi.org/10.1016/j.mtadv.2020.100120.

29. Chen, W.Y.; Lai, S.N.; Yen, C.C.; Jiang, X.; Peroulis, D.; Stanciu, L.A. Surface functionalization of $Ti_3C_2T_x$ MXene with highly reliable superhydrophobic protection for volatile organic compounds sensing. *ACS Nano* **2020**, *14*, no. 9, 11490–11501, https://doi.org/10.1021/acsnano.0c03896.

30. Gong, K.; Zhou, K.; Qian, X.; Shi, C.; Yu, B. MXene as emerging nanofillers for high-performance polymer composites: A review. *Compos. B. Eng.* **2021**, https://doi.org/10.1016/j.compositesb.2021.108867.

31. Hart, J.L.; Hantanasirisakul, K.; Lang, A.C.; Anasori, B.; Pinto, D.; Pivak, Y.; van Omme, J.T.; May, S.J.; Gogotsi, Y.; Taheri, M.L. Control of MXenes' electronic properties through termination and intercalation. *Nat. Commun.* **2019**, *10*, 1–10.

32. Schultz, T.; Frey, N.C.; Hantanasirisakul, K.; Park, S.; May, S.J.; Shenoy, V.B.; Gogotsi, Y.; Koch, N. Surface termination dependent work function and electronic properties of Ti3C2T x MXene. *Chem. Mater.* **2019**, *31*, no. 17, 6590–6597.

33. Tu, S.; Jiang, Q.; Zhang, X.; Alshareef, H.N. Large dielectric constant enhancement in MXene percolative polymer composites. *ACS Nano*, **2018**, *12*, 3369–3377.

34. Ling, Z.; Ren, C.E.; Zhao, M.Q.; Yang, J.; Giammarco, J.M.; Qiu, J.; Barsoum, M.W.; Gogotsi, Y. Flexible and conductive MXene films and nanocomposites with high capacitance. *Proc. Natl. Acad. Sci. U. S. A.* **2014**, *111*, 16676–16681.

35. Tang, Q.; Zhou, Z.; Shen, P. Are MXenes promising anode materials for Li ion batteries? Computational studies on electronic properties and Li storage capability of Ti3C2 and Ti3C2X2 (X= F, OH) monolayer. *J. Am. Chem. Soc.* **2012**, *134*, 16909–16916.

36. Weng, H.; Ranjbar, A.; Liang, Y.; Song, Z.; Khazaei, M.; Yunoki, S.; Arai, M.; Kawazoe, Y.; Fang, Z.; Dai, X. Large-gap two-dimensional topological insulator in oxygen functionalized MXene. *Phys. Rev. B: Condens. Matter Mater. Phys.* **2015**, *92*, 075436.

37. Khazaei, M.; Ranjbar, A.; Arai, M.; Yunoki, S. Topological insulators in the ordered double transition metals M 2′ M ″C 2 MXenes (M′= Mo, W; M ″= Ti, Zr, Hf). *Phys. Rev. B.* **2016**, *94*, 125152.

38. Khazaei M.; Arai, M.; Sasaki, T.; Chung, C.Y.; Venkataramanan, N.S.; Estili, M.; Sakka, Y.; Kawazoe, Y. Novel electronic and magnetic properties of two-dimensional transition metal carbides and nitrides. *Adv. Funct. Mater.* **2013**, *23*, 2185–2192.

39. Ibragimova, R.; Erhart, P.; Rinke, P.; Komsa, H.P. Surface functionalization of 2D MXenes: Trends in distribution, composition, and electronic properties. *J. Phys. Chem. Lett.* **2021**, *12*, no. 9, 2377–2384. https://doi.org/10.1021/acs.jpclett.0c03710.

40. Norrish K. The swelling of montmorillonite. *Discuss Faraday Soc.* **1954**, *18*, no. 0, 120–134.
41. Bannov, A.; Anastasya, T.; Yusin, S.; Ksenya, D.; Maksimovskii, E.; Ukhina, A.; Shibaev, A. Synthesis of the exfoliated graphite from graphite oxide. *Adv. Mater. Res.* **2014**, *1085*, 171–175. https://doi.org/10.4028/www.scientific.net/AMR.1085.171.
42. Khazaei, M.; Arai, M.; Sasaki, T.; Ranjbar, A.; Liang, Y.; Yunoki, S. OH-terminated two-dimensional transition metal carbides and nitrides as ultralow work function materials. *Phys. Rev. B: Condens. Matter Mater. Phys.* **2015**, *92*, 075411.
43. Liu, Y.; Xiao, H.; Goddard, W.A. Schottky-barrier-free contacts with two-dimensional semiconductors by surface-engineered MXenes. *J. Am. Chem. Soc.* **2016**, *138*, 15853–15856.
44. Naguib, M.; Mashtalir, O.; Carle, J.; Presser, V.; Lu, J.; Hultman, L.; Gogotsi, Y.; Barsoum, M.W. Two-dimensional transition metal carbides. *ACS Nano* **2012**, *6*, 1322–1331.
45. Maleski, K.; Ren, C.E.; Zhao, M.Q.; Anasori, B.; Gogotsi, Y. Size-dependent physical and electrochemical properties of two-dimensional MXene flakes. *ACS Appl. Mater. Interfaces* **2018**, *10*, 24491–24498.
46. Mashtalir, O.; Naguib, M.; Mochalin, V.N.; Dall'Agnese, Y.; Heon, M.; Barsoum, M.W.; Gogotsi, Y. Intercalation and delamination of layered carbides and carbonitrides. *Nat. Commun.* **2013**, *4*, 1716.
47. Habib, T.; Zhao, X.; Shah, S.A.; Chen, Y.; Sun, W.; An, H.; Lutkenhaus, J.L.; Radovic, M.; Green, M.J. Oxidation stability of $Ti_3C_2T_x$ MXene nanosheets in solvents and composite films. *Npj 2D Mater. Appl.* **2019**, *3*, 1–6. https://doi.org/10.1038/s41699-019-0089-3.
48. Balach, J.M.; Giebeler, L. MXenes and the progress of Li–S battery development—A perspective. *J. Phys. Energy* **2020**, *3*, 021002, https://doi.org/10.1088/2515-7655/abd5c4.
49. Aslam, M.K.; AlGarni, T.S.; Javed, M.S.; Ahmad Shah, S.S.; Hussain, S.; Xu, M. 2D MXene materials for sodium ion batteries: A review on energy storage, *J. Energy Storage* **2021**, *37*, 102478, ISSN 2352 152X, https://doi.org/10.1016/j.est.2021.102478.
50. Peng, Y.Y.; Akuzum, B.; Kurra, N.; Zhao, M.Q.; Alhabeb, M.; Anasori, B.; Kumbur, E.C.; Alshareef, H.N.; Ger, M.D.; Gogotsi, Y. All-MXene (2D titanium carbide) solid-state microsupercapacitors for on-chip energy storage. *Energy Environ. Sci.* **2016**, *9*, 2847–2854, https://doi.org/10.1039/c6ee01717g.
51. Podsiadlo, P.; et al. Ultrastrong and stiff layered polymer nanocomposites. *Science* **2007**, *318*, no. 5847, 80–83.
52. Wang, J.; Cheng, Q.; Lin, L.; Jiang, L. Synergistic toughening of bioinspired poly (vinyl alcohol)-clay-nanofibrillar cellulose artificial nacre. *ACS Nano* **2014**, *8*, no. 3, 2739–2745.
53. Liu, L.; et al. High mechanical performance of layered graphene oxide/poly(vinyl alcohol) nanocomposite films. *Small* **2013**, *9*, no. 14, 2466–2472.
54. Lu, X.; Yu, M.; Wang, G.; Tong, Y.; Li, Y. Flexible solid-state supercapacitors: Design, fabrication and applications. *Energy Environ Sci.* **2014**, *7*, 2160–2181.
55. Ling, Z.; Ren, C.; Zhao, M.Q.; Yang, J.; Giammarco, J.; Qiu, J.; Barsoum, M.; Gogotsi, Y. Flexible and conductive MXene films and nanocomposites with high capacitance. *Proc. Natl. Acad. Sci. U. S. A.* **2014**, *111*, https://doi.org/10.1073/pnas.1414215111.
56. Liu, R.; Li, W. High-thermal-stability and high-thermal-conductivity $Ti_3C_2T_x$ MXene/Poly(vinyl alcohol) (PVA) composites. *ACS Omega.* **2018**, *3*, 2609–2617. https://doi.org/10.1021/acsomega.7b02001.
57. Heon, M.; et al. Continuous carbide-derived carbon films with high volumetric capacitance. *Energy Environ Sci.* **2011**, *4*, no. 1, 135–138.

58. Yang, X.; Cheng, C.; Wang, Y.; Qiu L.; Li, D. Liquid-mediated dense integration of graphene materials for compact capacitive energy storage. *Science* **2013**, *341*, no. 6145, 534–537.

59. Zhu Y.; et al. Carbon-based supercapacitors produced by activation of graphene. *Science* **2011**, *332*, no. 6037, 1537–1541.

60. Soboľčiak, P.; Ali, A.; Hassan, M.K.; Helal, M.I.; Tanvir, A.; Popelka, A.; et al. 2D $Ti_3C_2T_x$ (MXene)-reinforced polyvinyl alcohol (PVA) nanofibers with enhanced mechanical and electrical properties. *PLoS ONE* **2017**, *12*, no. 8, e0183705. https://doi.org/10.1371/journal.pone.0183705.

61. Ren, Y.; Zhu, J.; Wang, L.; Liu, H.; Liu, Y.; Wu, W.; Wang, F. Synthesis of polyaniline nanoparticles deposited on two-dimensional titanium carbide for high-performance supercapacitors. *Mater. Lett.* **2018**, *214*, 84–87. https://doi.org/10.1016/j.matlet.2017.11.060.

62. Lipatov, A.; Alhabeb, M.; Lukatskaya, M.R.; Boson, A.; Gogotsi, Y.; Sinitskii, A. Effect of synthesis on quality, electronic properties and environmental stability of individual monolayer Ti3C2 MXene flakes. *Adv. Electron. Mater.* **2016**, *2*, 1600255.

63. Alhabeb, M.; Maleski, K.; Anasori, B.; Lelyukh, P.; Clark, L.; Sin, S.; Gogotsi, Y. Guidelines for synthesis and processing of two-dimensional titanium carbide (Ti3C2T x MXene). *Chem. Mater.* **2017**, *29*, 7633–7644.

64. VahidMohammadi, A.; Moncada, J.; Chen, H.; Kayali, E.; Orangi, J.; Carrero, C.; Beidaghi, M. Thick and freestanding MXene/PANI pseudocapacitive electrodes with ultrahigh specific capacitance. *J. Mater. Chem. A* **2018**, *6*, https://doi.org/10.1039/C8TA05807E.

65. Huang, P.; Cao, C.Y.; Sun, Y.B.; Yang, S.L.; Wei, F.; Song, W.G. One-pot synthesis of sandwich-like reduced graphene oxide@ CoNiAl layered double hydroxide with excellent pseudocapacitive properties. *J. Mater. Chem. A* **2015**, *3*, 10858–10863.

66. Yan, J.; Ren, C.E.; Maleski, K.; Hatter, C.B.; Anasori, B.; Urbankowski, P.; Sarycheva, A.; Gogotsi, Y. Flexible MXene/graphene films for ultrafast supercapacitors with outstanding volumetric capacitance. *Adv. Funct. Mater.* **2017**, *27*, 1701264.

67. Zhao, M.Q.; Ren, C.E.; Ling, Z.; Lukatskaya, M.R.; Zhang, C.; Van Aken, K.L.; Barsoum, M.W.; Gogotsi, Y. Flexible MXene/carbon nanotube composite paper with high volumetric capacitance. *Adv. Mater.* **2015**, *27*, 339–345.

68. Boota, M.; Anasori, B.; Voigt, C.; Zhao, M.Q.; Barsoum, M.W.; Gogotsi, Y. Pseudocapacitive electrodes produced by oxidant-free polymerization of pyrrole between the layers of 2D titanium carbide (MXene). *Adv. Mater.* **2016**, *28*, 1517–1522.

69. XU, H.; Zheng, D.; Liu, F.; Li, W.; Lin, J.. Synthesis of MXene/polyaniline composite with excellent electrochemical properties. *J. Mater. Chem. A* **2020**, *8*, https://doi.org/10.1039/D0TA00572J.

70. Boota, M.; Anasori, B.; Voigt, C.; Zhao, M.Q.; Barsoum, M.; Gogotsi, Y. Pseudocapacitive Electrodes Produced by Oxidant-Free Polymerization of Pyrrole between the Layers of 2D Titanium Carbide (MXene). *Adv. Mater.* **2015**, *28*, https://doi.org/10.1002/adma.201504705.

71. Peng, C.; Hu, D.; Chen, G.Z. Theoretical specific capacitance based on charge storage mechanisms of conducting polymers: Comment on Vertically oriented arrays of polyaniline nanorods and their super electrochemical properties. *Chem. Commun.* **2011**, *47*, 4105.

72. Sekiguchi, K.; Atobe, M.; Fuchigami, T. Electropolymerization of pyrrole in 1-ethyl-3-methylimidazolium trifluoromethanesulfonate room temperature ionic liquid. *Electrochem. Commun.* **2002**, *4*, 881.

73. Liu, T.; Finn, L.; Yu, M.; Wang, H.; Zhai, T.; Lu, X.; Tong, Y.; Li, Y. Polyaniline and polypyrrole pseudocapacitor electrodes with excellent cycling stability. *Nano Lett.* **2014**, *14*, 2522.

74. Tang, H.; Wang, J.; Yin, H.; Zhao, H.; Wang, D.; Tang, Z. Growth of Polypyrrole Ultrathin Films on MoS2 Monolayers as High-Performance Supercapacitor Electrodes. *Adv. Mater.* **2015**, *27*, 1117.

75. Hughes, M.; Chen, G.Z.; Shaffer, M.S.P.; Fray, D.J.; Windle, A.H. Electrochemical capacitance of a nanoporous composite of carbon nanotubes and polypyrrole. *Chem. Mater.* **2002**, *14*, 1610.

76. Wu, W.; Wei, D.; Zhu, J.; Niu, D.; Wang, F.; Wang, L.; Yang, L.; Yang, P.; Wang, C. Enhanced electrochemical performances of organ-like Ti_3C_2 MXenes/polypyrrole composites as supercapacitors electrode materials. *Ceramics Int.* **2019**, *45*, 7328–7337. https://doi.org/10.1016/j.ceramint.2019.01.016.

77. Wang, Q.W.; Zhang, H.B.; Liu, J.; Zhao, S.; Xie, X.; Liu, L.; Yang, R.; Koratkar, N.; Yu, Z.Z. Multifunctional and water-resistant MXene-decorated polyester textiles with outstanding electromagnetic interference shielding and Joule heating performances. *Adv. Funct. Mater.* **2019**, *29*, 1806819.

78. Wu, C.; Huang, X.; Wang, G.; Lv, L.; Chen, G.; Li, G.; Jiang, P. Highly conductive nanocomposites with three-dimensional, compactly interconnected graphene networks via a self-assembly process. *Adv. Funct. Mater.* **2013**, *23*, 506–513.

79. Yan, D.X.; Pang, H.; Li, B.; Vajtai, R.; Xu, L.; Ren, P.G.; Wang, J.H.; Li, Z.M. Structured reduced graphene oxide/polymer composites for ultra-efficient electromagnetic interference shielding. *Adv. Funct. Mater.* **2015**, *25*, 559–566.

80. Zhang, H.B.; Yan, Q.; Zheng, W.G.; He, Z.; Yu, Z.Z. Tough graphene–polymer microcellular foams for electromagnetic interference shielding. *ACS Appl. Mater. Interfaces* **2011**, *3*, 918–924.

81. Xu, M.K.; Liu, J.; Zhang, H.B.; Zhang, Y.; Wu, X.; Deng, Z.; Yu, Z.Z. Electrically conductive $Ti_3C_2T_x$ MXene/polypropylene nanocomposites with an ultralow percolation threshold for efficient electromagnetic interference shielding. *Ind. Eng. Chem. Res.* **2021**, *60*, no. 11, 4342–4350. https://doi.org/10.1021/acs.iecr.1c00320.

82. Friedman, M. Chemistry, biochemistry, and safety of acrylamide. A review. *J. Agric. Food Chem.* **2003**, *51*, 4504–4526.

83. Taeymans, D.; Wood, J.; Ashby, P.; Blank, I.; Studer, A.; Stadler, R.; Gonde, P.; Eijck, P.; Lalljie, S.; Lingnert, H.; Lindblom, M.; Matissek, R.; Mueller, D.; Tallmadge, D.; O'Brien, J.; Thompson, S.; Silvani, D.; Whitmore, T. A review of acrylamide: an industry perspective on research, analysis, formation, and control. *Crit. Rev. Food Sci. Nutr.* **2004**, *44*, 323–347.

84. Naguib, M.; Saito, T.; Lai, S.; Rager, M.S.; Aytug, T.; Parans Paranthaman, M.; Zhao, M.Q.; Gogotsi, Y. $Ti_3C_2T_x$ (MXene)–polyacrylamide nanocomposite films. United States. https://doi.org/10.1039/C6RA10384G. https://www.osti.gov/servlets/purl/1286978.

85. Yang, J.; Tang, L.S.; Bao, R.Y.; Bai, L.; Liu, Z.Y.; Yang, W.; et al. Largely enhanced thermal conductivity of poly (ethylene glycol)/boron nitride composite phase change materials for solar-thermal-electric energy conversion and storage with very low content of graphene nanoplatelets. *Chem. Eng. J.* **2017**, *315*, 481–490.

86. Zhou, Y.; Sheng, D.; Liu, X.; Lin, C.; Ji, F.; Dong, L.; et al. Synthesis and properties of crosslinking halloysite nanotubes/polyurethane-based solid-solid phase change materials. *Sol. Energy Mater. Sol. Cells* **2018**, *174*, 84–93.

87. Atinafu, D.G.; Dong, W.; Huang, X.; Gao, H.; Wang, J.; Yang, M.; et al. One-pot synthesis of light-driven polymeric composite phase change materials based on N-doped porous carbon for enhanced latent heat storage capacity and thermal conductivity. *Sol. Energy Mater. Sol. Cells* **2018**; *179*, 392–400.

88. Lu, X.; Huang, H.; Zhang, X.; Lin, P.; Huang, J.; Sheng, X.; Zhang, L.; Qu, J.P. Novel light-driven and electro-driven polyethylene glycol/two-dimensional MXene form-stable phase change material with enhanced thermal conductivity and electrical conductivity for thermal energy storage, *Composites Part B: Engineering* **2019**, *177*, 107372, ISSN 1359-8368, https://doi.org/10.1016/j.compositesb.2019.107372.
89. Li, W.Y.; Song, Z.Q.; Zhong, J.M.; Qian, J.; Tan, Z.Y.; Wu, X.Y.; Chu, H.Y.; Nie, W.; Ran, X.H. Multilayer-structured transparent MXene/PVDF film with excellent dielectric and energy storage performance. *J. Mater. Chem. C* **2019**, *7*, 10371–10378.
90. Feng, Y.; Li, M.L.; Li, W.L.; Zhang, T.D.; Zhao, Y.; Fei, W.D. Polymer/metal multi-layers structured composites: A route to high dielectric constant and suppressed dielectric loss. *Appl. Phys. Lett.* **2018**, *112*, 022901.
91. Liu, F.; Li, Q.; Cui, J.; Li, Z.; Yang, G.; Liu, Y.; Dong, L.; Xiong, C.; Wang, H.; Wang, Q. High-energy-density dielectric polymer nanocomposites with trilayered architecture. *Adv. Funct. Mater.* **2017**, *27*, 1606292.
92. Li, Q.; Liu, F.; Yang, T.; Gadinski, M.R.; Zhang, G.; Chen, L.Q.; Wang, Q. Sandwich-structured polymer nanocomposites with high energy density and great charge–discharge efficiency at elevated temperatures. *Proc. Natl. Acad. Sci. U. S. A.* **2016**, *113*, 9995–10000.
93. Wu, W.; Zhao, W.; Sun, Q.; Yu, B.; Yin, X.; Cao, X.; Feng, Y.; Li, R.K.Y.; Qu, J. Surface treatment of two dimensional MXene for poly(vinylidene fluoride) nanocomposites with tunable dielectric permittivity. *Composites Communications*, **2021**, *23*, 100562, ISSN 2452-2139, https://doi.org/10.1016/j.coco.2020.100562.
94. Mannsfeld, S.C.B.; Tee, B.C.K.; Stoltenberg, R.M.; Chen, C.V.H.H.; Barman, S.; Muir, B.V.O.; Sokolov, A.N.; Reese, C.; Bao, Z. Highly sensitive flexible pressure sensors with micro structured rubber dielectric layers. *Nat. Mater.* **2010**, *9*, 859–864.
95. Hantanasirisakul, K.; Gogotsi, Y. Electronic and optical properties of 2D transition metal carbides and nitrides (MXenes). *Adv. Mater.* **2018**, *30*, 1804779.
96. Zhu, J.; Ha, E.; Zhao, G.; Zhou, Y.; Huang, D.; Yue, G.; Hu, L.; Sun, N.; Wang, Y.; Lee, L.Y.S.; Xu, C.; Wong, K.Y.; Astruc, D.; Zhao, P. Recent advance in MXenes: A promising 2D material for catalysis, sensor and chemical adsorption. *Coord. Chem. Rev.* **2017**, *352*, 306–327.
97. Li, L.; Fu, X.; Chen, S.; Uzun, S.; Levitt, A.S.; Shuck, C.E.; Han, W.; Gogotsi, Y. Hydrophobic and stable MXene–polymer pressure sensors for wearable electronics. *ACS Appl. Mater. Interfaces* **2020**, *12*, no. 13, 15362–15369. https://doi.org/10.1021/acsami.0c00255.

7 Electromagnetic Interference Shielding Behavior of MXenes

Theoretical and Experimental Perspectives

Poushali Das
School of Biomedical Engineering
McMaster University
Ontario, Canada

Seshasai Srinivasan and Amin Reza Rajabzadeh
School of Biomedical Engineering
and
W Booth School of Engineering Practice and Technology
McMaster University
Ontario, Canada

CONTENTS

DOI: 10.1201/9781003281511-7

7.1 INTRODUCTION

Currently, the speed of technological advancement in the industrialized world is accelerating, with fifth-generation (5G) technologies being commercialized in nations like South Korea, which is the one of the world leaders in information and communications technology (the first country to provide 5G services, in December 2018) [1]. The introduction of 5G technology is being followed by astonishing advancements in the mobile electronics and telecommunications sectors, which is further propelling the growth of the Internet of Things (IoT) and big data technologies [2]. In addition to the increasing interaction with technology, 5G networks are anticipated to expand the manufacturing of electronic devices and equipment significantly. Given that these systems receive, create, and/or transmit electromagnetic (EM) waves over a wide frequency range, the electromagnetic interference (EMI) generated by any electronic equipment that transmits, distributes, or utilizes electrical energy is likely to have a detrimental effect on the performance of the device as well as the environment [3–6]. With the shrinking size of the electrical components and their higher processing power, EMI increases, causing malfunctions and degradation of electronic services and security systems [7–11]. The growth in electromagnetic pollution can have a detrimental effect on health and the ecosystem if no regulations are in place [12–15].

A good EMI shielding material reduces unwanted emissions and protects the component from external signals. The primary mechanism of EMI shielding involves the reflection of radiation by deploying charge carriers that interact directly with electromagnetic fields [16]. The secondary mechanism involves the absorption of EM radiation as the material's electric and/or magnetic dipole interacts with the radiation. The primary component affecting the shield's reflection and absorption properties is the electrical conductivity [17]. The third mechanism, which involves multiple internal reflections, has received less attention, despite the fact that it contributes significantly to EMI shielding efficiency. Internal reflections originate at the scattering centers, interfaces, or defect sites within shielding material, causing EM wave scattering and then absorption [7, 18]. Two-dimentional (2D) materials like graphene and transition metal carbides, nitrides, or carbonitrides (MXenes) have exceptional properties for EMI shielding [19, 20]. However, due to the relatively high synthesis cost and lesser electrical conductivity of graphene, their prospective applications are restricted. On the other hand, the development of highly conductive MXenes is attracting researchers due to their interesting characteristics [21, 22]. They have the highest metallic conductivity of all of the synthetic 2D materials. This is due to their high electron density near the Fermi level [3, 23]. The excellent electrical conductivity, alongside better mechanical flexibility, surface functionalization, and suitable ease of processing, surpasses the benefits of competing shielding materials [24–26]. MXenes are capable of forming aqueous dispersions without the use of additives or surfactants, enabling the advancement of MXene polymeric composites. In this chapter, the EMI shielding efficiency (SE) of MXenes is explored from both an experimental and a theoretical standpoint, giving readers a comprehensive understanding of the topic.

7.2 FUNDAMENTAL THEORIES OF EMI SHIELDING

The term "electromagnetic shielding effectiveness" refers to a material's capacity to attenuate or reduce EM signals (SE). The effectiveness of shielding is defined as the ratio of impinging energy to residual energy. Absorption and reflection occur when an electromagnetic wave passes through a shield. The term "residual energy" refers to a portion of the remaining energy that is neither reflected nor absorbed but rather emerges through it. A magnetic field (H) and an electric field (E) are two basic components of all electromagnetic waves. These two fields are perpendicular to each other, and the wave propagation direction is perpendicular to the plane containing the two components. Wave impedance is defined as the ratio of E to H [27]. EMI shielding may be categorized into two subgroups: the near field and the far field shielding regions. In the far field shielding zone, the distance between the radiation source and shield is greater than $\lambda/2\pi$, whereas if the distance is smaller than $\lambda/2\pi$, it is considered as near field shielding (λ: wavelength of the source). In the far field zone, the electromagnetic plane wave theory is frequently employed for EMI shielding purposes. In the near field region, EMI shielding is accomplished using the contribution of electric and magnetic dipoles [28]. An EMI shielding material creates a barrier of electrically conductive materials that attenuates radiated or conducted EM waves through reflections and absorption. Figure 7.1 shows the mechanism of EMI shielding effectiveness. As seen in this figure, the incident energy is

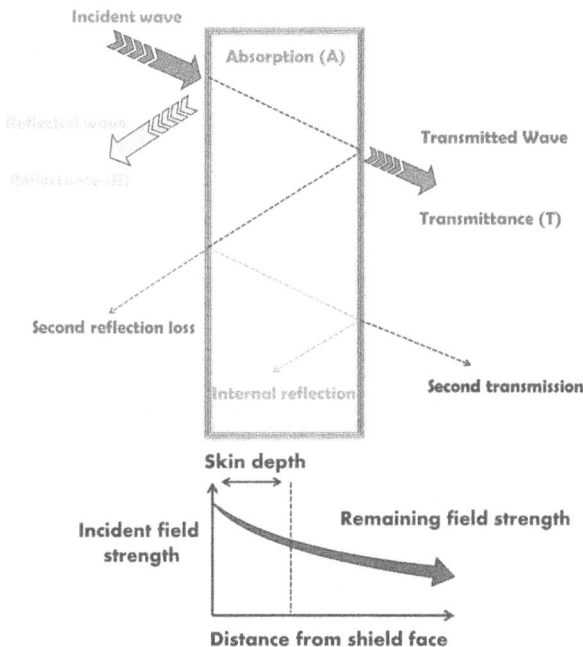

FIGURE 7.1 A schematic depiction of EMI shielding mechanism.

attenuated by the shielding material through reflections, such as direct and multiple reflections or second reflection and absorption. Thus, the attenuation of an electromagnetic wave is caused by three different mechanisms: Absorption (A), Reflection (R), and Multiple reflections (B). Shielding effectiveness can be designated as the sum of all these three terms, SE = A + R + B.

7.3 MECHANISM OF EMI SHIELDING

Many different methods for the interaction between electromagnetic radiation and the surface and interior of a compact shield film have been suggested. The incident power (P_I) of an electromagnetic wave interacts with the shield, and a portion of that power is reflected (P_R) on both the front and rear surfaces of the shield due to an impedance mismatch between the shield and the atmospheric air. The remaining power is absorbed and dissipated as heat energy within the shield owing to attenuation or transmission (P_T) [29]. The effectiveness of a shield against EM radiation is called EMI SE. It is defined as a logarithmic ratio of transmitted to incident power as shown here:

$$SE_T (dB) = 10 \log_{10} \frac{P_T}{P_I} = 20 \log_{10} \frac{E_T}{E_I} \tag{7.1}$$

where P and E signify power and electric field intensity, respectively, and subscripts I and T stand for the incident and transmitted EM waves, respectively.

Attenuation of EM radiation happens through a variety of methods, including reflection, absorption, and multiple reflections [30]. As per Schelkunoff's theory [31, 32], the total EMI shielding effectiveness (EMI SE_T) is the sum of the attenuation due to reflection (SE_R), absorption (SE_A), and multiple reflections (SE_M) [33–35], and is given as:

$$SE_T = SE_A + SE_R + SE_M \tag{7.2}$$

7.3.1 EMI SHIELDING: REFLECTION LOSS (SE_R)

Reflection is the primary EMI shielding mode achieved by the interface or surface between two propagation mediums with different impedances or refractive indices (for example, air and the shield). A simpler form of Fresnel's equation can be used to calculate the extent of reflection loss from the front to the rear of a shield surface. The equation is as follows:

$$SE_R (dB) = 20 \log \frac{(\eta + \eta_0)^2}{4 \eta \eta_0} = 39.5 + 10 \log \frac{\sigma}{2\pi f \mu} \tag{7.3}$$

where η and η_0 denote the impedances of the shield and air, respectively; σ and μ represent the electrical conductivity and magnetic permeability, respectively; and f is the frequency. With increasing conductivity, SE_R rises, suggesting that the material's electrical conductivity should be high in order to produce significant reflection loss. Furthermore, permeability and the frequency of EM waves also contribute to the reflection loss.

7.3.2   EMI SHIELDING: ABSORPTION LOSS (SE$_A$)

An EM wave is absorbed as it travels through a shielding material with an attenuation constant of α. The strength or amplitude (E) of the EM wave is exponentially reduced since $E = E_0 e^{-\alpha d}$ in a shield of the thickness of d. The attenuation constant of the material can be expressed as follows [29, 36]:

$$\alpha = \omega \sqrt{\frac{\mu\varepsilon}{2}\left[\sqrt{1+\left(\frac{\sigma}{\omega\varepsilon}\right)^2}-1\right]} \qquad (7.4)$$

where ω and ε are the angular frequency ($2\pi f$) and dielectric permittivity, respectively. The following criteria must be met in order to have a significant absorption loss: high electrical conductivity for ohmic loss [3], high dielectric permittivity for dielectric loss [37], and high magnetic permeability for magnetic loss [38]. Any absorbed energy will be dissipated as heat energy.

For non-magnetic and conducting shielding materials, the absorption loss (SE$_A$) can be calculated using the following equation:

$$SE_A(dB) = 20\log_{10} e^{\alpha d} = 20\left(\frac{d}{\delta}\right)\log_{10}e = 8.68\left(\frac{d}{\delta}\right) = 8.7d\sqrt{\pi f\mu 6} \qquad (7.5)$$

where δ denotes the skin depth and d is the distance traveled by the EM waves. The term "skin depth" represents the distance beneath the surface at which the intensity of the electric field decreases to 1 e^{-1} of the original incident wave intensity. The skin depth for a conductive shield is defined as $\delta = 1/\alpha = (\sqrt{\pi f\sigma\mu})^{-1}$. Thickness and electrical conductivity are the most important factors in absorption, whereas permittivity and permeability are the major elements in absorption loss.

7.3.3   EMI SHIELDING MULTIPLE REFLECTIONS (SE$_M$)

Due to multiple reflections, reflection from the rear surface of a thin shield impacts the final transmission. This is because the reflected radiation re-reflects at the front surface and contributes to a second transmission. This can be repeated indefinitely until the wave's energy has been entirely dissipated, as shown in Figure 7.1. Multiple reflections between the shield's front and rear surfaces contribute to minimal EMI SE. The loss due to the multiple reflections (SE$_M$) can be calculated as follows:

$$SE_M(dB) = 20\log_{10}\left(1-e^{-2\alpha d}\right) = 20\log_{10}\left(1-e^{\frac{-2d}{\delta}}\right) \qquad (7.6)$$

SE$_M$ is strongly dependent on the thickness, and becomes insignificant at a thickness close to or more than δ, or when the SE$_T$ is greater than 15 dB. It is necessary to take into account numerous reflections when evaluating shield performance, if the thickness is much lower than δ.

7.3.4 Internal Scattering (Internal Multiple Reflections)

The shield's ability to protect may be improved by increasing additional interfaces. Increased internal scattering, also known as internal multiple reflections, can occur at interfaces when the impedance characteristics are mismatched. Internal multiple reflections can result in increased absorption loss. Porous and segregated structures provide additional interfaces between the shield and EM radiation, which leads to considerable internal scattering. Internal scattering increases the propagation path length of an electromagnetic wave before transmission and improves the probability of interaction [39, 40]. As discussed earlier, internal scattering should be separated from SE_M. Internal scattering generated by the additional internal interfaces within the shield always contributes to greater absorption loss and total shielding effectiveness, whereas SE_M takes place between the front and rear surfaces of the shield, causing a reduction in shielding performance.

7.3.5 Absolute Effectiveness of Shielding

Lightweight materials are desirable in most applications, including aviation and smart electronics. The absolute shielding efficiency (SSE/t) has been introduced to measure a material's achievable shielding capability while incorporating density (ρ) and thickness (t). SSE/t can be determined by using the following equation, and a high SSE/t is desirable in numerous applications involving lightweight shielding materials:

$$\frac{SSE}{t} = SSE_t = \frac{EMI\,SE}{\dfrac{\rho}{t}} = \left(dB\,cm^2 g^{-1}\right) \tag{7.7}$$

7.4 PROPERTIES GOVERNING EMI SHIELDING MECHANISM

By analyzing the dielectric relative complex permittivity ($\varepsilon_r = \varepsilon'_r - j\varepsilon''_r$) and magnetic relative complex permittivity ($\mu_r = \mu'_r - j\mu''_r$) characteristics of an EMI shielding material, the mechanisms of EMI shielding can be comprehended [41]. The real parts refer to the charge storage (ε_r') and magnetic storage (μ_r') of the EM waves, whereas the imaginary parts indicate the dielectric loss (ε_r'') and magnetic loss (μ_r''), respectively. The magnitude of the loss can be determined from the tangent of dielectric loss and magnetic loss [42].

7.4.1 Dielectric Property

The dielectric loss is mostly controlled by ionic, orientational, electronic, and interfacial polarization. In the material, the bound charges are responsible for the ionic and orientational polarization. According to the Maxwell-Wagner-Sillars (MWS) theory, interfacial polarization emanates from the accumulation of space charges due to the dissimilarity in electrical conductivity/dielectric constant at the interface of the two different materials [43].

The ε_r' and ε_r'' can be related to the Cole-Cole equation shown here [44]:

$$\left(\varepsilon_r' - \frac{\varepsilon_S + \varepsilon_\infty}{2}\right)^2 + (\varepsilon_r'')^2 = \left(\frac{\varepsilon_S - \varepsilon_\infty}{2}\right)^2 \tag{7.8}$$

where ε_s and ε_∞ stand for the static dielectric constant and relative dielectric constant, respectively.

When the Cole-Cole plot is a semicircle, the semicircle is associated with the Debye relaxation process. It must be noted that the semicircle may not be noticed in high conducting materials, as the loss is mostly due to conduction loss, and can be given as:

$$\varepsilon_r'' = \frac{6}{2\pi\omega\varepsilon_0} \tag{7.9}$$

The ε_r' and ε_r'' values will drop as the frequency increases, as represented in Figure 7.2, and follow the relation given by:

$$\varepsilon_r'' = \sigma_{AC} - \sigma_{DC} \ (\varepsilon_r' \times \omega) \tag{7.10}$$

where σ_{AC} and σ_{DC} refer to the AC electrical conductivity and DC electrical conductivity, respectively, and ω is frequency. This is due to the reduction in the polarization of the space charge and the absence of dipole orientation when the field is varied at higher frequencies [45].

7.4.2 MAGNETIC PROPERTY

The magnetic loss is caused by a combination of several factors, namely, the domain wall loss, hysteresis loss, eddy current loss, and residual loss [47]. The hysteresis loss occurs due to hysteresis (i.e., the time interval between the magnetization vector M and the magnetic field vector H), during which the magnetic energy is released as heat. The eddy current loss [48] remains the same even if the frequency changes. The other types of losses are natural resonance and exchange resonance [49, 50]. The theory of ferromagnetic resonance states that the natural resonance is governed by the effective magnetic field (anisotropic energy), as indicated in the following equations [51]:

$$\omega_r = \frac{\gamma}{2\pi} H_e \tag{7.11}$$

$$H_e = \frac{4\kappa_1}{3\mu_0 \, M_S} \tag{7.12}$$

In the preceding equations, $\gamma/2\pi$ is the gyromagnetic ratio, H_e, κ_1, and $\mu_0 M_s$ represent the effective magnetic field, magnetic crystalline anisotropy co-efficient, and saturation magnetization, respectively. A larger effective magnetic field supports a

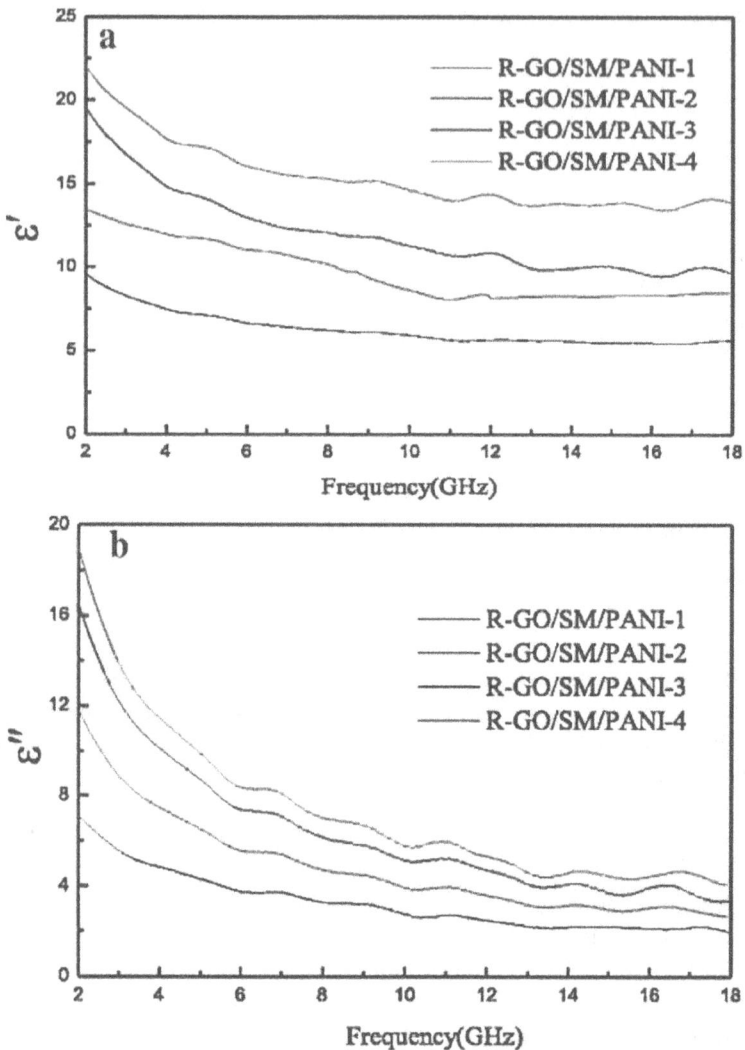

FIGURE 7.2 ε_r' and ε_r'' values of r-GO/Strontium ferrite/PANI composites vs. frequency in the range of 2–18 GHz.

(Reproduced with permission from Ref. [46].)

greater absorption of electromagnetic waves in higher frequencies [52]. For the EM wave absorption, a lower value of magnetic storage, μ_r', is more desirable than the magnetic loss, μ_r''.

The material's reflection loss varies with the charge carriers, whereas the absorption loss is dominated by electric and magnetic dipoles. Therefore, when the magnetic properties are absent, EMI shielding is solely reliant on dielectric properties, and vice versa [53].

7.4.3   EM Absorption

Since reflection loss contributes to secondary electromagnetic pollution, researchers have focused their attention on developing EMI shielding materials with high absorption. The SE_A, SE_R, and SE_M of EMI shielding materials, as a function of their electromagnetic characteristics, can be represented as follows [54]:

$$SE_A = 20d\sqrt{\frac{6_{AC}}{2}} \tag{7.13}$$

$$SE_R = 10\ log_{10}\frac{6_{AC}}{16.\ \varepsilon_0.\ \mu_r.\ \omega} \tag{7.14}$$

$$SE_M = 20\ log_{10}1 - e^{-\frac{2d}{\delta}}\ .\ e^{-j\frac{2}{\delta}} \tag{7.15}$$

The distance traveled by the EM waves within the shielding material is denoted by d. Equations 7.13–7.15 show that the reflection of a shielding material is related to the frequency of the signal and will drop as the frequency increases. On the other hand, the absorption loss will increase as the frequency increases. When the thickness of the shielding material is greater than δ, multiple internal reflections can be neglected [55]. Multiple reflections reduce the EMI SE for shielding materials when the thickness is less than the skin depth. Their EMI SE will increase as their thickness increases [56].

7.5   EMI SHIELDING OF MXenes

2D MXenes of various compositions exhibit a wide variety of electrical conductivities, ranging from 5–20,000 S cm^{-1} [57, 58]. Owing to the outstanding electrical conductivity of MXenes, they exhibit outstanding EMI shielding efficiency that is superior to that of synthetic materials and equivalent to that of typical metal, putting MXenes on the cutting edge of EMI shielding technology. Besides this, MXene films feature multilayered 2D sheets that distinguish them from traditional shielding materials. A multilayered MXene film has the advantage that each layer contributes to the attenuation of EM wave energy; as a result, effective shielding efficiency is attained with maximal absorption at a relatively lower thickness. Recently, Koo and coworkers studied the EMI shielding of a single MXene layer with a thickness of about 2.3 nm [59]. Figure 7.3a depicts multiple reflection mechanism in MXenes at very low thicknesses [60]. Moreover, this structure of MXenes can be altered with the introduction of dielectric domains or pores to increase the multiple reflections, thereby attenuating the energy of EM waves (Figure 7.3b) [60]. Furthermore, the surface of MXenes leads to an increase in the absorption of EM waves by various polarization losses. 2D MXenes also have unique properties such as large specific surface area, low density, great flexibility (for wearable devices), and ease of processability. These features make MXenes suitable for EMI shielding materials for intelligent next-generation electronics.

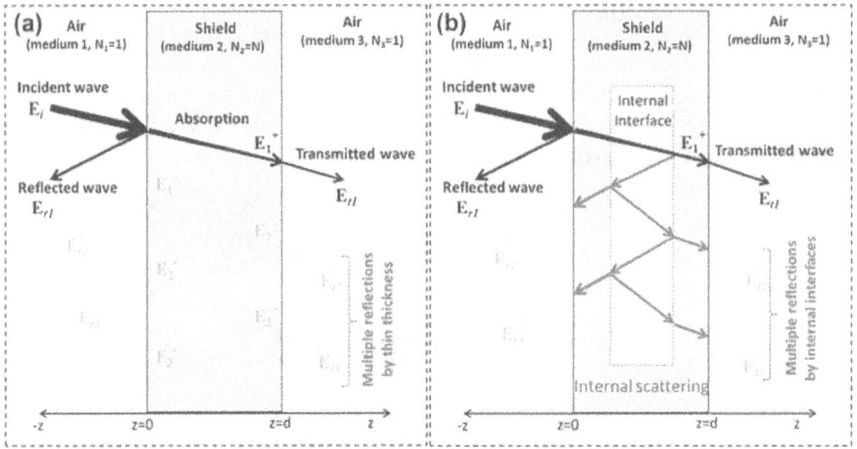

FIGURE 7.3 Mechanisms of EMI shielding of MXenes. (a) An ultrathin shield thickness and (b) induced inclusions or interfaces inside the shield playing as scattering sites within the shield.

(Reproduced with permission from Ref. [60].)

A modified clay approach known as the minimal intense layer delamination was used for the synthesis of $Ti_3C_2T_x$ MXene laminate films of μm thickness by Gogotsi et al. The MXenes showed remarkable EMI shielding capabilities of the films for the first time [3]. Besides this, these authors discussed EMI shielding of $Mo_2TiC_2T_x$ and $Mo_2Ti_2C_3T_x$ laminate films. Each MXene was prepared from its corresponding MAX phase. Compared to $Mo_2TiC_2T_x$ and $Mo_2Ti_2C_3T_x$, $Ti_3C_2T_x$ exhibited a much higher electrical conductivity of 4665 S cm^{-1}. In the X-band, the total EMI SE values of $Ti_3C_2T_x$, $Mo_2TiC_2T_x$, and $Mo_2Ti_2C_3T_x$ were 54, 20, and 24 dB, respectively, at a thickness of just 2.5 μm. The EMI SE of $Ti_3C_2T_x$ improved with the material thickness and achieved 92 dB at 45 μm. The outstanding EMI SE of $Ti_3C_2T_x$ was attributed to its higher electrical conductivity and laminar-like morphology. When an EM wave strikes the MXene's surface, a part of the wave is reflected due to the impedance mismatch between the MXene, which includes an abundance of charge carriers, and the surrounding atmosphere (air). Through the interfacial assembly of monolayer MXene flakes, Yun and colleagues were able to manufacture $Ti_3C_2T_x$ MXene films with thickness less than the skin depth [59]. For the multiple reflections, generally, a lower SE_T is obtained, along with a lower SE_R, and a higher SE_A. As the total EMI SE $\geq$ 15 dB or the thickness of the shielding material is more than the skin depth, the role of multiple reflections is assumed to be negligible. Nanometer thickness multilayer films were synthesized through a repeated collection of assembled monolayer films. An assembled monolayer $Ti_3C_2T_x$ MXene film (2.3 nm thickness) exhibited transparency of 90%. The authors reported that MXene film (thickness 55 nm) assembled with 24 layers showed EMI SE of 20 dB, which meets the minimum requirement for commercial applications. Recently, Han et al. demonstrated the importance of multiple reflections in thin films (nm size) that were manufactured by spray coating and spin casting. Furthermore, these scientists described the EMI shielding capabilities of 16 different MXene laminates, which were achieved by synthesizing films utilizing the vacuum filtration technique [57]. As a part of this study, MXenes were classified according to the number of atomic layers, the presence of single or double transition metal layers, and their presence in solid solutions to evaluate the effect of elemental composition, metal layer organization, and structure of MXenes in various environments. All the MXenes studied revealed an EMI SE of greater than 20 dB at submicrometer thickness. As shown in Figure 7.4a, depending on the composition and number of metal layers, an electrical conductivity of 5–8500 S cm^{-1} can be achieved. $Ti_3C_2T_x$ MXene has demonstrated the highest shielding effectiveness of 21 dB when examined for a 40 nm spray-coated film due to its good electrical conductivity. It is worth noting that the conductivity of MXenes depends greatly on the synthesis procedure and precursors. The electrical conductivity of $Ti_yNb_{2-y}CT_x$ and $Nb_yV_{2-y}CT_x$ films decreases with increasing Nb concentration for the solid solution MXenes, which is consistent with the change in EMI shielding ability of the materials. This indicates that the electrical conductivity and EMI shielding capabilities of MXenes can be tailored by their chemical composition. It can be seen from Figure 7.4b that MXenes with electrical conductivities greater than 100 S cm^{-1} were in close agreement with the simulated findings, as shown by continuousline, indicating a nonlinear growth in SE_T with increasing conductivity.

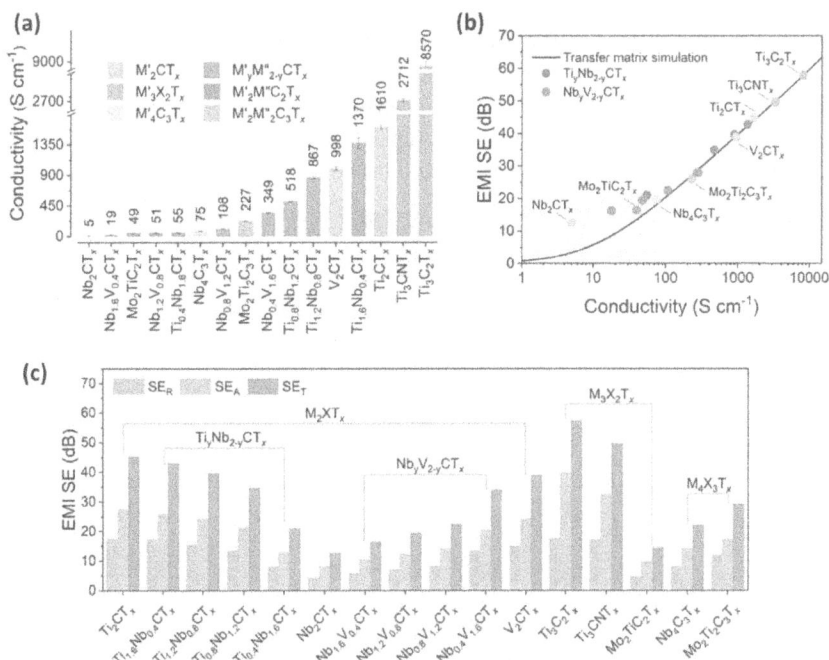

FIGURE 7.4 (a) Electrical conductivity of different MXenes; (b) comparison of the EMI SE of different MXene films, each of 5 μm of thickness; (c) average EMI SE values of different MXene films over the 8.2–12.4 GHz frequency range.

(Reproduced with permission from Ref. [57].)

MXenes with electrical conductivities less than 100 S cm^{-1} exhibit higher measured SE$_T$ values than expected. This difference between experimental data and theoretical predictions suggests that the electrical conductivity of MXenes is not the sole element influencing their EMI shielding activity. Figure 7.4c shows the EMI SE values in the X-band region for various MXenes of similar thicknesses (5±0.3 mm). In their report, Yin et al. studied the electromagnetic absorption of completely oxidized Ti$_3$C$_2$T$_x$ MXene across a wide frequency range [61]. Ti$_3$C$_2$T$_x$ was annealed in a CO$_2$ atmosphere at 800°C to convert the Ti layers into TiO$_2$ nanoparticles completely.

Figure 7.5a shows the structural evolution from Ti$_3$C$_2$T$_x$ to C/TiO$_2$ hybrids. The SEM images of the C/TiO$_2$ hybrids following a heat treatment (at 800°C) of Ti$_3$C$_2$T$_x$ in CO$_2$ environment are shown in Figure 7.5b–d. The lateral dimension reveals 2D carbon layers with encapsulated TiO$_2$ particles (Figure 7.5b). Interestingly, the aligned layers are so thin that the TiO$_2$ crystals under the carbon layer can be seen clearly, as evident from Figure 7.5c,d. The real and imaginary components of the permittivity improved as the MXene concentration increased (Figure 7.5e,f), which is ascribed to the large rutile TiO$_2$ particles. At 10.5 and 15.5 GHz, the composites demonstrated a minimum RL value of -36 dB for specimen thicknesses of 2.2 and 1.6 mm, respectively (Figure 7.5g). The quarter-wavelength effect was responsible for the shift in the peak frequency (f_m) to a lower value as the sample thickness was increased. The effective bandwidth encompassed the full Ku-band (12.4–18 GHz) at a thickness of 1.7 mm. The effective bandwidth of the X-band was 8.7 to 12.3 GHz at a matching thickness of 2.2 mm. Due to the adhesion of TiO$_2$ particles to carbon layers, the interface may be described as a resistor-capacitor circuit. As shown in Figure 7.5h, the dielectric TiO$_2$ particles support the development of minicapacitors, which enhance the absorption of EM waves. The significant number of capacitor-like arrangements at the interface causes increasing EM attenuation to rise and results in heat generation [62, 63].

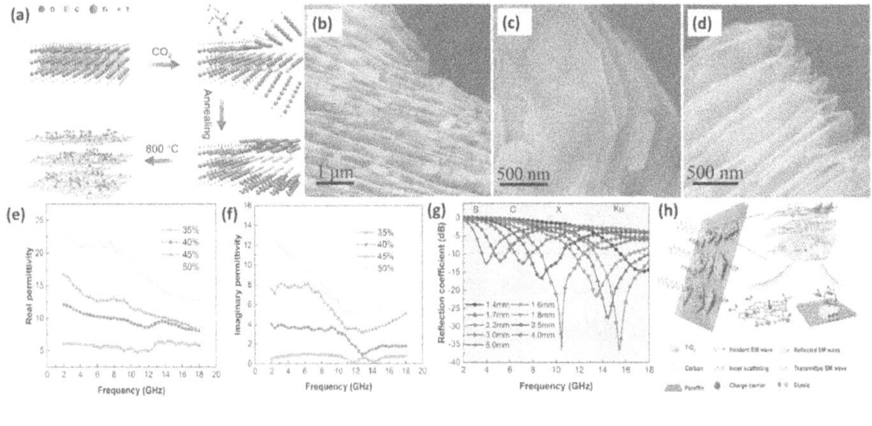

FIGURE 7.5   (a) Schematic presentation of the structural evolution from Ti$_3$C$_2$T$_x$ to C/TiO$_2$ hybrids; (b–d) SEM images of C/TiO$_2$ hybrids; (e) real (ε') and (f) imaginary (ε'') permittivity vs. frequency for composites with different loadings of C/TiO$_2$ hybrids; (g) RC curves vs. frequency and thickness of the sample with 45 wt% C/TiO$_2$ hybrids; (h) graphical depiction of EM wave absorption mechanisms for C/TiO$_2$ hybrids.

7.6   SUMMARY

MXenes have demonstrated a remarkable potential for EMI shielding applications due to their inherent features such as excellent metallic conductivity, tunable surface properties, two-dimensional sheet morphology, light weight, and ease of solution processing. $Ti_3C_2T_x$ MXene films with thicknesses ranging from nanometer to sub-micron demonstrated the highest shielding efficiency of all MXenes synthesized to date due to their high intrinsic electrical conductivities. However, there are several limitations, such as aggregation, storage, large-scale synthesis, and stability under sundry conditions in the MXenes system that should be taken into consideration in future investigations. Finally, even though more than 30 different MXene synthesis techniques have been reported, only a few MXenes have been examined, including the MXenes Ti_3C_2, Ti_2C, Mo_2TiC_2, $Mo_2Ti_2C_3$, and Ti_3CN. Furthermore, the advancement of magnetic MXenes with tailored chemical compositions can broaden the range of applications for MXenes. The magnetic MXene will cause an additional magnetic loss mechanism, which will improve the absorption of EM waves. Thus, MXenes are becoming one of the leading EMI shielding materials for a wide range of applications, such as smart and fast portable electronics, telecommunications devices, military hardware, radar systems, and medical appliances.

NOMENCLATURE

α	attenuation constant (dB m^{-1})
d	distance (mm)
E	electric field intensity (V m^{-1})
f	frequency (Hz)
σ	electrical conductivity (S m^{-1})
ε	dielectric permittivity (F m^{-1})
η	impedances of the shield (ohm)
η_0	impedances of the air (ohm)
μ	magnetic permeability (N A^{-2})
ω	angular frequency (rad s^{-1})
δ	skin depth (μm)
ρ	density (kg m^{-3})
ε'_r	charge storage
μ'_r	magnetic storage
ε''_r	dielectric loss
μ''_r	magnetic loss

REFERENCES

1. Kim, D.K., et al., 5G commercialization and trials in Korea. Communications of the ACM, 2020. **63**(4): p. 82–85.
2. Mudigonda, P. and SK. Abburi, A Survey: 5G in IoT is a Boon for Big Data Communication and Its Security, in ICDSMLA 2019. 2020, Springer. p. 318–327.
3. Shahzad, F., et al., Electromagnetic interference shielding with 2D transition metal carbides (MXenes). Science, 2016. **353**(6304): p. 1137–1140.

4. Das, P., et al., Acoustic green synthesis of graphene-gallium nanoparticles and PEDOT: PSS hybrid coating for textile to mitigate electromagnetic radiation pollution. ACS Applied Nano Materials, 2022. **5**(1): p. 1644–1655.

5. Ganguly, S., et al., Photopolymerized thin coating of polypyrrole/graphene nanofiber/iron oxide onto nonpolar plastic for flexible electromagnetic radiation shielding, strain sensing, and non-contact heating applications. Advanced Materials Interfaces, 2021. **8**(23): p. 2101255.

6. Ganguly, S., et al., Mussel-inspired polynorepinephrine/MXene-based magnetic nanohybrid for electromagnetic interference shielding in X-band and strain-sensing performance. Langmuir, 2022. **38**(12): p. 3936–3950.

7. Chen, Z., et al., Lightweight and flexible graphene foam composites for high-performance electromagnetic interference shielding. Advanced Materials, 2013. **25**(9): p. 1296–1300.

8. Ghosh, S., et al., Micro-computed tomography enhanced cross-linked carboxylated acrylonitrile butadiene rubber with the decoration of new generation conductive carbon black for high strain tolerant electromagnetic wave absorber. Materials Today Communications, 2020. **24**: p. 100989.

9. Ghosh, S., et al., 3D-enhanced, high-performing, super-hydrophobic and electromagnetic-interference shielding fabrics based on silver paint and their use in antibacterial applications. ChemistrySelect, 2019. **4**(40): p. 11748–11754.

10. Ghosh, S., et al., Fabrication of reduced graphene oxide/silver nanoparticles decorated conductive cotton fabric for high performing electromagnetic interference shielding and antibacterial application. Fibers and Polymers, 2019. **20**(6): p. 1161–1171.

11. Ghosh, S., et al., An approach to prepare mechanically robust full IPN strengthened conductive cotton fabric for high strain tolerant electromagnetic interference shielding. Chemical Engineering Journal, 2018. **344**: p. 138–154.

12. Mondal, S., et al., Low percolation threshold and electromagnetic shielding effectiveness of nano-structured carbon based ethylene methyl acrylate nanocomposites. Composites Part B: Engineering, 2017. **119**: p. 41–56.

13. Mondal, S., et al., Oxygen permeability properties of ethylene methyl acrylate/sepiolite clay composites with enhanced mechanical and thermal performance. J. Polym. Sci. Appl, 2017. **2**: p. 2.

14. Mondal, S. and S. Paria, Conductive carbon black filled polyvinyl alcohol composites for efficient electromagnetic interference shielding. 한국고분자학회 학술대회 연구논문 초록집, 2021. **46**(1): p. 102–102.

15. Shin, B., et al., Flexible thermoplastic polyurethane-carbon nanotube composites for electromagnetic interference shielding and thermal management. Chemical Engineering Journal, 2021. **418**: p. 129282.

16. Mondal, S., et al., High-performance carbon nanofiber coated cellulose filter paper for electromagnetic interference shielding. Cellulose, 2017. **24**(11): p. 5117–5131.

17. Das, N.C., et al., Single-walled carbon nanotube/poly (methyl methacrylate) composites for electromagnetic interference shielding. Polymer Engineering & Science, 2009. **49**(8): p. 1627–1634.

18. Al-Saleh, M.H., W.H. Saadeh, and U. Sundararaj, EMI shielding effectiveness of carbon based nanostructured polymeric materials: a comparative study. Carbon, 2013. **60**: p. 146–156.

19. Hong, S.K., et al., Electromagnetic interference shielding effectiveness of monolayer graphene. Nanotechnology, 2012. **23**(45): p. 455704.

20. Iqbal, A., P. Sambyal, and C.M. Koo, 2D MXenes for electromagnetic shielding: a review. Advanced Functional Materials, 2020. **30**(47): p. 2000883.

21. Kumar, P., et al., Ultrahigh electrically and thermally conductive self-aligned graphene/polymer composites using large-area reduced graphene oxides. Carbon, 2016. **101**: p. 120–128.

22. Kumar, P., et al., Large-area reduced graphene oxide thin film with excellent thermal conductivity and electromagnetic interference shielding effectiveness. Carbon, 2015. **94**: p. 494–500.
23. Zhang, J., et al., Scalable manufacturing of free-standing, strong $Ti_3C_2T_x$ MXene films with outstanding conductivity. Advanced Materials, 2020. **32**(23): p. 2001093.
24. Kim, D., et al., Nonpolar organic dispersion of 2D $Ti_3C_2T_x$ MXene flakes via simultaneous interfacial chemical grafting and phase transfer method. ACS Nano, 2019. **13**(12): p. 13818–13828.
25. Alhabeb, M., et al., Guidelines for synthesis and processing of two-dimensional titanium carbide ($Ti_3C_2T_x$ MXene). Chemistry of Materials, 2017. **29**(18): p. 7633–7644.
26. Ling, Z., et al., Flexible and conductive MXene films and nanocomposites with high capacitance. Proceedings of the National Academy of Sciences, 2014. **111**(47): p. 16676–16681.
27. Trivedi, D. and H. Nalwa, Handbook of Organic Conductive Molecules and Polymers. Wiley, New York, 1997. **2**: p. 505.
28. Wang, Y. and X. Jing, Intrinsically conducting polymers for electromagnetic interference shielding. Polymers for Advanced Technologies, 2005. **16**(4): p. 344–351.
29. Ott, H.W., Electromagnetic Compatibility Engineering. 2011, John Wiley & Sons.
30. Anasori, B. and Û.G. Gogotsi, 2D Metal Carbides and Nitrides (MXenes). Vol. 416. 2019, Springer.
31. Schelkunoff, S.A., Electromagnetic Waves. 1943.
32. Schulz, R.B., V. Plantz, and D. Brush, Shielding theory and practice. IEEE Transactions on Electromagnetic Compatibility, 1988. **30**(3): p. 187–201.
33. Modak, P. and D. Nandanwar, A review on graphene and its derivatives based polymer nanocomposites for electromagnetic interference shielding. Int. J. Adv. Sci. Eng. Technol., 2015. **1**: p. 212.
34. Joshi, A. and S. Datar, Carbon nanostructure composite for electromagnetic interference shielding. Pramana, 2015. **84**(6): p. 1099–1116.
35. Singh, A.P., et al., Encapsulation of γ-Fe_2O_3 decorated reduced graphene oxide in polyaniline core–shell tubes as an exceptional tracker for electromagnetic environmental pollution. Journal of Materials Chemistry A, 2014. **2**(10): p. 3581–3593.
36. Kaiser, K.L., Electromagnetic Shielding. 2005, CRC Press.
37. Sohi, N., M. Rahaman, and D. Khastgir, Dielectric property and electromagnetic interference shielding effectiveness of ethylene vinyl acetate-based conductive composites: effect of different type of carbon fillers. Polymer Composites, 2011. **32**(7): p. 1148–1154.
38. Kumar, R., et al., Carbon encapsulated nanoscale iron/iron-carbide/graphite particles for EMI shielding and microwave absorption. Physical Chemistry Chemical Physics, 2017. **19**(34): p. 23268–23279.
39. Han, M., et al., Anisotropic MXene aerogels with a mechanically tunable ratio of electromagnetic wave reflection to absorption. Advanced Optical Materials, 2019. **7**(10): p. 1900267.
40. Lee, S.H., et al., Density-tunable lightweight polymer composites with dual-functional ability of efficient EMI shielding and heat dissipation. Nanoscale, 2017. **9**(36): p. 13432–13440.
41. Cheng, C., et al., Tunable and weakly negative permittivity in carbon/silicon nitride composites with different carbonizing temperatures. Carbon, 2017. **125**: p. 103–112.
42. Wang, Z., et al., Ultralight, highly compressible and fire-retardant graphene aerogel with self-adjustable electromagnetic wave absorption. Carbon, 2018. **139**: p. 1126–1135.
43. Zhu, J., et al., Carbon nanostructure-derived polyaniline metacomposites: electrical, dielectric, and giant magnetoresistive properties. Langmuir, 2012. **28**(27): p. 10246–10255.

44. Sun, K., et al., Flexible polydimethylsiloxane/multi-walled carbon nanotubes membranous metacomposites with negative permittivity. Polymer, 2017. **125**: p. 50–57.
45. Maxwell, J.C., A Treatise on Electricity and Magnetism. Vol. 1. 1873, Clarendon Press.
46. Luo, J., et al., Synthesis, characterization, and microwave absorption properties of reduced graphene oxide/strontium ferrite/polyaniline nanocomposites. Nanoscale Research Letters, 2016. **11**(1): p. 1–14.
47. González, M., J. Pozuelo, and J. Baselga, Electromagnetic shielding materials in GHz range. The Chemical Record, 2018. **18**(7-8): p. 1000–1009.
48. Wu, T., et al., Facile hydrothermal synthesis of Fe_3O_4/C core–shell nanorings for efficient low-frequency microwave absorption. ACS Applied Materials & Interfaces, 2016. **8**(11): p. 7370–7380.
49. Toneguzzo, P., et al., Monodisperse ferromagnetic particles for microwave applications. Advanced Materials, 1998. **10**(13): p. 1032–1035.
50. Arias, R., P. Chu, and D. Mills, Dipole exchange spin waves and microwave response of ferromagnetic spheres. Physical Review B, 2005. **71**(22): p. 224410.
51. Kittel, C., On the theory of ferromagnetic resonance absorption. Physical Review, 1948. **73**(2): p. 155.
52. Liu, Y., et al., Low temperature preparation of highly fluorinated multiwalled carbon nanotubes activated by Fe_3O_4 to enhance microwave absorbing property. Nanotechnology, 2018. **29**(36): p. 365703.
53. Seng, L.Y., et al., EMI shielding based on MWCNTs/polyester composites. Applied Physics A, 2018. **124**(2): p. 1–7.
54. Ohlan, A., et al., Microwave absorption properties of conducting polymer composite with barium ferrite nanoparticles in 12.4–18 GHz. Applied Physics Letters, 2008. **93**(5): p. 053114.
55. Chen, M., et al., Highly conductive and flexible polymer composites with improved mechanical and electromagnetic interference shielding performances. Nanoscale, 2014. **6**(7): p. 3796–3803.
56. Zhang, K., et al., A facile approach to constructing efficiently segregated conductive networks in poly (lactic acid)/silver nanocomposites via silver plating on microfibers for electromagnetic interference shielding. Composites Science and Technology, 2018. **156**: p. 136–143.
57. Han, M., et al., Beyond $Ti_3C_2T_x$: MXenes for electromagnetic interference shielding. ACS Nano, 2020. **14**(4): p. 5008–5016.
58. Mathis, T.S., et al., Modified MAX phase synthesis for environmentally stable and highly conductive Ti_3C_2 MXene. ACS Nano, 2021. **15**(4): p. 6420–6429.
59. Yun, T., et al., Electromagnetic shielding of monolayer MXene assemblies. Advanced Materials, 2020. **32**(9): p. 1906769.
60. Iqbal, A., et al., MXenes for electromagnetic interference shielding: Experimental and theoretical perspectives. Materials Today Advances, 2021. **9**: p. 100124.
61. Han, M., et al., Laminated and two-dimensional carbon-supported microwave absorbers derived from MXenes. ACS Applied Materials & Interfaces, 2017. **9**(23): p. 20038–20045.
62. Liu, X., et al., Enhanced microwave absorption properties in GHz range of Fe_3O_4/C composite materials. Journal of Alloys and Compounds, 2015. **649**: p. 537–543.
63. Ma, Z., et al., Attractive microwave absorption and the impedance match effect in zinc oxide and carbonyl iron composite. Physica B: Condensed Matter, 2011. **406**(24): p. 4620–4624.

8 Role of Porous MXenes Foams and Aerogels in EMI Shielding

Sayan Ganguly
Bar-Ilan Institute for Nanotechnology and
Advanced Materials, Department of Chemistry
Bar-Ilan University
Ramat-Gan, Israel

CONTENTS

8.1 INTRODUCTION

Since graphene was discovered in 2004, two-dimensional (2D) layer-arranged materials have piqued the curiosity of researchers across the world owing to their distinct physical and chemical characteristics that distinguish them from their bulk corresponding item [1]. The most significant goals in this discipline are to utilize novel 2D materials with intriguing characteristics and to bring them into everyday uses as quickly as possible. The MXene family of 2D transition metal carbides and/or nitrides was recently discovered and named by Drexel University scientists. $M_{n+1}X_nT_x$ (n = 1–3) is the standard formula for MXenes, where M is an early transition metal (e.g., Ti, Zr, V, Nb, Ta, or Mo), X is carbon and/or nitrogen, and T_x is the number of atoms in the compound. This type of MXene is typically produced from layer-structured MAX-phase bulk ceramics by selective etching of A (generally Groups 3 and 4 elements) layers by fluoride-based compounds, and thus the basal faces of these MXenes are frequently terminated with surface moieties (T_x), such as a mixture of hydroxyl, oxygen, and fluorine. MXenes with highly adjustable

DOI: 10.1201/9781003281511-8

chemical and structural forms, which have a unique metallic conductivity and an abundance of surface functional groups, can be used as inherent dynamic materials and/or transporters of additional functionalities for a variety of uses, together with energy storage and transformation [2], electromagnetic interference shielding [3], sensors [4], biomedical imaging and therapy [5], water decontamination [6], gas separation [7], and catalysis [8, 9]. In order to be an effective EMI shielding material, MXenes must contain all of the key qualities necessary, which include strong electrical conductivity, large specific surface area, light weight, and, perhaps most significantly, simplicity of processing [10]. The creation of MXenes and composites with controlled structural designs, such as compact laminates, layer-by-layer assembly, porous and segregated structures, has been made possible by the programmable surface chemistry of these materials.

Although an MXene is susceptible to lamellar self-stacking in film formation, this decreases the number of active locations on its external areas and has an impact on its explicit characteristics, viz. loss of electromagnetic waves, obstructing ion transmission, and restricting the effective load of other functional materials, among other things. So far as we know, building an open-pored structure is a viable solution to the problem of self-stacking of 2D materials. Because of the remarkable features of MXene, it is predicted that MXene porous films would have higher performance and a wider range of applications than other porous films. The template sacrifice approach, the freeze-drying method, the induced pore creation method, and the coating of porous materials have all been used by researchers to create MXene porous films in recent years. These have significantly aided the development of film materials in the domains of EMI shielding, lithium/sodium ion batteries, pseudocapacitors, and biomedical research applications, among other things (Figure 8.1). A systematic evaluation of MXene porous films, on the other hand, has yet to be completed. In this context, it is necessary to summarize, elaborate, and forecast the functional applications, preparation methods, and internal mechanisms of MXene porous films in order to promote their further development. This chapter summarizes the advances made in the development of MXene porous films for EMI shielding. There is a strong emphasis on the "microstructure-macroscopic performance" of MXene porous films, as well as the pore-forming mechanism of the porous structure formed by the various preparation methods. The chapter also discusses the critical scientific and technical constraints that must be overcome immediately in the implementation of MXene porous films.

8.2 ELECTROMAGNETIC INTERFERENCE (EMI) SHIELDING BACKGROUNDS AND MATERIALS

Electronic systems have seen remarkable advancements in terms of technology and operating speeds as a result of the downsizing of contemporary electronic devices and circuits. When these miniature devices interact with one another, they generate unwanted electromagnetic interference (EMI), which can have a negative impact on the functioning of electronic systems. EMF radiation has been shown to be damaging to human health when exposed for an extended period of time. It has been

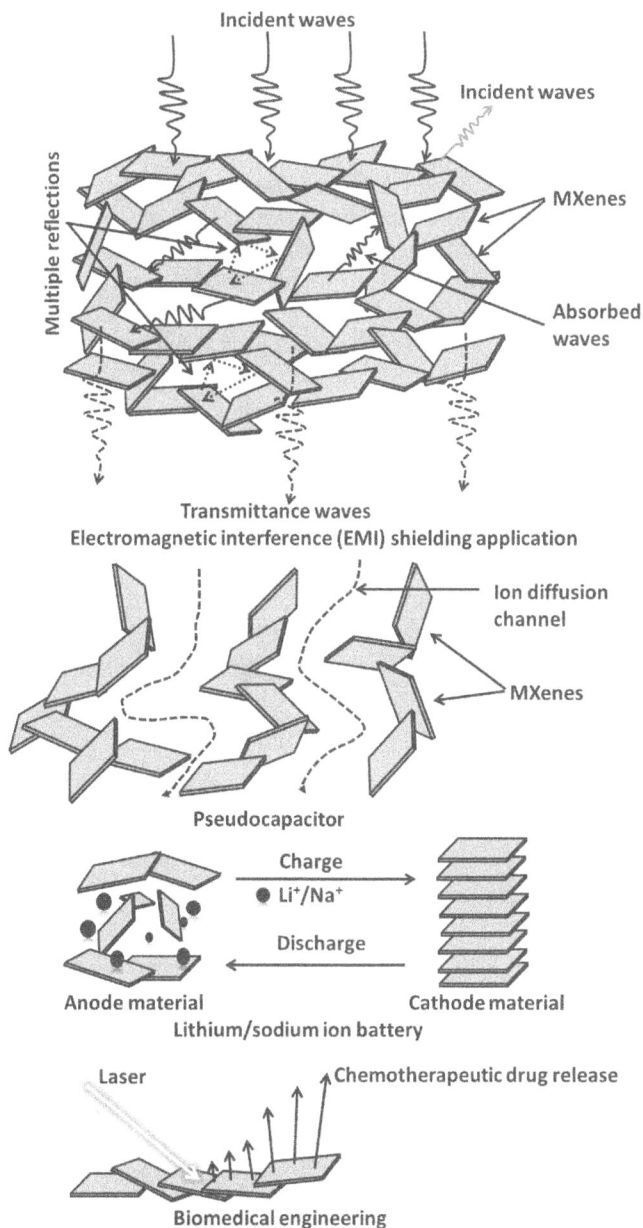

FIGURE 8.1   Various applications of MXenes-based porous structures.

linked to vomiting and diarrhea, migraines, eye issues, cancer, and damaging effects on newborn cognitive development. Medical implants and equipment (e.g., hearing aids, insulin pumps, and cardiac pacemakers) that operate in an alternating EM field are more prone to malfunctioning than those that do not. Modern warfare is also extremely sensitive to the effects of electromagnetic interference (EMI), and it is necessary to explore how to shield soldiers and equipment against EM pollution or attack [11]. As a result, the prevention or mitigation of unwanted electromagnetic radiations has emerged as a critical field in materials research. Metals such as copper, aluminum, silver, and stainless steel have been widely employed in the fight against electromagnetic pollution.

However, because of their high density, difficult processing ability, and high corrosion susceptibility, these metals have been restricted in their use in current mobile electronics that are highly integrated. To replace metals in EMI shielding applications, a variety of heterogeneous composites with conducting fillers have been developed. These include one-dimension (1D) fillers (e.g., carbon nanofibers [CNFs] and carbon nanotubes [CNTs]) and 2D fillers (e.g., expanded graphite, graphene, reduced graphene oxide, high-bandgap nitride [hBN], and MoS_2). These composites offer a number of favorable characteristics, including being lower in weight, having greater environmental resilience, and having superior anticorrosive capabilities. Their inadequate shielding ability, on the other hand, has prevented them from being widely used.

8.3 FABRICATION TECHNIQUES

It is generally known that porous 2D materials may be produced through the four approaches depicted in Figure 8.2: assembly, depositing or inserting, coating, and creating in-plane pores are some of the processes that might be used. Particular synthetic procedures for graphene and other 2D materials have been well-established for

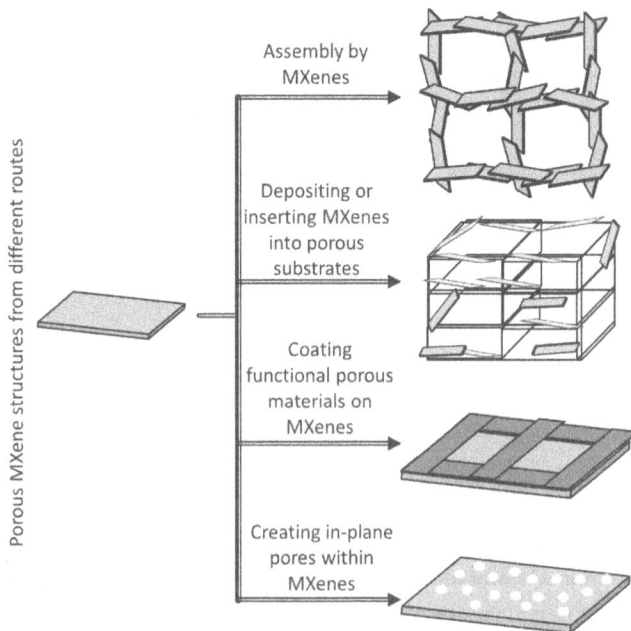

FIGURE 8.2 Various fabrication approaches for preparing porous MXenes.

TABLE 8.1

A Comparison of Porous MXenes Synthesized Using Different Processes and with Different Structural Features

Samples	Methods	Pore Structures	Ref.
MXenes assemblies	Freeze-dry	Meso/macro porous	[20]
MXenes aerogel	Freeze-dry	Meso/macro porous	[20]
MXenes lamella	Stirring and quenching	Vertically aligned meso/macro pores	[12]
MXenes/SiC nanowires	Bidirectional freezing	Parallel aligned meso/macro pores	[15]
MXenes/PI aerogel	Freeze-dry	Meso/macro pores (wide size distribution)	[16]
Fluffy MXenes microsphere	Spray dry	Meso/macro pores (wide size distribution)	[21]
3D macroporous MXenes film	Hard templating	Meso/macro pores (narrow size distribution)	[17]
Cellular MXenes foam	Hydrazine reduction	Meso/macro pores (wide size distribution)	[18]
MXenes insertion into 3D sponge	Dip coating	Macroporous (wide size distribution)	[19]
MXenes/RGO aerogel	Chemical reduction	Meso/macro pores (wide size distribution)	[22]
MXenes/RGO aerogel	Freeze-dry and calcination	Meso/macro pores (wide size distribution)	[23]
Cellulose/MXenes aerogel	Chemical crosslinking	Meso/macro pores (wide size distribution)	[24]
TiO_2 nanorod/MXenes/SnO_2	Self-assembly	Mesoporous (wide size distribution)	[24]
FeNi-LDH/MXenes	In situ growth	Mesoporous (wide size distribution)	[25]
Core-shell MXenes/SiO_2	Sol-gel	Vertical mesopore (narrow size distribution)	[26]
Porous MXenes flakes	Oxidative etching	Mesoporous (narrow size distribution)	[27]
Porous MXenes/divalent ion	Selective etching	Micro-meso pores (narrow size distribution)	[28]

the creation of various porous structures, including porous membranes. As shown in Table 8.1, porous MXenes derived from the four processes just listed may be synthesized using a variety of techniques that are based on unique pore-generating mechanisms. It is possible to tailor pore shapes ranging from sub-10 nm in-plane pores and limited superficial apertures to a few micrometers pervaded big openings, while the synthetic approaches can range from straightforward physical modifications to complex chemical procedures. These groundbreaking achievements are explored in detail in the following subsections, with the goal of providing an open window into the fabrication of porous MXenes and other developing 2D materials in general.

8.3.1 FABRICATION OF POROUS MXenes BY ASSEMBLY

In this scenario, the crucial thing is to drive the stratifying of MXene nanosheets in the presence of porogen materials in order to generate diffused frames, which is what is needed. Porogen agents include ice, water, sulfur, and polymer spheres, among other things [12]. Following the removal of the porogen agents using certain procedures, the organized scaffolds may be kept and porous MXene materials can be generated from the porogen agents. These approaches primarily include freeze-drying, spray-drying, hard templating, and chemical reduction, among other things.

Overall, this method may produce free-standing MXene structures with guest- and electroconductive 3D networks that can be exploited for a variety of applications immediately out of the gate. Consequently, it is the most promising and useful technique when compared to the other three options available. Freeze-drying has been extensively utilized to produce porous 2D materials by expelling the nanosheet precursors into the limits of ice crystals, a technique that is often used to construct perforated 3D composites [13]. If there are strong contacts between 2D nanosheets, this approach has the potential to create porous and resilient structures, according to theory. As a result of the robust hydrogen bonding communication and Van der Waal's force of interactions between MXene nanosheets, Wang et al. have used lyophilization or freeze-drying to produce a porous MXene with a huge surface area. The researchers have also used the product as a support to prepare porous MXene/ Fe_2O_3 composites [14]. It should be emphasized that by altering the freeze-drying procedure, the porosity structure may be easily modified to meet specific requirements. For a case in point, the Gogotsi research group has created free-standing perpendicularly associated porous MXene films by power-driven shearing of a discotic lamellar liquid-crystal phase of MXene nanosheets in a liquid-crystal environment [12]. This study introduces the use of a nonionic surfactant, hexaethylene glycol monododecyl ether ($C_{12}E_6$), which has a resilient hydrogen bonding interaction with $Ti_3C_2T_x$ and may be used to enhance the packing symmetry of $Ti_3C_2T_x$ by increasing its packing symmetry. $C_{12}E_6$ is incorporated between MXene nanosheets, resulting in the formation of a single discotic lamellar phase. This study represents a substantial advancement in the field and gives a guideline for the creation of porous MXenes with adaptable structural and functional features.

It is also possible to build porous MXene-based hybrids by incorporating additional functional materials into MXene dispersion prior to freeze-drying. It is a straightforward and generic method of modifying the structures and characteristics of porous MXenes, and it has the potential to be widely researched in order to broaden their uses. Several materials, such as SiC nanowires and polyimide (PI), have been included in $Ti_3C_2T_x$ aerogels, for instance [15]. The unsupported, ultralight, and well-ordered lamellar $Ti_3C_2T_x$/SiCnws composite foams have been created using electrostatic assembly and controlled bidirectional freezing procedures, along with regulated bidirectional freezing. As shown in Figure 8.3, $Ti_3C_2T_x$ nanosheets firmly wrap SiC nanowires inside each layer, increasing the mechanical characteristics of MXene aerogel while also introducing a large number of junction contacts and flaws that are helpful for its application [15]. According to $Ti_3C_2T_x$ surface terminations and polar PI chains, strong polar interfacial contacts between surface terminations and polar PI chains lead to the creation of one extremely resilient network in the MXene/PI aerogels. A significant improvement over the delicate and stiff pure MXene aerogel is the super-elasticity of the MXene/PI aerogels, which exhibit significant reversible compressibility as well as superior fatigue resistance and reversible stretchability [16]. By using capillary driven crumpling and self-assembly, spray drying may be used to create porous 2D materials that are otherwise difficult to produce. Qiu and colleagues have recently reported the spray drying synthesis of porous MXenes, which is a novel approach. The MXene slurry is first converted into aerosol droplets, and then it is immediately sintered at an elevated temperature

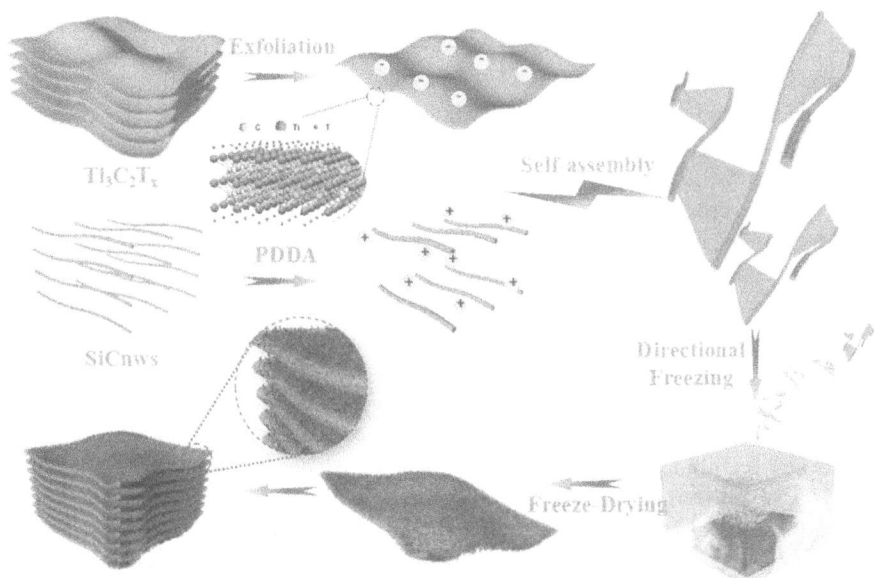

FIGURE 8.3 Diagram depicting the production of f-Ti$_3$C$_2$T$_X$/SiC nanowire hybrid foams in three dimensions.

(Reproduced with the permission from ref. [15], © 2018 American Chemical Society.)

to eliminate the mediators from the droplets. Following complete evaporation of the solvent, the inward suction force causes the isotropic compression and rapid assembly of MXene nanosheets into 3D structures with a fluffy form. It is the creation of this strong porous structure that effectively prevents the restacking of MXene nanosheets, which not only enhances the accessible surface area but also provides a framework that is kinetics-friendly, as previously stated. Also possible is the incorporation of transitional and main-group metal oxides (e.g., SnO2, Co3O4), metal phosphides (CoP), perovskite-type metal oxides (MnTiO3), and noble metals (e.g., Pt, Ag) into porous MXenes by the simple addition of metal salts to the dispersion of MXene. As a result of its versatility, this approach may be utilized to create novel porous MXene-based hybrids for a variety of different applications.

The hard templating technique is another successful way for fabricating porous materials. It is based on the pre-deposition of 2D materials on the surface of a hard template and the subsequent removal of the hard template from the workpiece. Using the bonds between the surface hydroxyl groups of poly(methyl methacrylate) (PMMA) sphere composites, Gogotsi and colleagues first manufactured MXene-wrapped poly(methyl methacrylate) (PMMA) spherical composites and then vacuum filtered them into free-standing films, as an example [17]. It is possible to generate hollow MXene spheres and 3D macroporous MXene films after removing the PMMA hard template. The 3D macroporous Ti$_3$C$_2$T$_x$ films are made up of hollow Ti$_3$C$_2$T$_x$ spheres that are linked to form a 3D structure. Because of their high conductivity and excellent contact between spheres, they are free-standing, flexible, hydrophobic, and highly conductive. PMMA spheres of various diameters may be used to alter the pore size of these 3D porous MXene films, and this approach is also used to fabricate 3D macroporous Mo2 CTx and V 2 CTX films. Furthermore, in this example, the pores are closer together than in other porous MXenes. The chemical reduction strategy can form holes inside thick MXene films by removing functional groups from the surface, resulting in foaming. By treating its thick films with hydrazine, Yu et al. created porous MXene foams [18]. An oxygen-containing group reacts with hydrazine, releasing gaseous species like CO_2 and H_2O that produce high pressure between MXene sheets to overcome the Van der Waals forces that keep them together, according to this study. Thus, dense hydrophilic $Ti_3C_2T_x$ films are changed into porous hydrophobic $Ti_3C_2T_x$ films that have a cellular structure, as opposed to dense hydrophilic $Ti_3C_2T_x$ films. It is important to note that the usage of hydrazine is required, and that H_2O, ascorbic acid, and ethanol will not cause the foaming process. It is probable that they will be unable to react with MXenes in order to create the required gases. However, the approaches described here are adaptable and may be utilized to produce porous MXenes with a variety of architectural configurations. Spray drying, for example, generates porous MXene particles, whereas other techniques produce free-standing films and aerogels (Figure 8.4a,b,c); freeze-drying creates porous MXenes with interpenetrating pores, and the hard templating approach provides porous MXenes with variable pore size. It may construct porous MXenes with customizable structures for specific purposes using the proper procedures, depending on the situation [19].

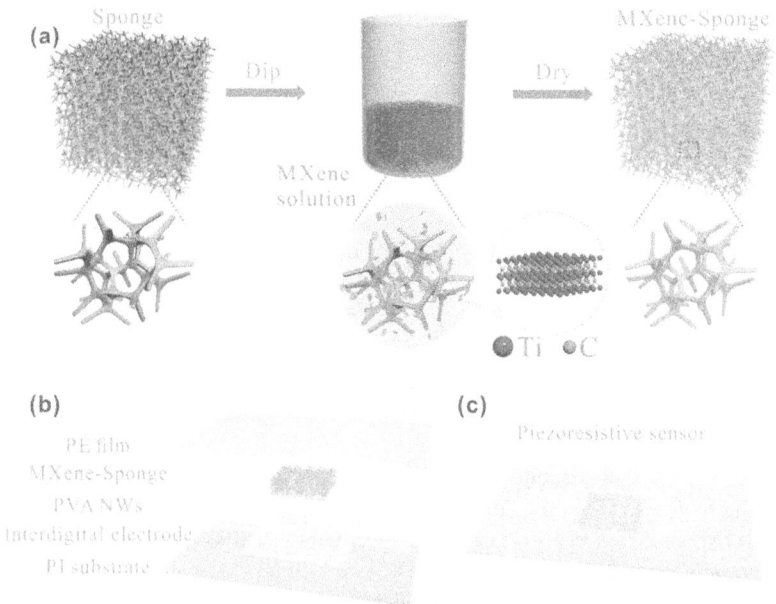

FIGURE 8.4 (a) Schematic depiction of the MXene-sponge manufacturing process; (b, c) schematic representation of the construction of a sensor based on MXene-Sponge/PVA NWs.

(Reproduced with the permission from Ref. [19], © 2018 Elsevier.)

8.3.2 MXenes Deposited onto Porous Substrate

Fundamental to these porous structures is to create porous substrates capable of supporting MXene nanosheets. Porous platforms can be manufactured in advance and also formed concurrently with the deposition and insertion of MXenes. Porous MXene-based hybrids may be produced using a simple dip-coating and drying procedure for the pre-synthesized substrates. For instance, Gao et al. constructed a 3D hybrid porous MXene-sponge network using a sponge as a substrate [19]. While the sponge's porosity structure is well preserved, $Ti_3C_2T_x$ nanosheets are densely deposited on the skeleton of the sponge, resulting in a strong bond between the two materials. Due to the fact that the deposition of MXene nanosheets has no effect on the creation of the porous structures in this situation, the loading mass of MXenes may be simply adjusted. Various porous substrates, including as metal foams and polymer sponges, can be used to develop novel porous MXene-based hybrids for specific applications. These porous substrates include metal foams, polymer sponges, and metal sponges with varying compositions. RGO aerogels are one kind of hydrogel/aerogels that may be used as porous substrates. Other types of hydrogels/aerogels can be used as porous substrates as well, as long as the introduction of MXene nanosheets does not significantly impede the production of hydrogels/aerogels. The availability of hydrogels/aerogels presents us with a plethora of options in a variety of fields [29, 30]. For example, using Ti_3C_2-cellulose dispersion as a precursor, Zhang and colleagues have created 3D porous cellulose/ Ti_3C_2 aerogels by a chemical cross-linking process [31]. It is widely known that the phase separation of cellulose-rich and cellulose-poor areas can result in the formation of porous aerogels from cellulose alone. The inclusion of Ti_3C_2 into the cellulose chains has no effect on the phase separation behavior of the cellulose chains, allowing the formation of cellulose/Ti_3C_2 aerogels. MXene-sponge has a porous structure that is similar to that of sponge, and $Ti_3C_2T_x$ nanosheets are densely coated on the sponge's skeleton to give it a crystalline appearance. In this instance, the deposition of MXene nanosheets has no effect on the creation of the porous structures, and the loading mass of MXenes may be simply adjusted without difficulty. Various porous substrates, including as metallic foams and polymer loofas, can be used to develop novel porous MXene-based composites for specific applications [32]. These porous substrates include metal foams, polymer sponges, and metal sponges with varying compositions. The latter situation is achieved by pre-mixing MXenes with the raw materials of a porous substrate, after which MXenes are placed and/or implanted on the porous substrate using procedures that are similar to those just described. The assembly of graphene oxide (GO) into porous reduced-GO (rGO) aerogel can be accomplished using a variety of methods, comprising undeviating chemical reduction, freeze-drying, and the subsequent thermal management [33]. rGO aerogels are good porous supports since they are compatible with both MXenes and GO. To give an example, Luo et al. [22] employed a direct chemical reduction approach to promote the self-assembly of GO in a $Ti_3C_2T_x$/GO dispersion to generate an MXene/rGO aerogel using a direct chemical reduction method. Gao and colleagues have utilized $Ti_3C_2T_x$/GO dispersion as a precursor to produce MXene/rGO aerogel by freeze-drying, and then used a thermal reduction approach to reduce the temperature of the aerogel [34]. rGO aerogels are one kind of hydrogel/aerogels that may be used as porous substrates. Other types of hydrogels/aerogels can be used as porous substrates as well, as long as the insertion of MXene nanosheets does not significantly impede the production of hydrogels/aerogels. The availability of hydrogels/aerogels presents us with a plethora of options in a variety of fields. According to Zhang et al., utilizing Ti_3C_2-cellulose dispersion as a precursor, they were able to construct 3D porous cellulose/Ti_3C_2 aerogels by a chemical cross-linking reaction using Ti_3C_2-cellulose dispersion [24]. It is widely known that the phase separation of cellulose-rich and cellulose-poor areas can result in the formation of porous aerogels from cellulose alone. The inclusion of Ti_3C_2 into the cellulose chains has no effect on the phase separation behavior of the cellulose chains, allowing the formation of cellulose/Ti_3C_2 aerogels.

8.3.3   FUNCTIONAL COATING ON MXenes

Mostly on surface of MXenes, various functional materials are organized to create interparticle voids, which are the spaces in between individual particles. Depending on their configuration, these holes can be chaotic or well-organized, and they can be created by a distinct arrangement of additional serviceable materials with MXenes or by in situ development of surplus functional materials on their external area. According to one recent study, TiO_2 nanorods and SnO_2 nanowires were initially synthesized by Xu et al., who then exploited Van der Waals interactions to accomplish the self-assembly of these transition metal oxides in the presence of $Ti_3C_2T_x$ [35]. In addition to preventing $Ti_3C_2T_x$ from restacking, these transition oxides are also well-dispersed on their surface, resulting in constricted and even continuous holes on the surface of the material. It has also been demonstrated that pyrite FeS_2 nanodots/$Ti_3C_2T_x$ composites may be made by dumping iron hydroxide on the surface of $Ti_3C_2T_x$ and then subjecting the composite to a sulfurization procedure [36]. Yu and colleagues have, on the other hand, developed spongy layered double hydroxide (LDH)/MXene composites by growing LDH in situ on $Ti_3C_2T_x$, a titanium alloy [25]. In this case, negatively charged -OH and -F groups on $Ti_3C_2T_x$ enhance the nucleation of LDH platelets, and the subsequent development of these LDH platelets results in the formation of a consistent spongy network on the MXene sheets. Various functional materials combined with MXenes not only allow for the integration of the advantages of both functional materials and MXenes, but it also frequently results in the emergence of unique features that are the result of the mutual interactions between these two materials. This opens up a plethora of possibilities for the development of novel MXene-based hybrids that deliver impressive performance. These porous MXene-based hybrids, like the highly developed porous graphene-based composites, are deserving of further investigation. Pores formed are often poor, and the pore diameters are not uniform but rather random and widely distributed, which is still unsuitable for several purposes. Yang et al. sought to coat one layer of mesoporous silica on the surface of MXenes, which were denoted as Ti_3C_2@mMSNs, using a basic and uncomplicated sol-gel process [26].

8.3.4 IN-PLANE PORE GENERATION IN MXenes

It should be emphasized that the pores in the three preceding examples are all out-of-plane holes, and constructing in-plane pores inside the 2D sheets might be a viable alternative method of fabricating porous MXenes. Dissimilar from out-of-plane pores, compactly dispersed in-plane pores have extra uncovered superficial positions and exhibit more effective ion/guest transport capabilities than out-of-plane pores do. Using metal ions catalyzed oxidative etching, for example, Gogotsi and colleagues have been able to manufacture porous MXenes using this approach [37]. Metal ions are used to catalyze the oxidation process of $Ti_3C_2T_x$ with O_2, which results in the conversion of certain $Ti_3C_2T_x$ sites into TiO_2. Porous $Ti_3C_2T_x$ ($P\text{-}Ti_3C_2$) may be produced by eliminating TiO_2 from the solution with HF. Mesopores with a diameter ranging from 4 to 6 nm are evenly dispersed on the surface of MXene nanosheets. Guo et al. [51] have also successfully synthesized porous MXenes by employing bimetal $Mo_{1.33}Sc_{0.67}AlC$ phase as a precursor. Sc and Al are removed from $Mo_{1.33}Sc_{0.67}AlC$ via a straightforward HF etching procedure, resulting in 2D $Mo_{1.33}C$ with divacancy ordering. It should be highlighted that nanosized in-plane holes inside $Mo_{1.33}C$ nanosheets can also be formed at the same time (Figure 8.5f), although this was not explicitly stated by the authors. Typically, nanopores and metal vacancies may be created during the production process of MXenes by using

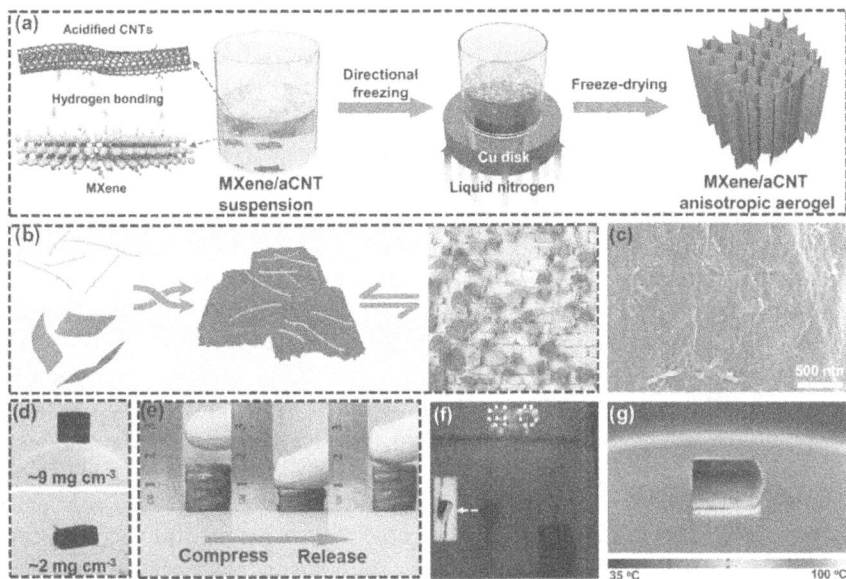

FIGURE 8.5 (a) Preparation of an MXene/aCNT aerogel; (b) MXene nanosheets and aCNTs bionic assembly; (c) MXene/aCNT microstructure in SEM picture of MCA; (d) MCA on feather (top) and strand of hair (bottom); (e) 50% strain reversible MCA-1 compression–release method; (f) electrically conductive MCA-1 LEDs; (g) thermographic picture of a 9 mm cuboid MCA-1 heated to 100°C.

(Reproduced with the permission from Ref. [39], © 2021 American Chemical Society.)

a simple HF etching procedure [38]. The presence of divacancy in this environment naturally enhances this process, resulting in the formation of visible holes on the $Mo_{1.33}C$ nanosheets. A novel synthetic approach for porous MXenes has been discovered using atomic and sub-10 nm scale structural design, which has the potential to transform and broaden the notion of property-tailoring in MXene materials.

8.4   MXenes AEROGELS/FOAMS FOR EMI SHIELDING APPLICATIONS

The ability to be lightweight is a requirement for effective EMI-shielding composites used in aerospace, military, and mobile electronics uses, among others. Because of their extremely low density, porous MXene foams and aerogels are particularly well suited for this use. $Ti_3C_2T_x$ MXene foams for lightweight and malleable EMI shielding materials were initially described by Liu et al. [18]. At 90°C, the compacted MXene sheets were submerged in a hydrazine solution. Cavity action allowed the hydrazine solution to permeate and extend the MXene layers, resulting in a low-density foam structure composed of MXene stacked sheets. As an example, a $Ti_3C_2T_x$ sheet with thicknesses of 1, 3, and 6 microns expanded to 6.2, 18, and 60 microns, respectively, when stretched. While the electrical conductivities of the porous MXene foams were lower than those of the bulk compacted MXene film (4000 S cm1), the electrical conductivities of the porous MXene foams were higher than those of the bulk compacted MXene film. An MXene-based aerogel with high superelasticity, light weight, and electrical conductivity has been developed by Deng et al. It is produced through the hybridization of 2D MXene sheets with 1D acidified carbon nanotubes (aCNTs) through directional freezing and subsequent freeze-drying [39]. Because they enhance the contacts among MXene nanosheets and help to form an unified network, carbon nanotubes (aCNTs) are critical in strengthening the flexibility of the MXene aerogel. An ultralow density of 9.1 mg cm^{-3} and a high conductivity of 447.2 S m^{-1} are achieved with an ultralow filler content of 0.30 vol% in the resulting MXene/acidified carbon nanotubes hybrid aerogel (MCA), which is obtained by combining MXene with acidified carbon nanotubes. In addition to the hydrogen bonding and Van der Waals' forces that exist among MXene nanosheets and aCNTs, the biomimetic design also possesses exceptional flexibility and compressibility. MXene aerogels with carbon nanotube reinforcement (aCNT) exhibit superior integrated characteristics when compared to conventional MXene aerogels, including super elasticity, increased fatigue resistance, and high electrical conductivity, as compared to conventional MXene aerogels. MCAs that are low specific gravity, squeezable, and conductive show promise for EMI shielding applications in aerospace and electrical equipment, according to the researchers.

Because of their distinctive ordered structures and high electrical conductivity (σ) values, renewable porous biochar and 2D MXenes have gotten a lot of attention in the high-end electromagnetic interference (EMI) shielding industries. As previously reported, the fabrication of a wood-derived porous carbon (WPC) skeleton from natural wood was used as a template [40]. Natural wood was first carbonized at a high temperature in order to produce a skeleton of wood-derived porous carbon (WPC), which was then processed further. In order to organize the MXene aerogel/WPC hybrids, the better-conducting and ultra-light spatial MXene aerogel was first produced, and then tested (Figure 8.5). As a membrane reactor, the WPC frame serves as a template, with highly organized honeycomb cells lining the inside of the WPC. The SE values, mechanical and thermal lining, and flame-retardant qualities of the materials were all evaluated in a systematic manner. Among other things, this research is expected to provide an easy synthesis for the preparation of ultra-lightweight, environmentally friendly, and highly effectual multioperational bio-carbon-based components, that could be applied in the high-end EMI shielding composite development areas of the aviation and governmental defense industries, among other things.

$Ti_3C_2T_x$ sheets were assembled into extremely conducting, permeable, and 3D structures using graphene oxide (GO)-aided hydrothermal association, tailed by directional freezing and successive freeze-drying, as demonstrated by Zhao et al. [41] (Figure 8.6). The $Ti_3C_2T_x$ MXene/RGO composite aerogel (MGA) with line up core–shell structure, when joined with the well-preserved fundamental arrangement of $Ti_3C_2T_x$, results in an epoxy nanocomposite with an exceptional electrical conductivity of 695.9 S m^{-1} and an EMI-shielding applicability greater than 0 dB at a tiny $Ti_3C_2T_x$ concentration of 0.74 vol%.We have discovered the sub-micron morphology formed by the assembly of $Ti_3C_2T_x$ and RGO sheets in the hybrid aerogel, as well as the processes that caused the assembly.

$Ti_3C_2T_x$ MXene/graphene fusion foams have been developed for use in electromagnetic interference shielding applications [42]. It was possible to manufacture hybrid foams with different MXene to GO ratios by the general vacuum drying of quenched solutions at 65°C tracked by thermal reduction at 300°C for 1 h. rGO

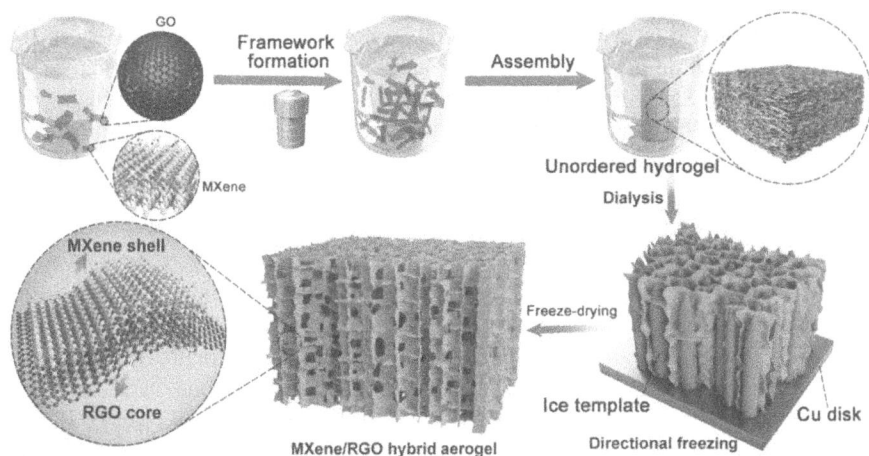

FIGURE 8.6 GO-assisted hydrothermal assemblies, directional freezing, and freeze-drying of a $Ti_3C_2T_x$ MXene/RGO hybrid aerogel.

(Reproduced with the permission from ref. [41], © 2018 American Chemical Society.)

foam with a pure rGO composition had a density of 3.1 mg/cm^3, but the density of a hybrid foam with an MXene to graphene ratio of 1:1 had a density of 7.2 mg/cm^3. As MXene (Ti$_3$C$_2$T$_x$) has a higher density than graphene, the inclusion of Ti$_3$C$_2$T$_x$ MXene resulted in an increase in the density of the hybrid foam. It was discovered that raising the graphene content to 1:2 and 1:3 lowered the density of the hybrid foam to 4.6 mg/cm^3 and 3.7 mg/cm^3, respectively. Likewise, the electrical conductivity of the rGO foam was greatly enhanced by the addition of MXene, increasing from 140 S cm^{-1} to 1250 S cm^{-1} with the addition of MXene. The conductive porous structure demonstrated improved EMI shielding efficiency, which rose when the MXene concentration in the structure was increased. EMI SE of 15 dB was observed in the pure rGO foam at a thickness of 1.5 mm, which doubled in the 1:1 MXene hybrid when the foam was thicker. Internal scattering in the highly conductive MXene network was shown to be responsible for the increased EMI shielding effectiveness. The EMI SE increased by more than 50 dB with an increase in thickness (3 mm) in the hybrid with an MXene to rGO ratio of 1:2 when the thickness was increased.

In this study, the EMI shielding capabilities of a hybrid foam composed of silver nanowire AgNW and titanium trioxide Ti$_3$C$_2$T$_x$ MXene were examined [43]. Three-dimensional triaxial compression of porous melamine formaldehyde (MF) substrate resulted in the formation of buckled melamine formaldehyde (BMF) foam. Through successive dip-coating, AgNW was consistently grown and uniformly sized. With an increasing number of dip-coating cycles, the electrical conductivity steadily increased in the coating. To create an irregular honeycomb MXene structure, the conductive foams were submerged in a Ti$_3$C$_2$T$_x$ MXene solution, followed by directed freezing and freeze-drying, and then dried in the same manner. The electrical conductivity and EMI-shielding capabilities of the BMF/AgNW/Ti$_3$C$_2$T$_x$ hybrid composite were greater than those of the individual BMF/AgNW and BMF/MXene composites. Because the AgNW and Ti$_3$C$_2$T$_x$ MXene worked together, they were able to achieve greater electrical conductivity in the porous design while also demonstrating superior EMI-shielding performance. In comparison, the EMI SE of the 2 mm thick BMF/AgNW/Ti$_3$C$_2$T$_x$ hybrid foam with a density of 49.5 mg/cm^3 was 52.6 dB, whereas the EMI SE of the 2 mm thick BMF/AgNW and BMF/MXene foams were both 40. The higher electrical conductivity was related to the greater shielding behavior, which resulted in an impedance mismatch between the hybrid and the unoccupied areas, according to the researchers. Furthermore, the internal scattering of incident EM waves from conductive porous surfaces was enhanced, resulting in absorption dominated EMI shielding as a result of the increased internal scattering [44]. A lightweight MXene/SA foam with 95% MXene has a density of 20 mg/cm^3, an electrical conductivity of 22.11 S cm^{-1}, and an outstanding EMI SE of 72 dB. It is 2 mm thick and has a density of 20 mg/cm^3 and an electrical conductivity of 22.11 S cm^{-1}. A thin layer of PDMS was added to the MXene/SA foam via vacuum filtering, resulting in a small reduction in EMI shielding effectiveness to 50 dB. When the porous design of the hybrid foam was compared to a basic mix of MXene/SA/PDMS and MXene/SA foam coated with a solid PDMS layer that covered all of the pores, the importance of the porous architecture in EM radiation attenuation was brought to light. Following the hybrid foam with a thin uniform PDMS coating and a thin MXene/SA/PDMS coating, the foam with a solid PDMS coating (50 dB) and the simple blended composite (50 dB) displayed the greatest EMI SE, followed by the simple blended composite (50 dB) (9 dB).

A unidirectional freeze-drying technique was used by Wang et al. to fabricate $Ti_3C_2T_x$ MXene-based hybrid aerogels with an anisotropic or orientated microporous structure using NCO-terminated waterborne polyurethane (WPU) prepolymer as a polymer matrix and cross-linker [45]. In this study, $NiFe_2O_4$ nanoparticles (NPs) are used as magnetic components in hybrid aerogels (Figure 8.7) because of their comparatively high permeability and permittivity. The NPs are incorporated into the aerogels by a liquid phase in situ synthesis procedure. A significant interfacial contact is formed when the NCO-terminated WPU prepolymer reacts with the hydroxyl functionalized MXene nanosheets and forms a tight interconnection between them. This results in the successful construction of WPU/MXene/$NiFe_2O_4$ hybrid aerogels with favorable flexibility, conductivity, and mechanical properties, even at low densities; in particular, the high compressive stresses of hybrid aerogels outperform those of other MXene- and WPU-based porous materials by a significant margin. For their part, WPU/MXene/NiFe2O4 hybrid aerogels have readily controllable micropore shape, porosity, and density. This, in combination with the changeable MXene concentration, results in an EMI SE that is highly adaptable to various application requirements.

Through the use of a sol-gel and thermal reduction process, a very structurally robust epoxy was infused into a $Ti_3C_2T_x$ MXene/carbon hybrid foam [46]. 10 mL of $Ti_3C_2T_x$

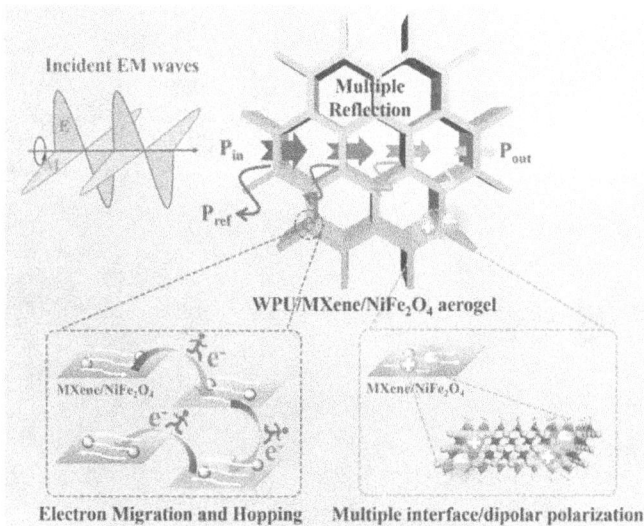

FIGURE 8.7 Schematic illustration of waterborne PU aerogel fabrication for EMI shielding applications.

(Reproduced with the permission from Ref. [45], © 2018 American Chemical Society.)

MXene dispersion were treated with resorcinol, formaldehyde, and a sodium carbon-ate catalyst before curing at 90°C for five hours under nitrogen and freeze-drying. The MXene/carbon hybrid foam was created by annealing the porous foam at 400°C for 2 h. Under vacuum, an epoxy solution was injected into the porous hybrid foam, which was then removed. MXene/carbon foam was created by combining 1.64 wt% MXene and 2.61 wt% carbon in an epoxy solution. At a thickness of 2 mm, the foam demonstrated electrical conductivity of 1.84 S cm^{-1} and EMI SE of 46 db throughout the whole X-band frequency range. When MXene concentration was raised, the hardness and Young's modulus of the composite increased to 0.31 GPa and 3.96 GPa, respectively, which were both greater than the values obtained from pure carbon/epoxy composites (0.28 GPa, and 3.51 GPa, respectively). Conductive 2D MXene flakes have a signifi-cant impact on the electrical, mechanical, and EMI-shielding characteristics of epoxy-based composites and hybrid foams, as well as on their overall performance.

Ti$_2$CT$_x$ MXene/PVA composite foams (Figure 8.8) with minimal reflection were investigated for EMI-shielding applications in the telecommunications industry [47]. The Ti$_2$CT$_x$/PVA foam was made via freeze-drying. Strong hydrogen bonds between MXene and PVA molecular chains provided added flexibility and mechani-cal strength. Figure 8.9 depicts a putative mechanism for the superior absorption-dominated EMI shielding performance of MXene/PVA composites.

This foam has a lower electrical conductivity than porous Ti$_3$C$_2$T$_x$ but a far greater conductivity than other carbon compounds [48]. 10.9 mg cm^{-3} of Ti$_2$CT$_x$/PVA

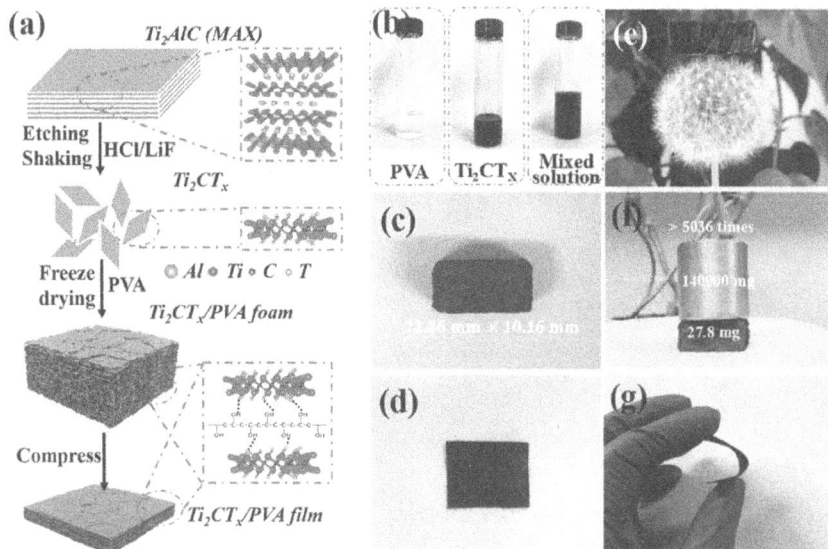

FIGURE 8.8   (a) Preparation of f-Ti$_2$CT$_x$/PVA composite foam and film; (b) bottles of PVA, f-Ti2CTx, and f-Ti2CTx/PVA solutions; (c) f-Ti2CTx/PVA foam; (d) typical f-Ti$_2$CT$_x$/PVA film picture; (e) an ultralow density f-Ti2CTx/PVA foam piece on a dandelion; (f) f-Ti$_2$CT$_x$/PVA foam sustains more than 5000 times its own weight; (g) f-Ti$_2$CT$_x$/PVA film photographed.

(Reproduced with the permission from Ref. [47], © 2019 American Chemical Society.)

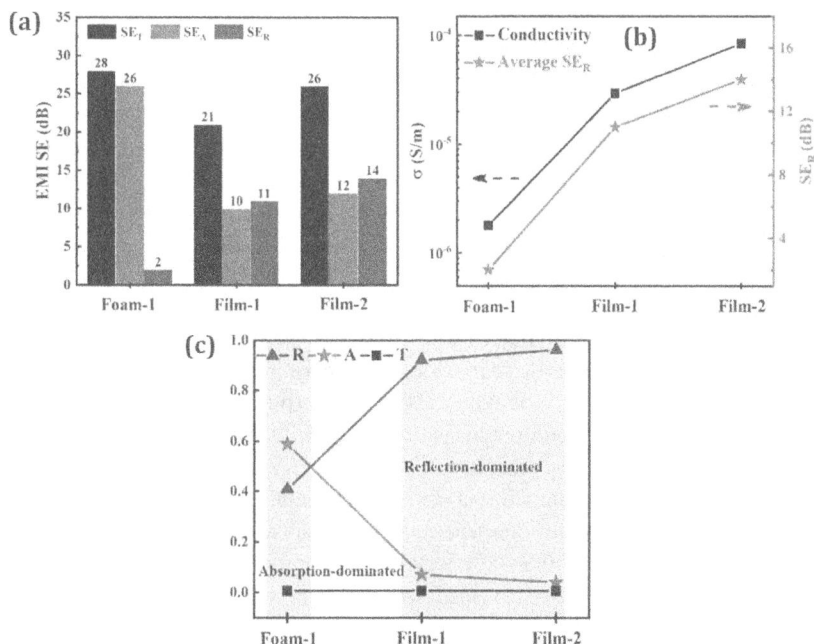

FIGURE 8.9 (a) EMI shielding values of f-Ti2CTx/PVA foam; (b) conductivity and shielding comparison of f-Ti2CTx/PVA foam; (c) composite f-Ti2CTx/PVA reflectance (R), absorption (A), and transmission (T).

(Reproduced with the permission from Ref. [47], © 2019 American Chemical Society.)

foam has an EMI SE of 26 to 33 dB. Low-density compressed foam structure led to 5136 dB cm^2 g^{-1} SSE/t at low Ti2CTx MXene content (0.15 vol%). Because of the porous foam's superior absorption efficiency, it outperformed a compact film construction, where EM waves entered the shield and high internal dispersion caused the dissipation of energy in the form of heat. In addition, the dielectric and interfacial polarization produced by an alternating EM field contributed to the absorption of electromagnetic radiation.

It was possible to create a $Ti_3C_2T_x$ MXene-based aerogel by employing traditional freeze-drying of MXene dispersions with varying concentrations [49]. Although the density of the aerogel was influenced by the concentration of the MXene solution used during the freeze-drying process, the shape of the pores was governed by the ice crystals generated during the freezing process. However, despite having a low density of 20.7 mg cm^{-3}, the $Ti_3C_2T_x$ MXene aerogels were dimensionally stable and possessed an electrical conductivity of 22 S cm^{-1} and an electromagnetic interference (EMI) shielding of 75 dB. It had an SSE/t of 18116 dB cm^2 g^{-1}.

Natural loofah sponge, which has been sustainably regenerated and is ecologically benign, has the potential to become a multifunctional bio-carbon material, particularly in the field of electromagnetic shielding, due to its unique hierarchical structure. A successful combination of carbonized loofah sponge and MXene aerogel allowed the complementing benefits of the two materials to be fully exploited while also compensating for their respective drawbacks [50]. Chemical vapor deposition (CVD) was used to coat a few layers of thick carbon onto the carbonized loofah sponge in order to correct tiny surface imperfections on the bio-carbon. This provided a suitable supporting substrate for the weak MXene aerogel, which was then fastened to the sponge. An MXene aerogel with a directionally arranged microporous MXene structure was then created to fill the macroscopic pores of the carbonized loofah sponge skeletons covered with CVD carbon. EMI shielding performance, electrical conductivity, rational mechanical strength, flame resistance, and thermal stability were all outstanding in the composites that were created, as predicted. In addition, the mechanism of anisotropy imparted to the composite by the directional alignment of the MXene aerogel was thoroughly investigated and understood. This research may pave the way for more effective development and application of aerogels and loofah sponges in the future.

Aerogels based on biomimetic materials with absorption-dominated EMI shielding have demonstrated higher promise in civilian and military applications than typical metal shields with reflection-dominated EMI shielding. Nanostructured cell walls and micrometer-sized holes combined with biomimetic cellular architecture have demonstrated enormous potential for constructing high-performance absorption membranes in recent years. Because of their low weight, facile processability, and customizable EMI shielding performance, EMI shielding materials have the potential to replace conventional reflection-dominated metal shields in a variety of applications. Using the unidirectional freezing process, Guo et al. were able to manufacture poly(vinyl alcohol) (PVA) aided transition metal carbides (MXene) aerogels with cellular morphology that were then tested [51]. The size of the pore channels and the thickness of the wall were customized by changing the freezing temperature. Interestingly, the EMI shielding performance improves while the compressible strength drops with increasing the pore channel size/wall thickness. It is necessary to achieve saturation of pores-induced multiple reflections and scattering in order to provide EMI shielding capability. As a result, the wall thickness-dominant absorption becomes the most important factor to consider. In terms of mechanical characteristics, the increase in stress concentration and the looser stacking of the PVA and MXene can be linked to the degradation in compressible strength. Finally, the lightweight PVA/MXene composite aerogels (approximately 33 mg/cm^3) can achieve a compressible strength at 60% strain of 127.3 kPa and an EMI shielding effectiveness of 40.6 dB at freezing temperatures, despite the fact that MXene is present in only 0.58% of the total volume of the composite aerogels.

A bidirectional freezing approach was used to create a 3D porous $Ti_3C_2T_x$/carbon nanotube (CNT) hybrid aerogel (Figure 8.10) for lightweight EMI shielding [52]. In the MXene/CNT hybrid aerogels, the synergy between their lamellar and porous structures contributed significantly to their high electrical conductivity (9.43 S cm^{-1}) and outstanding electromagnetic shielding effectiveness (EMI SE) value of 103.9 dB at 3 mm thickness at the X-band frequency; either one however is the best value documented for synthetic porous nanomaterials. MXene/CNT hybrid aerogels with CNT reinforcement demonstrated improved mechanical resilience and higher compressional modulus by 9661% compared to MXene aerogels without CNT reinforcement.

Freeze-drying and thermal annealing of $Ti_3C_2T_x$ MXene and the conductive polymer poly(3,4-ethylenedioxythiophene)-poly(styrenesulfonate) resulted in a novel $Ti_3C_2T_x$ MXene/PEDOT:PSS hybrid aerogel (PEDOT:PSS) (Figure 8.11) [53].

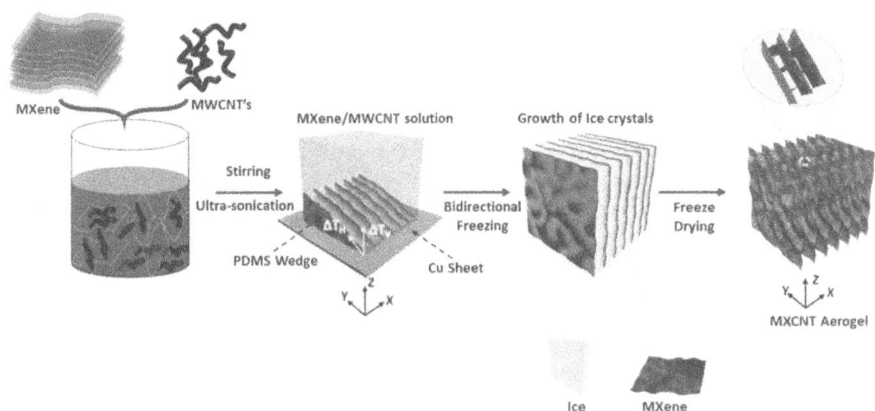

FIGURE 8.10 Bidirectional freezing method for $Ti_3C_2T_x$/CNT hybrid aerogel fabrication.

(Reproduced with the permission from Ref. [52], © 2019 American Chemical Society.)

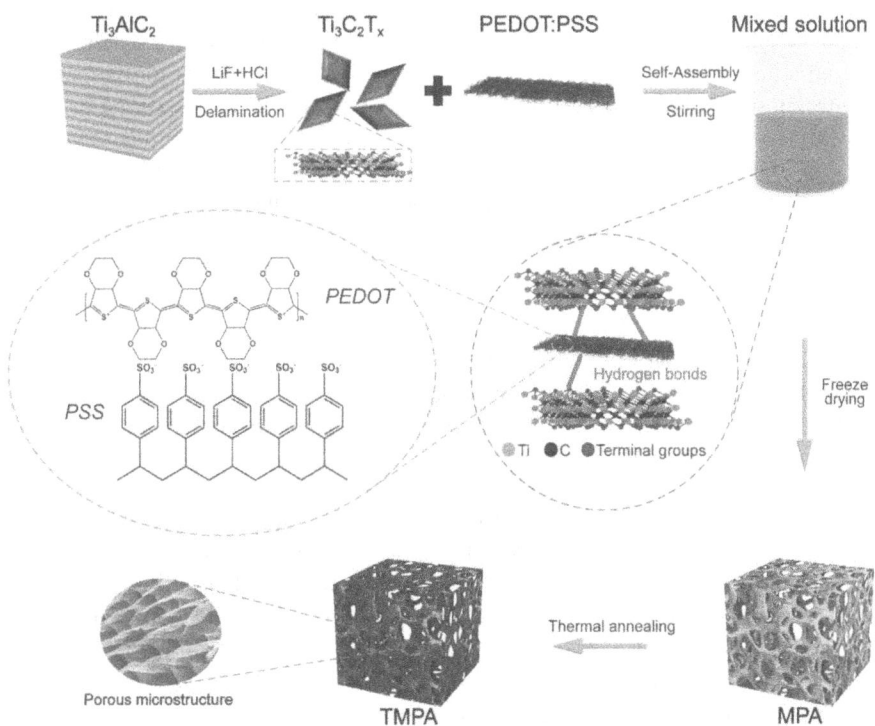

FIGURE 8.11 The diagram illustrates the preparation of $Ti_3C_2T_x$/PEDOT:PSS hybrid aerogel (MPA) and thermally treated MPA (TMPA).

(Reproduced with the permission from Ref. [53], © 2021 American Chemical Society.)

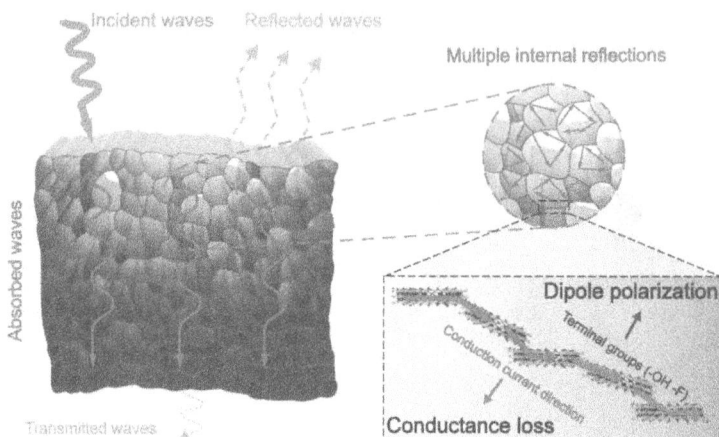

FIGURE 8.12 Hybrid aerogel's plausible EMI shielding mechanism.

(Reproduced with the permission from Ref. [53], © 2021 American Chemical Society.)

PEDOT:PSS not only increased the gelling ability of $Ti_3C_2T_x$, but it also succeeded in establishing a conductive bridge between MXene nanosheets, which was previously impossible. It was established by the experimental results that the hybrid aerogel displayed a distinct porosity microstructure, which was advantageous for the multiple scattering of electromagnetic waves inside the materials. When measured in terms of EMI shielding efficacy and specific shielding effectiveness, the results were up to 59 and 10,841 $dBcm^2g^{-1}$, respectively, while the SE_R/SE_T ratio value was just 0.05, showing that the material had exceptional wave absorption properties. Furthermore, the perfect impedance matching achieved by the composites as a result of the electrical conductance loss and polarization loss effects plays a crucial role in the outstanding wave absorption and EMI shielding performance of the materials. The results of the study demonstrate that the electrical conductance loss effect, interfacial polarization, dipole polarization, and excellent impedance matching of the EMI shielding aerogel with a porous microstructure are favorable to the EM wave absorption of the EMI shielding aerogel. As a result of this research, a novel concept for the design and preparation of novel EMI shielding materials with an absorption-dominated mechanism (Figure 8.12) has been developed.

Specifically, natural wood was delignified in constructing the wood aerogel as a porous framework, and then f-$Ti_3C_2T_x$ nanosheets were assembled into the wood aerogel to produce a novel ultralight, highly compressible, and anisotropic MXene@Wood (M@W) nanocomposite aerogel (0.108 $g/cm3$) that exhibits both EMI shielding and electromagnetic absorbing characteristics in multiple directions [54]. By virtue of their anisotropic wood aerogel skeleton, the M@W aerogels have both the channel-like microstructure and perfect structural load-bearing capacity that come with vertical growth as well as the layered microstructure and high compressibility that come with this growth orientation. A remarkable 72 dB EMI shielding efficacy in the parallel development direction and an expanded effective absorption bandwidth of 8.2–12.4 GHz in the vertical growth direction may both be achieved by controlling the loading of MXene with f-Ti$_3$C$_2$T$_x$. A simple alternative technique for generating wood-derived anisotropic MXene@Wood nanocomposite aerogels that exhibit both EMI shielding and EM absorption capabilities in various directions is shown in this study.

8.5 SUMMARY

MXenes' strong metallic conductivity, which surpassed all other conducting materials, and their composites at equivalent thickness values, allows Ti$_3$C$_2$T$_x$ to provide state-of-the-art EMI shielding performance,. Nonetheless, there are several limitations in the MXenes system that should be taken into consideration for future investigation. Furthermore, because MXenes are a big family, it is conceivable to synthesis and study MXenes with novel atomic compositions or new crystal structures, such as M$_5$X$_4$, in the future. Most of the time, the highly conductive MXenes contribute significantly to EMI shielding, which may result in secondary pollution in some cases. The use of porous foams and aerogels as well as segregated structures have been proposed to combat this problem; however, it is still necessary to develop MXene-shielding materials that have different structural factors, such as meta-structure, that can significantly improve the absorption of the incident EM waves. The research of MXenes is still in its infancy, and there is a lot of room to learn more about the new MXenes and their shielding capabilities in the future. In this detailed assessment, we aim to have provided an insight into future difficulties, as well as guidance for discovering material solutions for next-generation shielding applications.

REFERENCES

1. Shuck, C.E., et al., *Scalable synthesis of Ti$_3$C$_2$T$_x$ mxene*. Advanced Engineering Materials, 2020. **22**(3): p. 1901241.
2. Anasori, B., M.R. Lukatskaya, and Y. Gogotsi, *2D metal carbides and nitrides (MXenes) for energy storage*. Nature Reviews Materials, 2017. **2**(2): p. 1–17.
3. Iqbal, A., P. Sambyal, and C.M. Koo, *2D MXenes for electromagnetic shielding: a review*. Advanced Functional Materials, 2020. **30**(47): p. 2000883.
4. Riazi, H., G. Taghizadeh, and M. Soroush, *MXene-based nanocomposite sensors*. ACS Omega, 2021. **6**(17): p. 11103–11112.
5. Jamalipour Soufi, G., et al., *MXenes and MXene-based materials with cancer diagnostic applications: challenges and opportunities*. Comments on Inorganic Chemistry, 2021: p. 1–34.
6. Yu, S., et al., *MXenes as emerging nanomaterials in water purification and environmental remediation*. Science of The Total Environment, 2021: p. 152280.
7. Petukhov, D., et al., *MXene-based gas separation membranes with sorption type selectivity*. Journal of Membrane Science, 2021. **621**: p. 118994.
8. Li, Z. and Y. Wu, *2D early transition metal carbides (MXenes) for catalysis*. Small, 2019. **15**(29): p. 1804736.
9. Das, P., et al., *Carbon-dots-initiated photopolymerization: An in situ synthetic approach for MXene/poly (norepinephrine)/copper hybrid and its application for mitigating water pollution*. ACS Applied Materials & Interfaces, 2021. **13**(26): p. 31038–31050.

10. Ganguly, S., et al., *Mussel-inspired polynorepinephrine/MXene-based magnetic nano-hybrid for electromagnetic interference shielding in X-band and strain-sensing performance.* Langmuir, 2022. **38**(12): p. 3936–3950.

11. Ganguly, S., et al., *Polymer nanocomposites for electromagnetic interference shielding: a review.* Journal of Nanoscience and Nanotechnology, 2018. **18**(11): p. 7641–7669.

12. Xia, Y., et al., *Thickness-independent capacitance of vertically aligned liquid-crystalline MXenes.* Nature, 2018. **557**(7705): p. 409–412.

13. Gutiérrez, M.C., M.L. Ferrer, and F. del Monte, *Ice-templated materials: Sophisticated structures exhibiting enhanced functionalities obtained after unidirectional freezing and ice-segregation-induced self-assembly.* Chemistry of Materials, 2008. **20**(3): p. 634–648.

14. Tang, X., et al., *A novel lithium-ion hybrid capacitor based on an aerogel-like MXene wrapped Fe 2 O 3 nanosphere anode and a 3D nitrogen sulphur dual-doped porous carbon cathode.* Materials Chemistry Frontiers, 2018. **2**(10): p. 1811–1821.

15. Li, X., et al., *Ultralight MXene-coated, interconnected SiCnws three-dimensional lamellar foams for efficient microwave absorption in the X-band.* ACS Applied Materials & Interfaces, 2018. **10**(40): p. 34524–34533.

16. Liu, J., et al., *Multifunctional, superelastic, and lightweight MXene/polyimide aerogels.* Small, 2018. **14**(45): p. 1802479.

17. Zhao, M.Q., et al., *Hollow MXene spheres and 3D macroporous MXene frameworks for Na-ion storage.* Advanced Materials, 2017. **29**(37): p. 1702410.

18. Liu, J., et al., *Hydrophobic, flexible, and lightweight MXene foams for high-performance electromagnetic-interference shielding.* Advanced Materials, 2017. **29**(38): p. 1702367.

19. Yue, Y., et al., *3D hybrid porous MXene-sponge network and its application in piezoresistive sensor.* Nano Energy, 2018. **50**: p. 79–87.

20. Bao, W., et al., *Porous cryo-dried MXene for efficient capacitive deionization.* Joule, 2018. **2**(4): p. 778–787.

21. Xiu, L., et al., *Aggregation-resistant 3D MXene-based architecture as efficient bifunctional electrocatalyst for overall water splitting.* ACS Nano, 2018. **12**(8): p. 8017–8028.

22. Zhang, X., et al., *MXene aerogel scaffolds for high-rate lithium metal anodes.* Angewandte Chemie, 2018. **130**(46): p. 15248–15253.

23. Ma, Y., et al., *3D synergistical MXene/reduced graphene oxide aerogel for a piezoresistive sensor.* ACS Nano, 2018. **12**(4): p. 3209–3216.

24. Xing, C., et al., *Two-dimensional MXene (Ti3C2)-integrated cellulose hydrogels: toward smart three-dimensional network nanoplatforms exhibiting light-induced swelling and bimodal photothermal/chemotherapy anticancer activity.* ACS Applied Materials & Interfaces, 2018. **10**(33): p. 27631–27643.

25. Yu, M., et al., *Boosting electrocatalytic oxygen evolution by synergistically coupling layered double hydroxide with MXene.* Nano Energy, 2018. **44**: p. 181–190.

26. Li, Z., et al., *Surface nanopore engineering of 2D MXenes for targeted and synergistic multitherapies of hepatocellular carcinoma.* Advanced Materials, 2018. **30**(25): p. 1706981.

27. Ren, C.E., et al., *Porous two-dimensional transition metal carbide (MXene) flakes for high-performance Li-ion storage.* ChemElectroChem, 2016. **3**(5): p. 689–693.

28. Tao, Q., et al., *Two-dimensional Mo1. 33C MXene with divacancy ordering prepared from parent 3D laminate with in-plane chemical ordering.* Nature Communications, 2017. **8**(1): p. 1–7.

29. Ganguly, S. and S. Margel, *3D printed magnetic polymer composite hydrogels for hyperthermia and magnetic field driven structural manipulation.* Progress in Polymer Science, 2022. p. 101574.

30. Ganguly, S. and S. Margel, *Design of magnetic hydrogels for hyperthermia and drug Delivery.* Polymers, 2021. **13**(23): p. 4259.

31. Wang, Q., et al., *Modified Ti3C2TX (MXene) nanosheet-catalyzed self-assembled, anti-aggregated, ultra-stretchable, conductive hydrogels for wearable bioelectronics.* Chemical Engineering Journal, 2020. **401**: p. 126129.
32. Hu, M., et al., *Self-assembled Ti 3 C 2 T x MXene film with high gravimetric capacitance.* Chemical Communications, 2015. **51**(70): p. 13531–13533.
33. Bu, F., I. Shakir, and Y. Xu, *3D graphene composites for efficient electrochemical energy storage.* Advanced Materials Interfaces, 2018. **5**(15): p. 1800468.
34. Yan, J., et al., *Flexible and high-sensitivity piezoresistive sensor based on MXene composite with wrinkle structure.* Ceramics International, 2020. **46**(15): p. 23592–23598.
35. Liu, Y.T., et al., *Self-assembly of transition metal oxide nanostructures on MXene nanosheets for fast and stable lithium storage.* Advanced Materials, 2018. **30**(23): p. 1707334.
36. Du, C.-F., et al., *Porous MXene frameworks support pyrite nanodots toward high-rate pseudocapacitive Li/Na-ion storage.* ACS Applied Materials & Interfaces, 2018. **10**(40): p. 33779–33784.
37. Zhao, M.-Q., et al., *2D titanium carbide and transition metal oxides hybrid electrodes for Li-ion storage.* Nano Energy, 2016. **30**: p. 603–613.
38. Sang, X., et al., *Atomic defects in monolayer titanium carbide (Ti3C2T x) MXene.* ACS Nano, 2016. **10**(10): p. 9193–9200.
39. Deng, Z., et al., *Superelastic, ultralight, and conductive $Ti_3C_2T_x$ MXene/acidified carbon nanotube anisotropic aerogels for electromagnetic interference shielding.* ACS Applied Materials & Interfaces, 2021. **13**(17): p. 20539–20547.
40. Liang, C., et al., *Ultra-light MXene aerogel/wood-derived porous carbon composites with wall-like "mortar/brick" structures for electromagnetic interference shielding.* Science Bulletin, 2020. **65**(8): p. 616–622.
41. Zhao, S., et al., *Highly electrically conductive three-dimensional Ti3C2T x MXene/reduced graphene oxide hybrid aerogels with excellent electromagnetic interference shielding performances.* ACS Nano, 2018. **12**(11): p. 11193–11202.
42. Fan, Z., et al., *A lightweight and conductive MXene/graphene hybrid foam for superior electromagnetic interference shielding.* Chemical Engineering Journal, 2020. **381**: p. 122696.
43. Weng, C., et al., *Buckled AgNW/MXene hybrid hierarchical sponges for high-performance electromagnetic interference shielding.* Nanoscale, 2019. **11**(47): p. 22804–22812.
44. Wu, X., et al., *Compressible, durable and conductive polydimethylsiloxane-coated MXene foams for high-performance electromagnetic interference shielding.* Chemical Engineering Journal, 2020. **381**: p. 122622.
45. Wang, Y., et al., *Flexible, ultralight, and mechanically robust waterborne polyurethane/Ti3C2T x MXene/nickel ferrite hybrid aerogels for high-performance electromagnetic interference shielding.* ACS Applied Materials & Interfaces, 2021. **13**(18): p. 21831–21843.
46. Wang, L., et al., *3D $Ti_3C_2T_x$ MXene/C hybrid foam/epoxy nanocomposites with superior electromagnetic interference shielding performances and robust mechanical properties.* Composites Part A: Applied Science and Manufacturing, 2019. **123**: p. 293–300.
47. Xu, H., et al., *Lightweight Ti_2CT_x MXene/poly (vinyl alcohol) composite foams for electromagnetic wave shielding with absorption-dominated feature.* ACS Applied Materials & Interfaces, 2019. **11**(10): p. 10198–10207.
48. Li, X.-H., et al., *Thermally annealed anisotropic graphene aerogels and their electrically conductive epoxy composites with excellent electromagnetic interference shielding efficiencies.* ACS Applied Materials & Interfaces, 2016. **8**(48): p. 33230–33239.
49. Bian, R., et al., *Ultralight MXene-based aerogels with high electromagnetic interference shielding performance.* Journal of Materials Chemistry C, 2019. **7**(3): p. 474–478.

50. Li, S., et al., *CVD carbon-coated carbonized loofah sponge loaded with a direction-ally arrayed MXene aerogel for electromagnetic interference shielding.* Journal of Materials Chemistry A, 2021. **9**(1): p. 358–370.
51. Guo, Z., et al., *Poly (vinyl alcohol)/MXene biomimetic aerogels with tunable mechani-cal properties and electromagnetic interference shielding performance controlled by pore structure.* Polymer, 2021. **230**: p. 124101.
52. Sambyal, P., et al., *Ultralight and mechanically robust Ti3C2T x hybrid aerogel rein-forced by carbon nanotubes for electromagnetic interference shielding.* ACS Applied Materials & Interfaces, 2019. **11**(41): p. 38046–38054.
53. Yang, G.-Y., et al., *Ultralight, conductive Ti$_3$C$_2$T$_x$ MXene/PEDOT: PSS hybrid aerogels for electromagnetic interference shielding dominated by the absorption mechanism.* ACS Applied Materials & Interfaces, 2021. **13**(48): p. 57521–57531.
54. Zhu, M., et al., *Ultralight, compressible, and anisotropic MXene@ Wood nanocom-posite aerogel with excellent electromagnetic wave shielding and absorbing properties at different directions.* Carbon, 2021. **182**: p. 806–814.

9 Role of MXene-Based Conductive Polymer Composites in EMI Shielding

Bona Elizebath Baby
Post Graduate and Research Department of
Chemistry, Government College for Women
University of Kerala
Trivandrum, India

Anand Krishnamoorthy
Department of Basic Sciences
Amal Jyothi College of Engineering
Kanjirappally, Kerala, India
and
Apcotex Industries Limited
MIDC Industrial Area
Taloja, Kerala, India

CONTENTS

DOI: 10.1201/9781003281511-9

9.1 INTRODUCTION

In 2004, graphene was discovered [1] with its unusual properties, which paved the way for scientists to show an interest towards the 2D materials that extend up to MAX phases. They are in the form of layered, hexagonal carbides and nitrides with the general formula $M_{n+1}AX_n$ (MAX) [2], where n = 1 to 4, M is an early transition metal, A is an A-group (mostly IIIA and IVA or groups 13 and 14) element and X is either carbon and/or nitrogen. The selective etching of parent MAX phases, by eliminating the A layers from it, produces MXenes [3, 4]. In the MAX phase, the different types of bonds, such as covalent, ionic and metallic, exist simultaneously [5]. M–X is a combination of different bonds such as covalent bonds, ionic and metallic bonds. The more metallic bond components are possessed by M–A and A–A bonds [6], whereas the bond strength of M–X bond is relatively higher than that of the prior bonds [7]. Hence, the A layered atoms are peeled off more easily than other layers with the capacity of highest reactivity [8–9]. They are graphene-like structures having two-dimensional transition metal carbide, nitride or carbonitride with hydrophilic surface, which makes them different from that of graphene and most of the 2D materials having a hydrophobic nature. Both MAX phases and MXenes have unparalleled properties and applications to demand. The layered structure and the mixed metallic-covalent bonding nature of the MAX phase, which makes the M–X bonds stronger and where the M–A bonds weaker. They are excellent conductors of electricity and are oxidation resistant with ultra-low friction and have many other properties that arise in them as a new class of solids. Similarly, MXenes, with rare combinations of electronic conductivity [10–11] and hydrophilicity, provide a wide range of applications in nanocomposites as fillers, EMI shielding, energy storage and in electronic devices (Figure 9.1).

Though silver-, copper-, iron- and nickel-based conventional metals and their alloys exhibit EMI shielding properties, their broader applications in electronic products are restricted owing to their high density and tendency for corrosion [12–13]. Conductive polymer composites are a type of multi-phase composite and are obtained by the addition of electrically conductive fillers into the polymer matrix by different methods, and their EMI shielding performance is then studied [14–17]. The carbon materials, metals and materials having intrinsic conductivity are mainly used as the electrically conductive fillers in CPC [18–19]. The good mechanical properties, light weight characteristics in various forms of allotropes of carbon-based materials, allow them to be widely used in conductive polymer composites [20–24].

Compared to carbon-based materials, the 2D layered transition metal carbides, nitrides or carbonitrides offer excellent conductivity, good water affinity and chemically active [25–26]. The 3D structures have longer conductive pathways due to the complexity in their interfaces which enables several reflection and scattering results and better shielding property against the electromagnetic waves. But the development of self-standing 3D structures with superior mechanical properties through self-assembly of MXenes is difficult, as it has weak interaction and also poor gelation property [27–29]. In addition to electromagnetic interference shielding property, they are widely used in polymer matrix which can offer good mechanical property [30–32]. This chapter mainly discusses the applications of MXene [33] based conductive polymer hybrids in electromagnetic interference shielding.

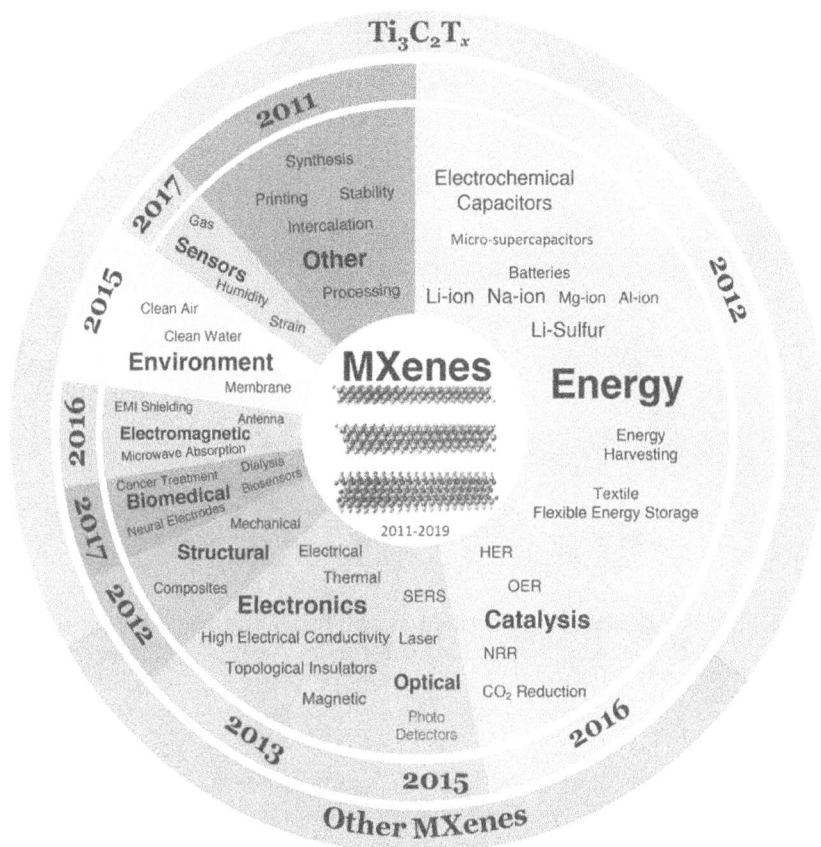

FIGURE 9.1 The properties and applications of MXenes. The pie diagram shows the research publications related to MXenes in various domain applications from 2011 to 2019. The outer ring explains the relation between the publications only on $Ti_3C_2T_x$ MXenes compared to other MXene compositions (M_2XT_x, $M_3X_2T_x$, $M_4X_3T_x$).

(Ref. 78, © 2019, American Chemical Society.)

9.2 MXene-BASED CONDUCTIVE POLYMER COMPOSITES

MXene–polymer composites with superior EMI shielding properties are increasing in popularity, particularly in smart, wearable electronic devices. The key requirements in the development of EMI shielding materials are flexibility, low density, light weight nature and good mechanical stability [34–36]. Owing to their properties (Figure 9.2) with excellent intrinsic electrical conductivity, MXenes can meet the latest trends of electronic components manufacturing [37–39]. MXenes have achieved much attention towards EMI shielding applications and are in the top list after carbon nanotubes and graphene [40, 41].

The electrically conductive 3D networks can be developed in the polymer by the introduction of MXene, which has excellent EMI shielding performance, outstanding σ,

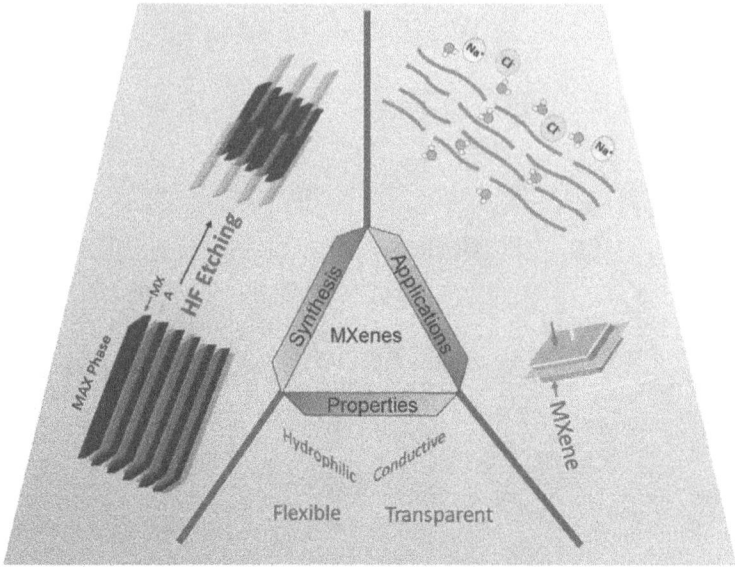

FIGURE 9.2 The different stages involved in fabrication of MXenes from its precursor MAX phase, which also includes the properties and various applications of MXene.

(Ref. 80, © 2019, Elsevier.)

is light weight and has superior mechanical properties, even at lower MXene loading [42–45]. Thus, these types of MXene/polymer matrix can be prepared by some common synthesis methods, including physical or mechanical blending, lyophilization, pre-support molding and vacuum-assisted filtration.

9.2.1 SYNTHESIS METHODS

9.2.1.1 Physical Blending

When MXenes are added into a polymer matrix and mechanically blended, polymer-MXene composites will be obtained. This process is called physical blending. In this process, there will not be any chemical crosslinking between the polymer and MXene. Rajavel et al. [46] fabricated $Ti_3C_2T_x$/(PVDF) composite (Figure 9.3a) by using a technique called solvent-assisted mixed compaction.

The EMI shielding capacity of the composite was 48.47 dB at which the amount of Ti3C2Tx was 22.55 vol% and a 2 mm thickness. Sun et al. prepared Ti_3C_2Tx/ polystyrene composites (Figure 9.3b) made up of isolated structure by techniques such as electrostatic assembly and press molding [47]. The Ti_3C_2Tx/PS had electrical conductivity of 1081 S/m and EMI shielding effectiveness (SE) of 54 dB, obtained when the amount of Ti_3C_2Tx was 1.9 vol% and the thickness was 2 mm. Industry demanding epoxy resins are heavily used in the field of aerospace, electronics and lot more because of its superior chemical stability and wonderful mechanical properties [48]. The impact of Ti_3C_2Tx annealing done thermally on the performance of EMI

(a)

(b)

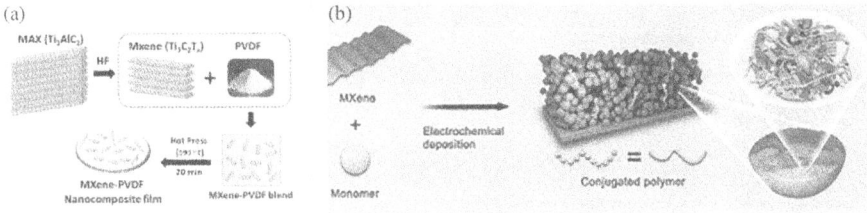

FIGURE 9.3   (a) Schematic representation of 2D bulk MXene facial exfoliation and MXene/ PVDF nanocomposite film preparation using the solvent-assisted mixed compaction method. (Reproduced with the permission from Ref. 46, © 2020, Elsevier.) (b) Schematic illustration of the electrochemical polymerization of conjugated polymer-MXene composite nanosphere. (Ref. 47, © 2022, Elsevier.)

shielding of Ti_3C_2Tx/epoxy composites is studied by Gu et al. [49]. Based on the preceding, it was identified that the amount of Ti_3C_2Tx was 15 wt% and thickness was 3 mm, where the Ti_3C_2Tx/epoxy composites had optimal electrical conductivity of 105 S/m and shielding effectiveness of 41 dB and improved by 176% and 37%, as compared to thermal annealing done before.

The physical blending method is not bound to any polymer matrix and MXenes, which is what makes this method simple and applicable. But for this reason the MXene/polymer EMI shielding composites behave as having very poor mechanical properties. In X-band the EMI shielding effectiveness (SE) of the blended composites is very small. Thus, to attain an ideal EMI SE, a filling amount which is higher is needed that significantly affects the MXene/polymer composites process ability and mechanical strength [50].

9.2.1.2   Freeze-Drying

Freeze-drying/lyophilization is considered the commonly used method for the synthesis of MXene/polymer composite foams for EMI shielding. In this technique the suspension is frozen after the self-assembly for the promotion to the formation of solid molecules (organic), which then is sublimated to ice in order to escape through vacuum drying. Xu et al. [51] synthesized the porous PVA/MXene foams for EMI shielding by freeze-drying (Figure 9.4a).

The EMI shielding performance (SE) of PVA/MXene composite foam was about 28 dB and the specific shielding effectiveness (SSE) was 5136 dBcm²g⁻¹ at the time when the quantity of MXene was 0.15 vol%, and the thickness was 2 mm. Dip-coating and directional freeze-drying combination was used for preparation of silver nanowire/MXene/MF sponges for EMI shielding [52]. Herein, melamine formaldehyde (MF) was used as the coating template (Figure 9.4b). When they attained the density of 49.5 mgcm⁻³ with a thickness of 2mm with excellent electromagnetic interference shielding performances of 52.6 dB, Wu et al. [53] prepared the 3D structured MXene/sodium alginate (SA) by drying using the method of directional freeze, which is later coated with thin layers of polydimethylsiloxane (PDMS) to make MXene/SA/PDMS EMI shielding composite foams (Figure 9.4c). In this case, when the quantity of MXene was 95 wt% with a thickness of 2mm, the resulting

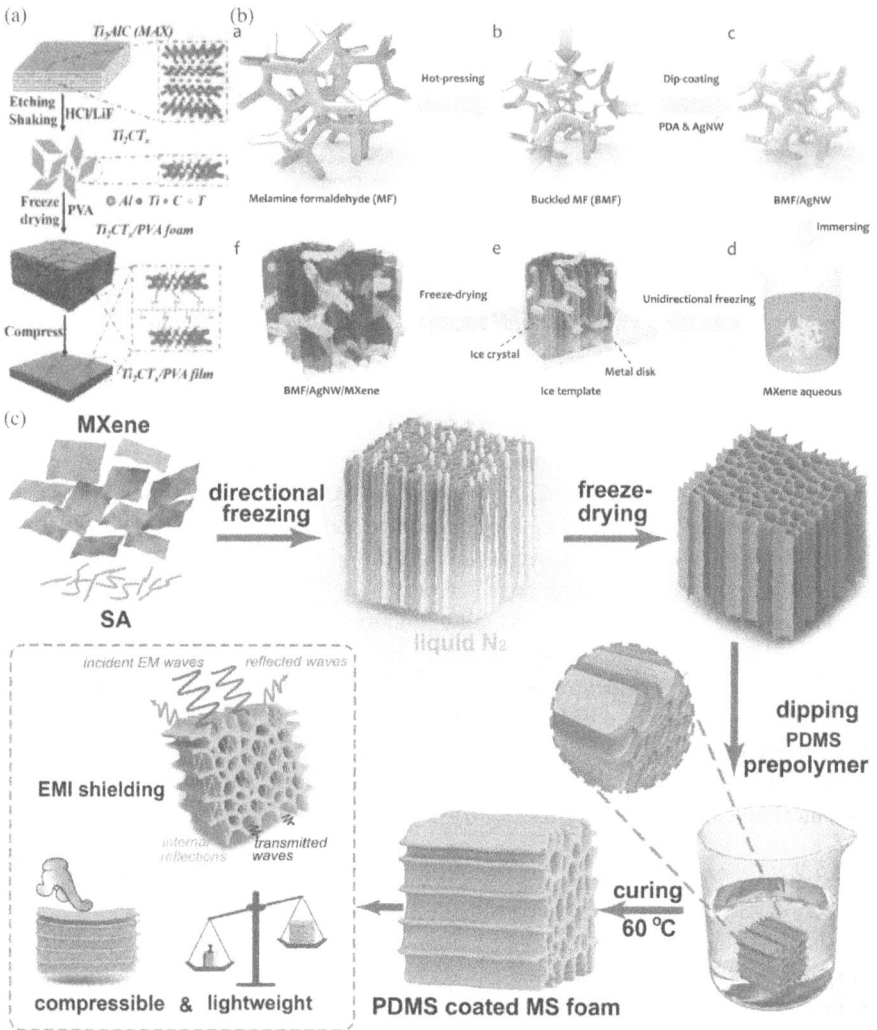

FIGURE 9.4 (a) Illustration of f-Ti$_2$CTx/PVA composite foams and films by the method of freeze-drying. (Ref. 51, © 2019, American Chemical Society.) (b) Diagram schematically represents the BMF/AgNW/MXene hybrid sponges by freeze-drying. In this method melamine formaldehyde sponge used in the form of coating template for the preparation of the silver nanowire/MXene/MF EMI shielding composite sponges by combined directional freeze-drying technique and dip coating. (Ref. 52, © 2009, Royal Society of Chemistry.) (c) The fabrication of MS hybrid aerogels and PDMS coated with MS foams in the presence of liquid nitrogen under freeze-drying technique, the 3D MXene/sodium alginate aerogel structure, prepared by the directional freeze-drying and then PDMS coated to make it more compressible and electrically conductive MXene/SA/PDMS EMI shielding composite foams. (Ref. 53, © 2020, Elsevier.)

foams had the excellent electrical conductivity of 2211 S/m and SE of 70.5 dB. The method of freeze-drying ensures the porosity of materials which are 3D to a greater level. Further, there won't be any disruption in its internal structure. It is evident that MXenes/polymer composites having high porosity, controllable pore size and distribution can be developed by this technique. However, there is a common problem for MXene/polymer foam materials for EMI shielding applications, as they are having very poor air stability because of the high specific surface area of the foam structure and the easy oxidizing nature of MXenes.

9.2.1.3 Pre-Support Molding

In pre-support molding, MXene/polymer composites are prepared by a molding method in which the polymer matrix are backfilled into the MXene that forms a network of 3D electrically conductive (Zhao et al. [54]) 3D $Ti_3C_2T_x$ structures which are porous, which is then finally backfilled by epoxy resins (Figure 9.5b). The $Ti_3C_2T_x$/epoxy composites have its best σ of 695.9 S/m and EMI SE of 50 dB with the amount of $Ti_3C_2T_x$ is 0.74 vol% and 2 mm thickness. Gu et al. [55] prepared MCF/epoxy composites in which the epoxy resins are backfilled to the MXene/C hybrid foam (MCF) that is formed using a sol-gel method (Figure 9.5a). The MCF quantity of 4.25 wt% with thickness of 3 mm showed σ of 184 S/m and the EMI SE as 46 dB for MCF/epoxy composites.

First, the rGO-MXene structure similar to honeycomb (rGMH) was synthesized using a method called template induction and electrostatic self-assembly, which is then backfilled with the epoxy resins to get the rGMH/epoxy EMI shielding composites (Figure 9.5c) [56]. In these types of composites it shows the highest

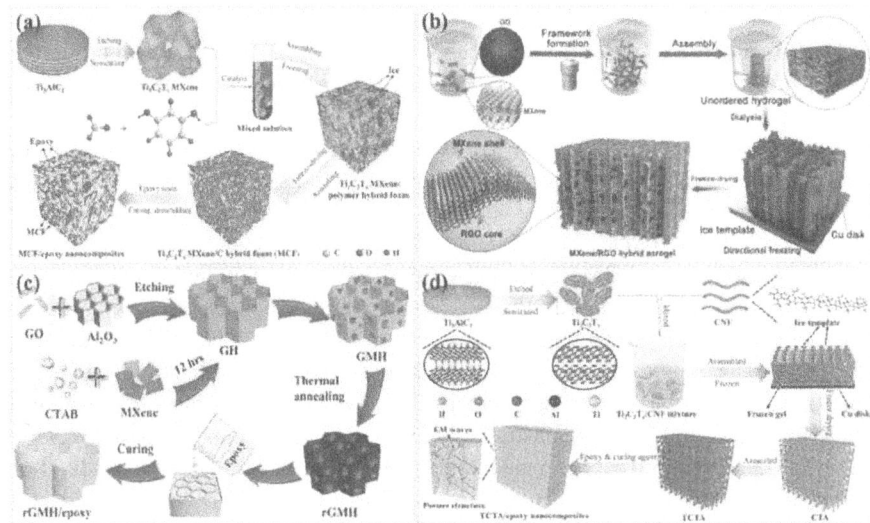

FIGURE 9.5   (a) The fabrication for the MCF/epoxy EMI composites. (Ref. 55, © 2019, Elsevier.) (b) Schematic diagram showing the preparation of $Ti_3C_2T_x$/RGO hybrid aerogels. (Ref. 54, © 2018, American Chemical Society.) (c) Fabrication of rGMH/epoxy nanocomposites. (Ref. 56, © 2020, Elsevier.) (d) The fabrication of MXene/CNF/epoxy nanocomposites [57].

σ of 387.1 S/m and EMI SE of 55 dB at which concentration of rGO and MXene were 1.2 wt% and 3.3 wt% and thickness was 3 mm. Despite, Gu et al. [57] synthesized the MXene/CNF/epoxy composites (Figure 9.5d) for EMI shielding applications. In involves two steps: First, using directional freeze-drying technology for the preparation of highly electrically conductive 3D MXene/cellulose nanofiber (CNF) aerogels; the second step involves the impregnation with epoxy resins. The electrical conductivity, shielding effectiveness (SE) and SE divided by the thickness (SE/d) of MXene/CNF/epoxy composites was 1672 S/m, 74 dB and 37 dB/mm respectively. The amount of MXene present was 1.38 vol%. The high σ and EMI SE with low MXenes loading is obtained in the pre-support molding without damaging conductive networks. Since MXenes are highly oxidizing in nature, this method helps to completely cover the MXenes by polymer matrix so as to neglect MXenes and air contact. However, mechanical properties shown by the MXenes/ polymer composites, due to no chemical bonding between the polymer matrix and the MXenes, are considered as inferior.

9.2.1.4 Vacuum-Assisted Filtration

In the vacuum-assisted filtration method, the negative pressure caused by the suction is used to remove liquid uniformly from mixed MXenes, and the matrix which is polymer is accelerated. Cao et al. [58] prepared flexible and thin d-$Ti_3C_2T_x$/CNF (d-$Ti_3C_2T_x$ = delaminated $Ti_3C_2T_x$ & CNF = cellulose nanofibers) EMI shielding composite papers through this method (Figure 9.6a). The quantity of d-$Ti_3C_2T_x$ was 80 wt% and thickness was about 47 μm, then it recorded a σ of 739.4 S/m, EMI SE of 25.8 dB and then the outstanding EMI SSE of 2647 dBcm^{-2}g^{-1} for d-$Ti_3C_2T_x$/CNF EMI shielding composite papers.

Xie et al. [59] synthesized d-$Ti_3C_2T_x$/aramid nanofiber (ANF) composite papers through vacuum-assisted filtration, which possesses excellent mechanical properties and σ, in which it shows an EMI SE of 28 dB at which d-Ti_3C_2Tx was 60 wt% and 17 μm thickness. While, Zhan et al. [60] prepared ultra-thin MXene/TOCNF composite films for shielding, that is, EMI through vacuum-assisted filtration, by using TEMPO oxidized cellulose nanofiber called TOCNF. Its EMI SE was 39.6 dB, for which the amount of $Ti_3C_2T_x$ is 50 wt% and of 38 μm thickness.

Hu et al. [61] introduced the MXene/CNF composite papers by the preceding technique and brought PDMS of thin layer for the protection of the MXene networks from oxidation (Figure 9.6b). In those types of composite paper it shows σ of 2756 S/m and EMI SE of 43 dB when the amount of MXene is 1.89 vol% and 200 μm thickness. The retentivity of EMI SE of MXene/CNF EMI shielding composite papers was above 90% after 2000 bending-release cycles. Another discovery was made by Luo et al. [62] who constructed highly electrically conductive and flexible $Ti_3C_2T_x$ MXene/natural rubber nanocomposite films. They exhibited a σ of 1400 S/m and shielding effectiveness (SE) of 53.6 dB. The amount of MXene and thickness were 6.71 vol% and 251 μm, respectively.

Xin et al. [63] synthesized MXene/CNF/silver composites for EMI shielding applications which gave much higher EMI SE as compared to MXene/CNF film (14.9 dB) and is 50.7 dB, for which silver nanoparticles used is 32.54 wt% and 46 μm thickness. The method of vacuum-assisted filtration is used by Zhou et al. [64]

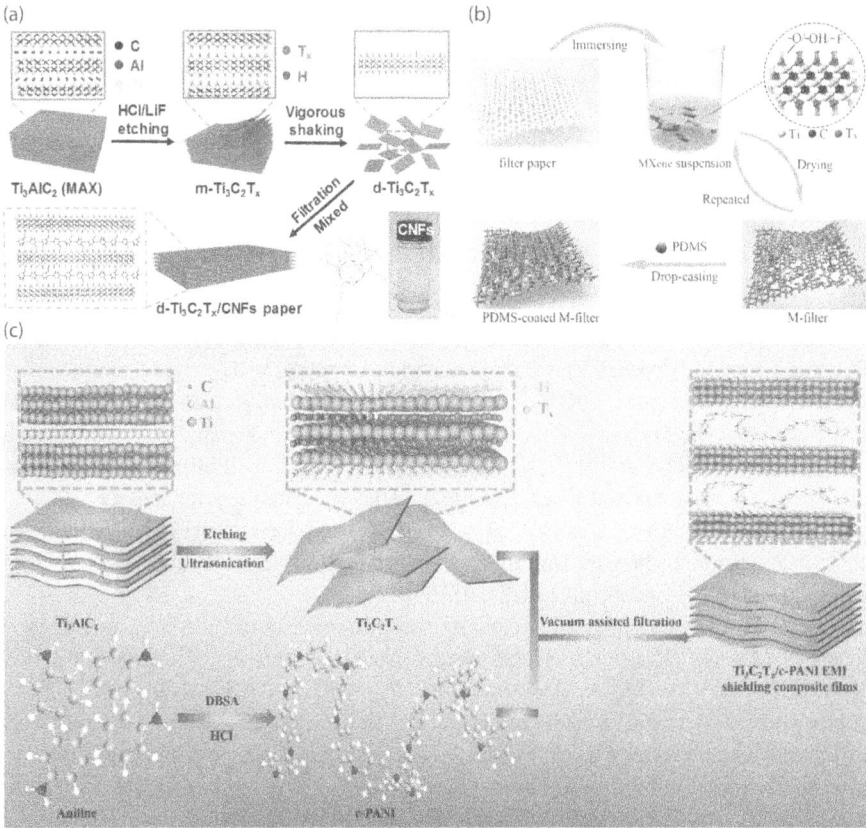

FIGURE 9.6 (a) Represents the fabrication of the d-$Ti_3C_2T_x$/CNF composite papers prepared by the self-assembly by the filtration which vacuum assisted. (Ref. 58, © 2018, American Chemical Society.) (b) Illustration of the preparation of freestanding nanocomposites process. (c) Synthesis of $Ti_3C_2T_x$/PANI composite films. (Ref. 66, © 2019, Elsevier.)

for preparing $Ti_3C_2T_x$/calcium alginate (CA) spongy films. This gives excellent EMI shielding effectiveness (SE) of 54.3 dB, and EMI-specific shielding effectiveness (SSE) of 17,586 $dBcm^{-2}g^{-1}$ when the amount MXene was 90 wt% and 26 μm thickness. Similarly, Shahzad et al. [65] used sodium alginate (SA) and MXenes including the $Ti_3C_2T_x$, Mo_2TiC_2 and $Mo_2Ti_2C_3T_x$ for the preparation of MXene/SA composite films. These films have extraordinary flexibility as well as pre-eminent σ through the vacuum-assisted filtration and they provide EMI SE of 57 dB when the amount of MXene was 90 wt% and the thickness was 8μm. Gu et al. [66] used dodecyl benzene sulfonic acid along with HCl co-doped PANI and thin layered $Ti_3C_2T_x$ for the synthesis of $Ti_3C_2T_x$/PANI composites (Figure 9.6c). Thus, it is observed that the electrical conductivity and EMI SE of $Ti_3C_2T_x$/C-PANI composite films will be 24.4 S/cm and 36 dB when the mass ratio of $Ti_3C_2T_x$ to PANI was 7:1 and 40 μm thickness.

The advantages of the vacuum-assisted filtration process are they are less complex and controlling and also easy. Thus, obtained MXene/polymer EMI shielding composite films have high electrical conductivity (σ) and EMI SE value, as they can make uniform electrically conductive networks. In those type of composites, a hydrogen bond is seen in between the MXenes and polymer matrix, where it's bonding force is limited. Hence, it should be identified that the technological/ mechanical properties of MXene/polymer EMI shielding composite require more improvement.

9.2.2   POLYVINYL BUTYRAL (PVB) COMPOSITE OF MXene

Polyvinyl butyral (Figure 9.7) is considered to be an acetal and formed from the reaction of polyvinyl alcohol with butyraldehyde. They are highly transparent materials and due to this they are extensively used to fabricate the laminated glass sheets as a resin glue in the glass industry and used in high-rise glass buildings and in automobiles, as they can withstand shock and break. They play an important safety role, and are applied between two sheets of glass. They protect the travelers against injuries caused by chipped glass during any accident. PVB are of different types ranging from glass panels to decorative glass [67, 68].

The adhesive properties exhibited by the polyvinyl butyral helped in the synthesis of MXene/PVB polymer composites. Over a broad range of frequency, EM wave-absorption properties are improved for MXene/PVB/Ba$_3$Co$_2$Fe$_{24}$O$_{41}$/Ti$_3$C$_2$ composites as observed by Yang et al. in 2017. For the preparation of Ti$_3$C$_2$ MXene, at first Al is etched, and after its successful etching, Co$_2$Z powders are added to methyl-ethyl-ketone (40 vol%) and ethanol (60 vol%). Triethyl phosphate dispersant and MEK were ball milled for 4 hours.

The mixture ball-milled for another 4 hours, in which PVB is acting as a binder and PEG/diethyl phthalate as a plasticizer. The layered polymer composites are produced from the slurry that is obtained after the ball milling process was used to tape cast a thin sheet of thickness of just 200 μm. The plate-like structure is possessed by the Co$_2$Z with the particle size ranging from 200–250 μm, while the MXene resembles the layered structure of graphene (Figure 9.8). And this multilayered composite exhibits successful attenuation and efficient absorption of the EM wave. At 5.8 GHz the RL$_{max}$ value of PVB/Co$_2$Z/Ti$_3$C$_2$ MXene composites is 46.3 dB and its absorption bandwidth at 1.6 GHz is loss below -10dB. In the current low observable aircrafts, this composite is the best choice.

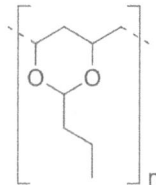

FIGURE 9.7   Structure of polyvinyl butyral (PVB). It is a clear, colorless, amorphous thermoplastic resin with the general formula (C$_8$H$_{14}$O$_2$)$_n$.

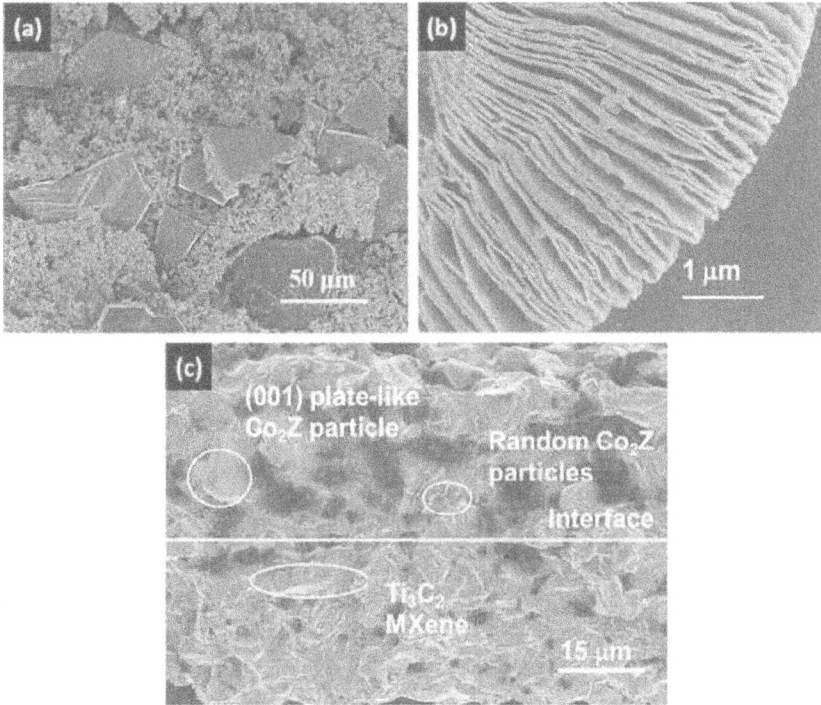

FIGURE 9.8   SEM images of (a) Co_2Z powder, (b) Co_2Z plate-like oriented powder blend Ti_3C_2 MXene; and (c) synthesized $PVB/Co_2Z/Ti_3C_2$ composite.

(Ref. 69, © 2017, Elsevier.)

9.2.3   POLYACRYLAMIDE (PAM) COMPOSITE OF MXene

In 1962, polyacrylamide was produced from the co-polymerization method of acrylamide and N, N-methylene bisacrylamide. The special characteristics of this polymer include hydrophilicity, biocompatibility and also protein resistance [70, 71]. M. Naguib et al. synthesized the nanocomposites that possess the enhanced flexibility and conductivity by the combination of PAM and MXene [72]. A glaring rise in the conductivity was observed at 3.3×10^{-2} Sm^{-1} after the addition of 6 wt% MXene onto the membrane. Experimentation in the field of MXene-polymer composites paved a wide way to new learnings and future discoveries.

9.2.4   POLYANILINE (PANI) COMPOSITE OF MXene

Polyaniline polymer was discovered 150 years ago and they have high conductivity. For the production of intelligent and multifunctional yarns, PANIs are used in the conductive coatings of yarns. Since the polymerization mechanism and the oxidative nature of PANIs is intricate, they can be easily prepared by adjusting the pH [73, 74]. Huawei et al. synthesized $Ti_3C_2T_x$ MXene/(PANI) composites with sandwich intercalated structure to study microwave absorption properties.

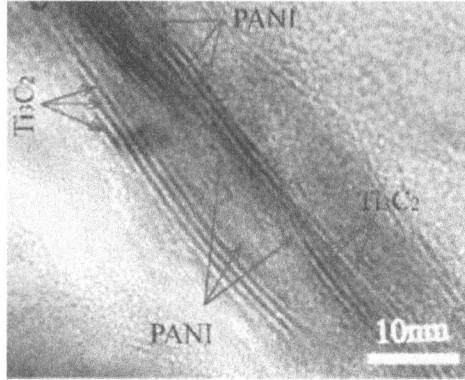

FIGURE 9.9 $Ti_3C_2T_x$ MXene/PANI composites TEM image.
(Ref. 75, © 2019, Elsevier.)

Various PANI concentrations are made by blending MXene and aniline monomers suspension which is later incorporated with multiple MXene/PANI composites.
Oxidative polymerization after the addition of APS, the mixture was agitated for
hours in an ice water bath. The different mass ratios of the composite can be obtained
after hydrochloric acid is used to wash the precipitate and then by distilled water.
After polymerization, tiny particles were found to be spread across the surface. The
amount of aniline used can be used to control the change in morphology and the
particle size of the sample. This relation is directly related to its dielectric properties.
The multiple internal reflections of the composite were due to the reflected surfaces
of its multiple layers. The uneven charge distribution and the attenuation which is
electromagnetic at each reflection site of MXene/PANI (Figure 9.9) resembles the
resistor-capacitor circuit model. The 1:3 sample with ratio exhibited good microwave absorption. A reflection loss (RL) of -56.3 dB is observed at 13.80 GHz with
an 99.9999% absorption efficiency because of the increment that is observed in the
conductive tracks of MXene [75]. The high rate of the absorption efficiency makes
this considered to be trailblazing [76].

9.3 MECHANISM FOR ELECTROMAGNETIC INTERFERENCE SHIELDING

The EM interference shielding is a type of process in which the transmission of electromagnetic waves is minimized by using a shield to reflect or absorb interface that
falls between two media [77]. EM waves can transmit, reflect or absorb in a medium.
It is known to all that the EM radiation component that is electric and magnetic field
are perpendicular to each other as well as to the propagation direction (Figure 9.10).

Electric and magnetic fields that are associated with the EM waves are described
in a phasor form by Equations 9.1 and 9.2 [78, 79]:

$$\vec{E} = E\ e^{-\gamma z}\hat{a}_x = E\ e^{-\alpha z}e^{-j\beta z}\hat{a}_x = E\ e^{-\alpha z}\left(\cos \beta_z - j \sin \beta_z\right)\hat{a}_x \qquad (9.1)$$

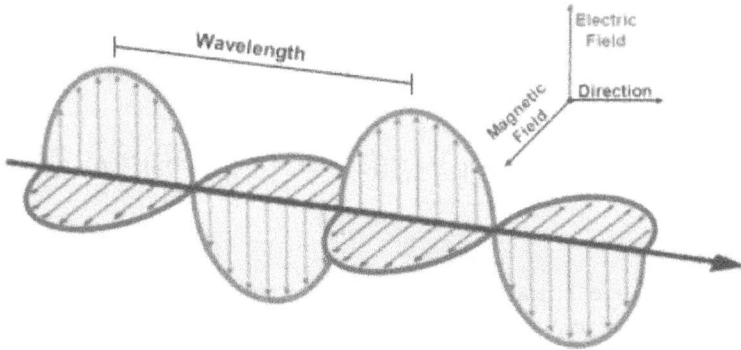

FIGURE 9.10   Schematic representation of EM wave.

(Ref. 77, © 2018, Elsevier.)

$$\rightarrow H = H\ e^{-\gamma z} \hat{a}_y = H\ e^{-\alpha z} e^{-j\beta z} \hat{a}_y = H\ e^{-\alpha z} \left(\cos \beta_z - j \sin \beta_z \right) \hat{a}_y \qquad (9.2)$$

where γ is the propagation constant of the medium; α and β are the attenuation and phase shift constants respectively; E and H are the amplitudes of the electric and the magnetic fields. An EM wave usually propagated without the reflection in a single medium. However, when an EM wave is incident at the interface of the two different media, reflection may occur. Equations 9.3 and 9.4 represents the reflection coefficient (RC: R_{12}) and the transmission coefficient (TC: T_{12}) at the interface, which are determined by the impedance of the media as follows:

$$R_{12} = E_r/E_i = \left(\eta_2 - \eta_1 \right)/\left(\eta_2 + \eta_1 \right) \qquad (9.3)$$

$$T_{12} = E_t/E_i = 2\eta_{2}/\left(\eta_2 + \eta_1 \right) \qquad (9.4)$$

where E_i and E_r are the amplitudes of the incident and the reflected electric fields respectively, while η_1 and η_2 are the impedance of medium 1 and medium 2 respectively. Wave impedance is defined as the ratio of the electric and the magnetic fields. Intrinsic impedance of a medium can be expressed as a function of the angular frequency (ω), magnetic permeability (μ), electrical conductivity (σ) and permittivity (ε):

$$\eta = |E|/|H| = \sqrt{\left[(j\ \omega\ \mu)/(\sigma + j\ \omega\ \varepsilon) \right]} \qquad (9.5)$$

When an EM wave propagates in a lossy medium or a shield that has a non-zero α, the lossy medium generally absorbs the wave. Hence, the strength or the amplitude of the field diminishes exponentially as it increases the penetration depth (E = $E_0 e^{-\alpha z}$). The attenuation constant (α) of the material can be expressed as a function of the ω, σ, μ and ε of the shield:

$$\alpha = \omega \sqrt{\left\{ (\mu\ \varepsilon/2) \left[\left(\sqrt{1 + (\sigma/\omega\ \varepsilon)^2} \right) - 1 \right] \right\}} \qquad (9.6)$$

Thus, an EM wave passing through a shield undergoes the reflection at the two interfaces and the absorption by the shield. The ratio of the total transmission field relative to the incident field can be expressed as follows [80]:

$$E_t/E_i = T_{12}e^{-\alpha d}T_{23} = \left(2\eta/(\eta+\eta_0)\right)e^{-\alpha d}\left(2\eta_0/(\eta+\eta_0)\right) = \left[4\eta\eta_0/(\eta+\eta_0)^2\right]e^{-\alpha d}$$

$$(9.7)$$

where T_{12} and T_{23} are the transmission coefficients at the front and the back surfaces of the shield respectively, and α and d are the attenuation constant and the thickness of the shield respectively.

The shielding effectiveness (SE) of a thick shield is defined in dB as follows:

$$SE_T = 20 \log\left(E_i/E_t\right) \tag{9.8}$$

$$SE_T = SE_R + SE_A = 20 \log\left((\eta+\eta_0)^2/4\eta\eta_0\right) + 20 \log e^{\alpha d} \tag{9.9}$$

where SE_T is the total SE, and SE_R and SE_A represent the shielding due to the reflection and the absorption respectively. SE_R is induced by the reflection on the surface of the shield and SE_A represents the absorption of the EM wave by the shield. It should be noted that although the wave reflected from the second interface can contribute to the wave transmission after being re-reflected by the first interface, this value is usually negligible for a thick shield with an SE_T larger than 15 dB. For electrically conductive shield materials that have $\sigma \gg \omega\,\varepsilon$, the attenuation constant (α) and impedance (η) can be represented as follows: $\alpha = \sqrt{(\pi f \mu \sigma)}$ and $\eta = \sqrt{(j \omega \mu/\sigma)} = \sqrt{(j2\pi f \mu/\sigma)}$ whereas, for electrically insulating media such as air, $\sigma_0 = 0$, $\alpha_0 = 0$ and $\eta_0 = \sqrt{(\mu_0/\varepsilon_0)} = 377\ \Omega$. Here f is the frequency.

Thus, SE_R and SE_A can be rearranged as follows:

$$SER = 39.5 + 10 \log\left(\sigma/2\pi f\,\mu\right) \tag{9.10}$$

$$SEA = 8.7\alpha d = 8.7(d/\delta) = 8.7d\sqrt{(\pi\,f\,\mu\,\sigma)} \tag{9.11}$$

where δ is the skin or penetration depth, which is a useful term for shielding and which indicates the distance beneath the surface at which the intensity of the electric field decreases to the 1/e of the original incident wave intensity. For a conductive shield, the skin depth is expressed as $\delta = 1/\alpha = (\sqrt{\pi f \sigma \mu})^{-1}$.

Equations 9.10 and 9.11 clearly depict that the electrical conductivity is the major factor contributing to the reflection and the absorption. Permittivity and the permeability will also play a significant role in increasing the absorption contribution, thus increasing the overall electromagnetic interference shielding effect (EMI SE) (77, 78).

9.4 FUTURE PROSPECTS

By the discovery of MXene in 2016, it makes sense among the scientific community based on their application in electromagnetic interference shielding. Several published reports on the applications of MXene-based conductive polymer composites and their EMI shielding properties are available to the scientific community. The use and discovery of MXene hybrids are at its starting stage and hence there is much more development in this field to be discovered. Flexible and light weight conductive polymer composites-based MXene is the future of the science and pave a great vista ahead. Next generation telecommunication can be made possible by fewer electromagnetic emissions by making use of MXene-based transparent films as they have as high a transparency as that of graphene. Since the discovery of MXene is at its infancy, it should be noted that a large family of MXene is to be identified for EMI applications.

ACKNOWLEDGEMENTS

The authors are thankful to Lakshmi Narayan, Amity University, for editing and revising the chapter.

REFERENCES

1. K.S. Novoselov, A.K. Geim, S.V Morozov, D. Jiang, Y. Zhang, S.V Dubonos, I.V Grigorieva, A.A. Firsov, Electric field effect in atomically thin carbon films, Science 306 (2004) 666–669.
2. M.W. Barsoum, MAX phases: properties of machinable ternary carbides and nitrides. Weinheim: Wiley, 2013.
3. M. Naguib, et al. Two-dimensional nanocrystals produced by exfoliation of Ti_3AlC_2, Adv. Mater. 23 (2011) 4248–4253.
4. B. Dai, B. Zhao, X. Xie, T. Su, B. Fan, R. Zhang, R. Yang, Novel two-dimensional $Ti_3C_2T_x$ MXenes/nano-carbon sphere hybrids for high-performance microwave absorption, J. Mater. Chem. C 6 (2018) 5690–5697, https://doi.org/10.1039/C8TC01404C.
5. Z. Sun, D. Music, R. Ahuja, S. Li, J.M. Schneider, Bonding and classification of nano-layered ternary carbides, Phys. Rev. B. 70 (2004), 092102, https://doi.org/10.1103/PhysRevB.70.092102.
6. Y. Du, X. Kan, F. Yang, L. Gan, U. Schwingenschlogl, MXene/graphene heterostructures as high-performance electrodes for Li-ion batteries, ACS Appl. Mater. Interfaces 10 (2018) 32867–32873, https://doi.org/10.1021/acsami.8b10729.
7. J. Halim, K. Cook, M. Naguib, P. Eklund, Y. Gogotsi, J. Rosen, M.W. Barsoum, Xray photoelectron spectroscopy of select multi-layered transition metal carbides (MXenes), Appl. Surf. Sci. 362 (2016) 406–417, https://doi.org/10.1016/j.apsusc.2015.11.089.
8. Q. Tang, Z. Zhou, P. Shen, Are MXenes promising anode materials for Li Ion batteries? Computational studies on electronic properties and Li storage capability of Ti_3C_2 and $Ti_3C_2X_2$ (X=F, OH) monolayer, J. Am. Chem. Soc. 134 (2012) 16909–16916, https://doi.org/10.1021/ja308463r.
9. L. Gao, C. Li, W. Huang, S. Mei, H. Lin, Q. Ou, Y. Zhang, J. Guo, F. Zhang, S. Xu, H. Zhang, MXene/polymer membranes: synthesis, properties and emerging applications, Chem. Mater. 32 (2020) 1703–1747, https://doi.org/10.1021/acs.chemmater.9b04408.
10. A. Tanvir, P. Sobolciak, A. Popelka, M. Mrlik, Z. Spitalsky, I. Krupa, M. Micusik, J. Prokes, Electrically conductive transparent polymeric nanocomposites modified by 2D $Ti_3C_2T_x$ (MXene), Polymers 2019,11–1272, https://doi.org/10.3390/polym11081272.

11. Wang Z, Cheng Z, Fang C, Hou X, Xie L, Recent advances in MXenes composites for electromagnetic interference shielding and microwave absorption, Composites: Part A (2020), https://doi.org/10.1016/j.compositesa.2020.105956.

12. C. Liang, P. Song, A. Ma, X. Shi, H. Gu, L. Wang, H. Qiu, J. Kong, J. Gu, Highly oriented three-dimensional structures of Fe_3O_4 decorated CNTs/reduced graphene oxide foam/epoxy nanocomposites against electromagnetic pollution, Compos. Sci. Technol. 181 (2019) 107683, https://doi.org/10.1016/j.compscitech.2019.107683.

13. J.M. Thomassin, C. Pagnoulle, L. Bednarz, I. Huynen, R. Jerome, C. Detrembleur, Foams of polycaprolactone/MWNT nanocomposites for efficient EMI reduction, J. Mater. Chem. 18 (2008) 792–796, https://doi.org/10.1039/B709864B.

14. Y. Huangfu, K. Ruan, H. Qiu, Y. Lu, C. Liang, J. Kong, J. Gu, Fabrication and investigation on the PANI/MWCNT/thermally annealed graphene aerogel/epoxy electromagnetic interference shielding nanocomposites, Composer Part A Appl. S 121 (2019) 265–272, https://doi.org/10.1016/j.compositesa.2019.03.041.

15. N.A. Aal, F. El-Tantawy, A. Al-Hajry, M. Bououdina, New antistatic charge and electromagnetic shielding effectiveness from conductive epoxy resin/plasticized carbon black composites, Polym. Compos. 29 (2008) 125–132, https://doi.org/10.1002/pc.20334.

16. L. Wang, H. Qiu, C. Liang, P. Song, Y. Han, Y. Han, J. Gu, J. Kong, D. Pan, Z. Guo, Electromagnetic interference shielding MWCNT-Fe_3O_4@Ag/epoxy nanocomposites with satisfactory thermal conductivity and high thermal stability, Carbon 141 (2019) 506–514, https://doi.org/10.1016/j.carbon.2018.10.003.

17. J. Wu, D.D.L. Chung, Increasing the electromagnetic interference shielding effectiveness of carbon fiber polymer-matrix composite by using activated carbon fibers, Carbon 40 (2002) 445–467, https://doi.org/10.1016/S0008-6223(01)00133-6.

18. Y. Yang, M.C. Gupta, K.L. Dudley, R.W. Lawrence, Novel carbon nanotube polystyrene foam composites for electromagnetic interference shielding, Nano Lett. 5 (2005) 2131–2134, https://doi.org/10.1021/nl051375r.

19. C. Liang, H. Qiu, Y. Han, H. Gu, P. Song, L. Wang, J. Kong, D. Cao, J. Gu, Superior electromagnetic interference shielding 3D graphene nanoplatelets/reduced graphene oxide foam/epoxy nanocomposites with high thermal conductivity, J. Mater. Chem. C 7 (2019) 2725–2733, https://doi.org/10.1039/C8TC05955A.

20. D.D.L. Chung, Electromagnetic interference shielding effectiveness of carbon materials, Carbon 39 (2001) 279–285, https://doi.org/10.1016/S0008-6223(00)00184-6.

21. H. Zhang, G. Zhang, M. Tang, L. Zhou, J. Li, X. Fan, X. Shi, J. Qin, Synergistic effect of carbon nanotube and graphene nanoplates on the mechanical, electrical and electromagnetic interference shielding properties of polymer composites and polymer composite foams, Chem. Eng. J. 353 (2018) 381–393, https://doi.org/10.1016/j.cej.2018.07.144.

22. B. Yao, W. Hong, T. Chen, Z. Han, X. Xu, R. Hu, J. Hao, C. Li, H. Li, S.E. Perini, M. T. Lanagan, S. Zhang, Q. Wang, H. Wang, Highly stretchable polymer composite with strain-enhanced electromagnetic interference shielding effectiveness, Adv. Mater. 32 (2020) 1907499, https://doi.org/10.1002/adma.201907499.

23. J. Liang, Y. Wang, Y. Huang, Y. Ma, Z. Liu, J. Cai, C. Zhang, H. Gao, Y. Chen, Electromagnetic interference shielding of graphene/epoxy composites, Carbon 47 (2009) 922–925, https://doi.org/10.1016/j.carbon.2008.12.038.

24. M. Naguib, M.W. Barsoum, Y. Gogotsi, Ten years of progress in the synthesis and development of MXenes, Adv. Mater. (2021), 2103393, https://doi.org/10.1002/adma.202103393.

25. M. Naguib, V. Mochalin, M. Barsoum, Y. Gogotsi, 25th anniversary article: MXenes: a new family of two-dimensional materials, Adv. Mater. 26 (2014) 992–1005, https://doi.org/10.1002/adma.201304138.

26. A. Lipatov, M. Alhabeb, M.R. Lukatskaya, A. Boson, Y. Gogotsi, A. Sinitskii, Effect of synthesis on quality, electronic properties and environmental stability of individual monolayer Ti_3C_2 MXene flakes, Adv. Electron. Mater. 2 (2016) 1600255, https://doi.org/10.1002/aelm.201600255.

27. J. Liu, H.B. Zhang, X. Xie, R. Yang, Z. Liu, Y. Liu, Z.Z. Yu, Multifunctional, super elastic, and lightweight MXene/polyimide aerogels, Small 14 (2018) 1802479, https://doi.org/10.1002/smll.201802479.

28. P. Sambyal, A. Iqbal, J. Hong, H. Kim, M.K. Kim, S.M. Hong, M. Han, Y. Gogotsi, C.M. Koo, Ultralight and mechanically robust $Ti_3C_2T_x$ hybrid aerogel reinforced by carbon nanotubes for electromagnetic interference shielding, ACS Appl. Mater. Interfaces 11 (2019) 38046–38054, https://doi.org/10.1021/acsami.9b12550.

29. T. Shang, Z. Lin, C. Qi, X. Liu, P. Li, Y. Tao, Z. Wu, D. Li, P. Simon, Q.H. Yang, 3D Macroscopic architectures from self-assembled MXene hydrogels, Adv. Funct. Mater. 29 (2019) 1903960, https://doi.org/10.1002/adfm.201903960.

30. M. Cao, Y. Cai, P. He, J. Shu, W. Cao, J. Yuan, 2D MXenes: electromagnetic property for microwave absorption and electromagnetic interference shielding, Chem. Eng. J. 359 (2019) 1265–1302, https://doi.org/10.1016/j.cej.2018.11.051.

31. K. Li, M. Liang, H. Wang, X. Wang, Y. Huang, J. Coelho, S. Pinilla, Y. Zhang, F. Qi, V. Nicolosi, Y. Xu, 3D MXene architectures for efficient energy storage and conversion, Adv. Funct. Mater. 30 (2020) 2000842, https://doi.org/10.1002/adfm.202000842.

32. M. Han, C.E. Shuck, R. Rakhmanov, D. Parchment, B. Anasori, C.M. Koo, G. Friedman, Y. Gogotsi, Beyond $Ti_3C_2T_x$: MXenes for electromagnetic interference shielding, ACS Nano 14 (2020) 5008–5016, https://doi.org/10.1021/acsnano.0c01312.

33. P. Song, B. Liu, H. Qiu, X. Shi, D. Cao, J. Gu, MXenes for polymer matrix electromagnetic interference shielding composites: a review, Compos. Commun. 24 (2021) 100653, 2452–2139, https://doi.org/10.1016/j.coco.2021.100653.

34. C. Liang, K. Ruan, Y. Zhang, J. Gu, Multifunctional flexible electromagnetic interference shielding silver nanowires/cellulose films with excellent thermal management and joule heating performances, ACS Appl. Mater. Interfaces 12 (2020) 18023–18031, https://doi.org/10.1021/acsami.0c04482.

35. L. Jia, D. Yan, X. Liu, R. Ma, H. Wu, Z. Li, Highly efficient and reliable transparent electromagnetic interference shielding film, ACS Appl. Mater. Interfaces 10 (2018) 11941–11949, https://doi.org/10.1021/acsami.8b00492.

36. Y. Sun, S. Luo, H. Sun, W. Zeng, C. Ling, D. Chen, V. Chan, K. Liao, Engineering closed-cell structure in lightweight and flexible carbon foam composite for high efficient electromagnetic interference shielding, Carbon 136 (2018) 299–308, https://doi.org/10.1016/j.carbon.2018.04.084.

37. J. Fu, J. Yun, S. Wu, L. Li, L. Yu, K. Kim, Architecturally robust graphene encapsulated MXene Ti_2CT_x @ polyaniline composite for high-performance pouch type asymmetric supercapacitor, ACS Appl. Mater. Interfaces 10 (2018) 34212–34221, https://doi.org/10.1021/acsami.8b10195.

38. C. Liang, H. Qiu, P. Song, X. Shi, J. Kong, J. Gu, Ultra-light MXene aerogel/wood derived porous carbon composites with wall-like "mortar/brick" structures for electromagnetic interference shielding, Sci. Bull. 65 (2020) 616–622, https://doi.org/10.1016/j.scib.2020.02.009.

39. J. Liu, H. Zhang, R. Sun, Y. Liu, Z. Liu, A. Zhou, Z. Yu, Hydrophobic, flexible, and lightweight MXene foams for high-performance electromagnetic-interference shielding, Adv. Mater. 29 (2017) 1702367, https://doi.org/10.1002/adma.201702367.

40. W. Yuan, J. Yang, F. Yin, Y. Li, Y. Yuan, Flexible and stretchable MXene/polyurethane fabrics with delicate wrinkle structure design for effective electromagnetic interference shielding at a dynamic stretching process, Compos. Commun. 19 (2020) 90–98, https://doi.org/10.1016/j.coco.2020.03.003.

41. P. Song, C. Liang, L. Wang, H. Qiu, H. Gu, J. Kong, J. Gu, Obviously improved electromagnetic interference shielding performances for epoxy composites via constructing honeycomb structural reduced graphene oxide, Compos. Sci. Technol. 181 (2019) 107698, https://doi.org/10.1016/j.compscitech.2019.107698.

42. Z. Liu, Y. Zhang, H.B. Zhang, Y. Dai, J. Liu, X. Li, Z. Z. Yu, Electrically conductive aluminum ion-reinforced MXene films for efficient electromagnetic interference shielding, J. Mater. Chem. C 8 (2020) 1673–1678, https://doi.org/10.1039/C9TC06304H.

43. J. Liu, Z. Liu, H.B. Zhang, W. Chen, Z. Zhao, Q.W. Wang, Z.Z. Yu, Ultra strong and highly conductive MXene-based films for high-performance electromagnetic interference shielding, Adv. Electron. Mater 6 (2019) 1901094, https://doi.org/10.1002/aelm.201901094.

44. Z. Fan, D. Wang, Y. Yuan, Y. Wang, Z. Cheng, Y. Liu, Z. Xie, A lightweight and conductive MXene/graphene hybrid foam for superior electromagnetic interference shielding, Chem. Eng. J. 381 (2020) 122696, https://doi.org/10.1016/j.cej.2019.122696.

45. C. Liang, P. Song, H. Qiu, Y. Zhang, X. Ma, F. Qi, H. Gu, J. Kong, D. Cao, J. Gu, Constructing interconnected spherical hollow conductive networks in silver platelets/reduced graphene oxide foam/epoxy nanocomposites for superior electromagnetic interference shielding effectiveness, Nanoscale 11 (2019) 22590–22598, https://doi.org/10.1039/C9NR06022G.

46. K. Rajavel, S. Luo, Y. Wan, X. Yu, Y. Hu, P. Zhu, R. Sun, C. Wong, 2D $Ti_3C_2T_x$ MXene/polyvinylidene fluoride (PVDF) nanocomposites for attenuation of electromagnetic radiation with excellent heat dissipation, Composer Part A Appl. S 129 (2020) 105693, https://doi.org/10.1016/j.compositesa.2019.105693.

47. L. Qin, Q. Tao, X. Liu, M. Fahlman, J. Halim, P.O.A Persson, J. Rosen, F. Zhang, Electrochemical polymerization of conjugated polymer MXene composite nanosphere. Nano Energy 60 (2019) 734–742, https://doi.org/10.1016/j.nanoen.2019.04.002.

48. J. Gu, Q. Zhang, H. Li, Y. Tang, J. Kong, J. Dang, Study on preparation of SiO_2/epoxy resin hybrid materials by means of sol-gel, Polym.-Plast. Technol. 46 (2007) 1129–1134, https://doi.org/10.1080/03602550701558033.

49. L. Wang, L. Chen, P. Song, C. Liang, Y. Lu, H. Qiu, Y. Zhang, J. Kong, J. Gu, Fabrication on the annealed $Ti_3C_2T_x$ MXene/Epoxy nanocomposites for electromagnetic interference shielding application, Compos. B Eng. 171 (2019) 111–118, https://doi.org/10.1016/j.compositesb.2019.04.050.

50. B. Anasori, Y. Xie, M. Beidaghi, J. Lu, B. Hosler, L. Hultman, P.R. Kent, Y. Gogotsi, M.W. Barsoum, Two-dimensional, ordered, double transition metals carbides (MXenes), ACS Nano 9 (2015) 9507–9516, https://doi.org/10.1021/acsnano.5b03591.

51. H. Xu, X. Yin, X. Li, M. Li, S. Liang, L. Zhang, L. Cheng, Lightweight Ti_2CT_x MXene/poly(vinyl alcohol) composite foams for electromagnetic wave shielding with absorption-dominated feature, ACS Appl. Mater. Interfaces 11 (2019) 10198–10207, https://doi.org/10.1021/acsami.8b21671.

52. C. Weng, G. Wang, Z. Dai, Y. Pei, L. Liu, Z. Zhang, Buckled AgNW/MXene hybrid hierarchical sponges for high-performance electromagnetic interference shielding, Nanoscale 11 (2019) 22804–22812, https://doi.org/10.1039/C9NR07988B.

53. X. Wu, B. Han, H. Zhang, H. Xie, T. Tu, Y. Zhang, Y. Dai, R. Yang, Z. Yu, Compressible, durable and conductive polydimethylsiloxane-coated MXene foams for high-performance electromagnetic interference shielding, Chem. Eng. J. 381 (2020) 122622, https://doi.org/10.1016/j.cej.2019.122622.

54. S. Zhao, H.B. Zhang, J.Q. Luo, Q.W. Wang, B. Xu, S. Hong, Z.Z. Yu, Highly electrically conductive three-dimensional $Ti_3C_2T_x$ MXene/reduced graphene oxide hybrid aerogels with excellent electromagnetic interference shielding performances, ACS Nano 12 (2018) 11193–11202, https://doi.org/10.1021/acsnano.8b05739.

55. L. Wang, H. Qiu, P. Song, Y. Zhang, Y. Lu, C. Liang, J. Kong, L. Chen, J. Gu, 3D $Ti_3C_2T_x$ MXene/C hybrid foam/epoxy nanocomposites with superior electromagnetic interference shielding performances and robust mechanical properties, Composer Part A Appl. S 123 (2019) 293–300, https://doi.org/10.1016/j.compositesa.2019.05.030.

56. P. Song, H. Qiu, L. Wang, X. Liu, Y. Zhang, J. Zhang, J. Kong, J. Gu, Honeycomb structural rGO-MXene/epoxy nanocomposites for superior electromagnetic interference shielding performance, Sustain. Mater. Technol. 24 (2020) e00153, https://doi.org/10.1016/j.susmat.2020.e00153.

57. L. Wang, P. Song, C. Lin, J. Kong, J. Gu, 3D shapeable, superior electrically conductive cellulose nanofibers/$Ti_3C_2T_x$ MXene aerogels/epoxy nanocomposites for promising EMI shielding, Research 2020 (2020) 4093732, https://doi.org/10.34133/2020/4093732.

58. W. Cao, F. Chen, Y. Zhu, Y. Zhang, Y. Jiang, M. Ma, F. Chen, Binary strengthening and toughening of MXene/cellulose nanofiber composite paper with nacre inspired structure and superior electromagnetic interference shielding properties, ACS Nano 12 (2018) 4583–4593, https://doi.org/10.1021/acsnano.8b00997.

59. F. Xie, F. Jia, L. Zhuo, Z. Lu, L. Si, J. Huang, M. Zhang, Q. Ma, Ultrathin MXene/ aramid nanofiber composite paper with excellent mechanical properties for efficient electromagnetic interference shielding, Nanoscale 11 (2019) 23382–23391, https://doi.org/10.1039/C9NR07331K.

60. Z. Zhan, Q. Song, Z. Zhou, C. Lu, Ultra strong and conductive MXene/cellulose nanofiber films enhanced by hierarchical nano-architecture and interfacial interaction for flexible electromagnetic interference shielding, J. Mater. Chem. C 7 (2019) 9820–9829, https://doi.org/10.1039/C9TC03309B.

61. D. Hu, X. Huang, S. Li, P. Jiang, Flexible and durable cellulose/MXene nanocomposite paper for efficient electromagnetic interference shielding, Compos. Sci. Technol. 188 (2020) 107995, https://doi.org/10.1016/j.compscitech.2020.107995.

62. J. Luo, S. Zhao, H. Zhang, Z. Deng, L. Li, Z. Yu, Flexible, stretchable and electrically conductive MXene/natural rubber nanocomposite films for efficient electromagnetic interference shielding, Compos. Sci. Technol. 182 (2019) 107754, https://doi.org/10.1016/j.compscitech.2019.107754.

63. W. Xin, G. Xi, W. Cao, C. Ma, T. Liu, M. Ma, J. Bian, Lightweight and flexible MXene/ CNF/silver composite membranes with a brick-like structure and high performance electromagnetic-interference shielding, RSC Adv. 9 (2019) 29636–29644, https://doi.org/10.1039/C9RA06399D.

64. Z. Zhou, J. Liu, X. Zhang, D. Tian, Z. Zhan, C. Lu, Ultrathin MXene/calcium alginate aerogel film for high-performance electromagnetic interference shielding, Adv. Mater. Interfaces 6 (2019) 1802040, https://doi.org/10.1002/admi.201802040.

65. F. Shahzad, A. Mohamed, C.B. Hatter, B. Anasori, S.M. Hong, C.M. Koo, Y. Gogotsi, Electromagnetic interference shielding with 2D transition metal carbides (MXenes), Science 353 (2016) 1137–1140, https://doi.org/10.1126/science.aag2421.

66. J. Gu, Y. Zhang, L. Wang, J. Zhang, P. Song, Z. Xiao, C. Liang, H. Qiu, J. Kong, Fabrication and investigation on the ultra-thin and flexible $Ti_3C_2T_x$/co-doped polyaniline electromagnetic interference shielding composite films, Compos. Sci. Technol. 183 (2019) 107833, https://doi.org/10.1016/j.compscitech.2019.107833.

67. I. Glass, Building decorative glass, Build. Decor. Mater. (2011) 139–168. https://doi.org/10.1533/9780857092588.139.

68. L.W. McKeen, L.W. McKeen, 9 – Polyolefins, polyvinyls, and acrylics, Permeability Properties of Plastics and Elastomers, 2017, 157–207. doi:10.1016/B978-0-323-50859-9.00009-9.

69. H. Yang, J. Dai, X. Liu, Y. Lin, J. Wang, L. Wang, F. Wang, Layered PVB/$Ba_3Co_2Fe_{24}O_{41}$/ Ti_3C_2 MXene composite: enhanced electromagnetic wave absorption properties with high impedance match in a wide frequency range, Mater. Chem. Phys. 200 (2017) 179–186. doi:10.1016/j.matchemphys.2017.05.057.

70. N. Hadjesfandiari, A. Parambath, Stealth coatings for nanoparticles: Polyethylene glycol alternatives, Elsevier Ltd, 2018. doi:10.1016/B978-0-08-101750-0.00013-1.
71. G.T. Hermanson, Immobilization of ligands on chromatography supports, Bioconjugate Techniques, 2013. doi:10.1016/b978-0-12-382239-0.00015-7.
72. M. Naguib, T. Saito, S. Lai, M.S. Rager, T. Aytug, M. Parans Paranthaman, M.Q. Zhao, Y. Gogotsi, $Ti_3C_2T_x$ (MXene)–polyacrylamide nanocomposite films, RSC Adv. 6 (2016) 72069–72073. doi:10.1039/C6RA10384G.
73. A. Jalal Uddin, Coatings for technical textile yarns, Tech. Text. Yarns. (2010) 140–184. doi:10.1533/9781845699475.1.140.
74. J.D Arrigo, Early work with aqueous carbohydrate gels, Stud. Interface Sci. 25 (2011) 29–44. doi:10.1016/B978-0-444-53798-0.00004-3.
75. H. Wei, J. Dong, X. Fang, W. Zheng, Y. Sun, Y. Qian, Z. Jiang, Y. Huang, $Ti_3C_2T_x$ MXene/polyaniline (PANI) sandwich intercalation structure composites constructed for microwave absorption, Compos. Sci. Technol. 169 (2019) 52–59, https://doi.org/10.1016/j.compscitech.2018.10.016.
76. J. Jimmy, B. Kandasubramanian, MXene functionalized polymer composites: synthesis and applications, Eur. Polym. J. (2019), https://doi.org/10.1016/j.eurpolymj.2019.109367.
77. C. Wang, V. Murugadoss, J. Kong, Z. He, X. Mai, Q. Shao, Y. Chen, L.Guo, C. Liu, S. Angaiah, Z. Guo, Overview of carbon nanostructures and nanocomposite for electromagnetic wave shielding. Carbon 140 (2018) 696–733, https://doi.org/10.1016/j.carbon.2018.09.006.
78. K. L. Kaiser, Electromagnetic shielding. Boca Raton: CRC Press, 2005.
79. H. W. Ott, Electromagnetic compatibility engineering. New York: Wiley, 2011.
80. O. Salim, K.A. Mahmoud, K.K. Pant, R.K. Joshi, Introduction to MXenes: synthesis and characteristics, Mater. Today Chem. 14 (2019) 100191, 2468–5194, https://doi.org/10.1016/j.mtchem.2019.08.010.

10 MXene-Polymer Nanocomposites for Biomedical Applications

Amandeep Singh
Department of Polymer Science and Technology
University of Calcutta
Kolkata, India

Sonam Gupta
Associate Scientific Writer
Indegene
Bangalore, India

K. Kumari
Department of Chemical Engineering
SLIET Longowal
Punjab, India

P. P. Kundu
Department of Polymer Science and Technology
University of Calcutta
Kolkata, India
and
Department of Chemical Engineering
Indian Institute of Technology
Roorkee, India

CONTENTS

DOI: 10.1201/9781003281511-10

10.1 INTRODUCTION

In the last couple of years, with countless developments in synthetic practices, many two-dimensional novel materials, other than graphene, are being magnificently synthesized. Out of these, a recently explored large class of two-dimensional materials belongs to the MXenes. These are electrically conductive, hydrophilic, layered, two-dimensional nanomaterials developed from carbides/nitrides/carbonitrides of transition metal with a-few-atom thickness. It has been reported that >30 various compositions of such MXenes have already been synthesized to date [1, 2]. Generally, chemical composition of MXenes is expressed as $M_{n+1}X_nT_x$, where M indicates an early transition metal (i.e., $_{22}Ti$), X signifies C and/or N, n signifies number of repeating units (an integer between 1 and 4), T indicates functional group present onto the surface (i.e., $-OH$, $-F$, and $-O-$), and x suggests counts of such functional groups present onto the surface of MXene. Presence of functional groups onto MXenes compounds makes them hydrophilic. These groups make MXenes reactive towards reactions and interactions with several hydrophilic polymers (i.e., polyvinyl alcohol (PVA), poly-diallyldimethylammonium chloride (PDDA), polyacrylic acid (PAA)) and many biopolymers. However, a stable suspension in water (or in other polar solvents) can be attained by probe or bath sonication of an MXene-containing suspension around neutral pH. Being a comparatively novel and exciting family of two-dimensional nanomaterials, MXenes consist of an extensive range of compositions and outstanding unique properties, for instances, ease of dispersibility and metallic conductivity, which render them to be a promising candidate for functional reinforcement in polymer nanocomposites. However, the most extensively explored MXene is Ti_3C_2, first reported in 2011 [3]. As far as its morphological points are considered, MXene resembles to graphene oxide. Thickness of a sheet of a single layer MXene is about 1 nm, whereas its lateral dimensions vary from a few hundred nanometres to tens of microns [4].

Polymers possess excellent properties in terms of tensile and impact strengths, fatigue, corrosion, abrasion, fracture resistance, and other bulk properties. Their solubility in organic solvents improves their compatibility with nanoparticles, which further enables their incorporation into two-dimensional material systems. Nanotechnology and nano-engineering have brought significant advancements in scientific and technological fields, including medicine and physiology. Hybrid composite materials concurrently take advantages from the properties of both polymers and nanoparticles. This fact has driven the incorporation of nanoparticles, that is, clay [5, 6] graphene [7], GO [8, 9], Ag NPs [10, 11], Al_2O_3 [12], TiO_2 [13], nanocellulose [14, 15], and so on, are extensively used as filler materials to prepare polymer nanocomposites [16, 17] through different methods including *in situ* polymerization, grafting, solution processing, melt blending, and several other methods consisting of covalent and non-covalent modification [18, 19] for biomedical applications.

Generally, the same kind of processing method is used for MXene and solution-processed graphene, as both of them possess almost the same kind of surface morphology. Though graphene and graphene oxide-based polymer nanocomposites show active adsorption of organic solvents within the graphene layers that unfavourably impacts the characteristics of the materials and also influences their mechanical strength, electronic properties, thermal properties, and surface chemical activity [20, 21], Ti_3C_2 MXene shows outstanding solvent stability along with tunable adsorption properties because of the presence of several surface functional groups. Therefore, the relative lead of MXenes-based polymer nanocomposites over graphene and graphene oxide-based polymer nanocomposites allows greater interaction between polymer matrix and nano-reinforcements. The application of MXenes as reinforcements enables preparation of highly robust polymer nanocomposite hybrids. Moreover, the excellent dispersion ability of MXenes in polymer matrix without the addition of dispersing agents additionally enables its suitability to be incorporated into polymer matrix. MXenes can be successfully incorporated within polymer matrix either during initial polymer synthesis steps or during an ongoing production line.

Mostly, the *in situ* polymerization process induces excellent interfacial interactions between graphene and polymer matrix, but it also dissembles the viscosity of premix that constrains further processing. However, like graphene, MXene can be introduced while *in situ* polymerization without making compromises with viscosity alteration, stabilizing agents, solution incompatibility, and active agglomeration. Another advantage of using MXenes is their top-down synthesis approaches, through which high yield can be obtained as compared to other two-dimensional nanofiller reinforcement materials. Furthermore, several other applications, for instance gas sensors and electronics, need low error-prone systems. Such systems are quite costly to develop using graphene through bulk processing. However, large area graphene sheets without defects can be prepared through different routes, that is, chemical vapor deposition, but such graphene sheets are considered to be less viable as far as their economics is concerned, especially for sensitive technological applications, for instance, electronics and energy storage [22]. However, these objectives can be served by using MXene.

MXene-incorporated polymer nanocomposites were first reported in 2014. An intensively explored MXene is $Ti_3C_2T_z$, and it was imparted into two hydrophilic polymers: PDDA and PVA [23]. $Ti_3C_2T_z$ MXene was synthesized by etching method. Ti_3AlC_2 was etched in 50% HF solution at room temperature for 18 hours under continuous stirring. Thereafter, MXene multilayers were washed several times with distilled water to maintain the pH level at 6. The obtained powder was kept for drying at room temperature for a day. Thereafter, this powder was added into dimethyl sulfoxide (DMSO) and stirred for 18 hours at room temperature and later isolated by adding water and followed by centrifugation. Water was added in a ratio of 300:1 to the isolated DMSO-intercalated multilayer MXene powder. This reaction mixture was then sonicated in an argon environment for 5 hours. Thereafter, colloidal supernatant was separated out. PDDA and PVA were added separately into obtained colloidal suspensions in different proportions. Several thin films were developed from MXene-polymer solution by vacuum-assisted filtration method by using variable weight percentages (0, 40, 60, 80, 90, and 100 wt%) of MXene content [24]. Soon after the first publication, synthesis of MXene-incorporated polymer nanocomposites attracted the interest of the scientist community. However, research on MXenes itself is at its high level as compared to research on MXene-embedded nanocomposites. Many polymer matrixes have been used so far for developing MXene-incorporated polymer nanocomposites including several hydrophilic polymers, for instance, PVA [23, 25–38], PAA [39, 40], biopolymers [41–49], polyethylene (PE) [50–52], polyethylene oxide (PEO) [53–55], polypropylene (PP) [56], polystyrene (PS) [57], polyamide (PA) [40, 58, 59], polyimide (PI) [60–65], acrylamide [66–69], acrylates [70, 71], urethanes [72–77], silicones [62, 78–82], epoxies [28, 66, 83–90], and many more [31, 33, 71, 91–110]. Apart from these, several conductive polymers like poly(aniline) [111–117], polyvinylpyrrolidone (PVP) [118], polypyrrole (PPy) [31, 73, 119–125], poly(3,4-ethylenedioxythiophene) (PEDOT) [123, 126–128], polystyrene sulfonate (PSS) [129–133], and fluoride-based polymers named polyvinylidene difluoride (PVDF) [134–140], poly(vinylidenefluoride-trifluoroethylene) (PVDF-TrFE) [141], as well as poly(vinylidene fluoride-trifluoroethylene-chlorofluoroethylene) (PVDF-TrFE-CFE) [142–144] were also used to develop MXene-based polymer composites. However, several factors, including the overall cost, availability, sustainability, and recyclability of the substance, are needed to be considered before choosing the right one in order to maintain circular economy [145]. Furthermore, the outstanding properties of MXene nanoparticles imbue them with abilities which are exploited in a variety of uses and also became a subject that need to be explored in each and every bit to find the possibilities to be used in 3D printing [146].

10.2 PREPARATION OF MXene-POLYMER NANOCOMPOSITES

MXenes have hydrophilic surfaces and possess excellent mechanical properties and metallic conductivity, thus incorporation of MXenes as reinforcements into polymer matrix improves the mechanical and thermal properties of produced nanocomposites. However, single-layer MXenes contain a comparably higher degree of usable surface hydrophilicity and excellent compatibility with polymers than multilayered MXenes. Therefore, it is recommended that MXenes should be delaminated before using as re-enrolments for developing polymer nanocomposites.

In order to optimize the properties of nanocomposite, exfoliation of layered reinforcement materials and their homogeneous dispersion within host polymer matrix are categorically important. Nevertheless, mechanical reinforcement should not essentially be optimized with monolayers as established by Gong et al. [147] in case of graphene-polymer nanocomposites, in which most of the modulus takes place within tri-layer graphene with a 1 nm layer thickness. Therefore, use of multilayer MXene may be useful for polymer nanocomposites preparation because exfoliation of MXene is reducible up to an optimized level. Consequently, in order to obtain exfoliation as well as dispersion at the same time, the method through which nanocomposites are fabricated basically influences their structure and properties.

There are several methods to prepare MXene-incorporated polymer nanocomposites. Some of them are mentioned in the following sections.

10.2.1 IN SITU POLYMERIZATION

Generally, the *in situ* polymerization process includes the addition of nanoparticle material into a solution of monomer and thereafter the polymerization of monomers takes place in the presence of dispersed nanomaterials to obtain a nanocomposite. Several researches about the *in situ* polymerization process established that the subsequent nanocomposites contain covalent linkages between nanomaterial and polymer matrix. Though, *in situ* polymerization is also used to develop noncovalent nanocomposites using different polymers such as PE and PMMA. This polymerization process produces nanocomposites containing a comparatively higher degree of dispersion of nanomaterials into polymer matrix than the melt mixing method. Therefore, *in situ* polymerization methods are sometimes mentioned as an intercalation polymerization approach.

Another advantage of this technique is that the nanomaterial reinforcements are penetrated throughout the maximum excluded volume of coiled polymer matrix. Since the polymer chains remained coiled to get maximum stability by reducing surface energy and entropy as well, the maximum volume of polymer matrix remains available for inclusion of nanomaterials. *In situ* polymerizations to obtain MXene-polymer nanocomposites are widely found to prepare the curing systems using PDMS [47, 85, 86], epoxies [89, 95], etc. In such nanocomposites, one part of the amount of MXene is imparted into resin and another part of the amount of MXene is added into the curing agent, and then polymerization reaction is initiated. However, a solvent (acetone or DMF) is also used during this process to get better dispersion of nanomaterials into resin and curing agents. Several studies showed that *in situ* polymerizations can be carried out in an aqueous solution in case of PAM [69], PPy [132, 134–136], PEDOT [140, 141], polyaniline [120–126], and PAM/PVA [70]. Even though, most reports revealed the involvement of a solvent. Carey et al. [58] have performed *in situ* bulk polymerization of nylon-6 wherein 12-ALA-modified multilayer $Ti_3C_2T_z$ was intercalated with monomer ε-caprolactam without using any solvent. In this approach, 12-ALA-modified multilayer $Ti_3C_2T_z$ and ε-caprolactam monomers mixture was heated under inert atmosphere up to melting point of ε-caprolactam monomer (100°C). Thereafter, mixture was heated up to 250°C under inert conditions to carry out ring-opening polymerization of ε-caprolactam to produce MXene incorporated nylon-6.

10.2.2 SOLVENT PROCESSING

Another method to prepare MXene-polymer nanocomposites is the solvent processing method. In this method, integration of MXene into a polymer matrix is achieved by using a suitable solvent. Colloidal MXene materials are directly used in this method. Nevertheless, in this method, choice of solvent remained very limited as only polar (protic or aprotic) solvents are suitable. MXenes have inherent compatibility with polar solvents with a Hildebrand solubility parameter (d) of ~27 MPa$^{1/2}$ [60]. Therefore, the solvent processing approach is widely accepted to develop MXenes-polymer nanocomposites. However, surface modification of MXene can make it stable towards non-polar solvents [53]. Likewise, appropriate polymers suitable for the solvent processing approach are also limited, because only polar solvent can be used. Usually, MXene in the form of aqueous colloid is added into a polymer solution and thereafter solvent is separated-out either through vacuum filtration, evaporation, or sometimes precipitated to a non-solvent. Numerous MXene-incorporated nanocomposites have been developed through water as a solvent using several hydrophilic polymers, including PU [76, 112], PVA [23, 26–39, 54, 148], PAA [40], PDDA [23, 109], sulfonated polyether ether ketone [114], PEO [54–56], PA [61], PI [63], polydopamine [46], chitosan [41, 50, 114], cellulose nanofibrils [43, 44], poly(acrylamide) (PAM) [68], natural rubber [106], hyperbranched polyethyleneimine [107], polyallylamine hydrochloride [109], PEDOT:PSS [125, 147, 149, 150], polyether-polyamide block copolymer [112, 151, 152], and so on.

However, MXene-incorporated nanocomposites have also been synthesized using polar organic solvents for different polymers like PU [78, 79], polysulfone [117], epoxy [90, 153, 154], PVDF-TrFE-CFE [156–158], PVDF-CTFE-DB [150], PVDF-TrFE [155], PVDF [148, 149, 151, 153, 154], PVP [127, 155], PVA [111], and LLDPE [53], poly(2,2'-m-(phenylene)-5,5'-bibenzimidazole) [119], PI [65], acrylic resin elastomers [73], polyacrylonitrile [101, 153], polycaprolactone [153], PPy [127], and so on.

In case of graphene-incorporated polymer nanocomposites, an excellent dispersion of graphene can be accomplished through solvent-based processing techniques [156], but evidences for better dispersibility of MXene into polymer solution is yet to be explored, as there is no direct evaluations of solution blending approaches with other techniques of reinforcement integration. The effect of dispersibility of MXenes into polymer solution in solvent processing is among the domains of research which are yet to explored, and a study of dispersion in a model polymer by different processing approaches can bring a substantial involvement to the field of MXene polymer nanocomposites.

10.2.3 Melt Processing

In this method, MXenes nanomaterial is integrated into polymer matrix by melting the polymer host. The melt processing approach includes other techniques such as injection moulding, extrusion, and hot pressing; those are employed to incorporation of alignment of MXene nanomaterials into molten polymer. Such techniques are predominantly appropriate after optimization for large-scale production of MXene-polymer nanocomposites. The melt processing nanocomposite production approach seems advantageous for MXenes, as it has higher bulk density (~4.26 g/cm^3) than graphene, thus feeding of MXene nanomaterial into a hopper is easier and safer than low-bulk density graphene [156]. However, several disadvantages are also associated with the melt processing method. For instance, this method is usually restricted to only thermoplastic polymers, as they can be treated by melting. Also, the use of a high temperature (up to melting point of polymer) may result in the oxidation or degradation of MXene nanomaterial.

In recent times, the melt processing method has been employed to develop MXene-based polymer nanocomposites with thermoplastic polyurethane (TPU) [77, 78], ultrahigh-molecular-weight polyethylene (UHMW-PE) [51], nylon-6 [40, 60], PS [58], phthalonitrile [104], LLDPE [52, 53], and so on.

Generally, MXenes nanomaterial is used as a reinforcement filler without any surface modification. However, some reports revealed that the surface modification improves the compatibility of MXene with polymer host. For instance, Zhang et al. [50] have modified the surface of $Ti_3C_2T_z$ using (isopropyl dioleic(dioctyl phosphate) titanate) (DN-101) that was blended with multilayered powders in petroleum ether solvent for 2 hours at room temperature. Thereafter, surface-modified MXene nanopowder was mixed with UHMW-PE polymer host in a high-speed reactor at 2000 rpm for 1 hour. After some time, the reaction mixture was moulded in a press vulcanizer and kept at 220°C temperature under 10 MPa pressure for 30 minutes to obtain the final nanocomposite sheet. It has been found that the percentage of crystallinity (χ) increases almost linearly with the amount of incorporation of surface-modified MXene. A sample of unfilled UHMW-PE had χ value as ~45%, whereas a sample containing 2 wt% of surface-modified $Ti_3C_2T_z$ showed χ value as ~53%. The result shows that Ti_3C_2 nanoparticles have appeared as nucleation centres for UHMW-PE chains to form many small-sized spherulites. It establishes a heterogeneous nucleation mechanism for crystallization of such nanocomposites. Nevertheless, surface-modified Ti_3C_2 nanomaterials were not exfoliated and UHMW-PE chains were not intercalated into multilayer MXene stacks. In spite of this, organotitanate-modified MXene stacks were found to be dispersed in a uniform manner with UHMW-PE host matrix without any agglomeration, even for the highest MXenes loading. It has been found that mechanical properties, i.e., tensile strength, hardness, creep or recovery behavior, and coefficient of friction of UHMW-PE, get enhanced on the incorporation of organotitanate-modified MXene [51]. Apart from poor exfoliation, this method is capable as it does depend upon application of any restricted scale parameters, however, included weight from titanium-containing DN-101 can be a possible limiting parameter for organotitanate-based surface modification.

In another research, Sheng et al. [74] employed another approach in which colloidal $Ti_3C_2T_z$ MXene was treated with polyethylene glycol (PEG) for 1 hour, taking 1:2 weight ration of PEG and MXene. Thereafter, the mixture was freeze-dried for 96 hours at −60°C. PEG-pretreated $Ti_3C_2T_z$ was mixed using a Brabender Plasticorder with TPU at 180°C and 60 rpm for 6 minutes. Later on, the attained mixture was hot compression moulded at 180°C under a pressure of 10 MPa for 10 minutes to get compressed sheets of thickness 1 mm. Results exhibit that tensile strength of nanocomposites improves by up to 47.1%, strain-to-failure by 17.5%, and storage modulus by 2482% at room temperature. Dispersion of $Ti_3C_2T_z$ was found to be excellent throughout the TPU matrix due to the pre-treatment of MXenes which improved the d-spacing of MXenes up to 19.7 Å from 15.4 Å. Also, thermal properties were found to be significantly improved in the case of PEG pre-treated MXenes, which is attributed to the barrier effect of MXenes.

However, results have dependence to colloidal MXene as well as freeze-drying limiting factors, but results also suggest the importance of exfoliation of MXene sheets. In that order, an exclusive method in order to produce simultaneously exfoliated as well as functionalized $Ti_3C_2T_z$ sheets was established by He et al. [75] to develop $Ti_3C_2T_z$-TPU nanocomposites. This approach was attained through ball milling of multilayered $Ti_3C_2T_z$ into an argon-flowed solution of 20 wt% of PDDA at 500 rpm for 2 hours. After centrifugation, the obtained material was washed by deionized water many times through vacuum filtration. Thereafter, obtained delaminated and functionalized MXene (f-Ti_3C_2) was dried in a vacuum at 80°C for 36 hours. Also, for the comparison purpose, delaminated $Ti_3C_2T_z$ (d-Ti_3C_2) was synthesized by the same method without using PDDA. It was noticeable that the f-Ti_3C_2 possesses a slightly improved d_{002} basal spacing (Δd_{002} = 0.108 nm) as compared to d-Ti_3C_2, which suggests the intercalation of PDDA. However, the increase in interlayer spacing may be higher in case of intercalation of any high-molecular-weight (say MW 100,000–200,000) polymers. Therefore, it can be stated, after looking over obtained results, that polymer did not intercalate the multilayers of MXenes in this particular case. Thereafter, obtained powder was nicely blended with TPU in DMF solvent followed by flocculation in deionized water. The mixture was then filtered and dried at 60°C for 36 hours. After drying, it finally was hot pressed under a pressure of 12 MPa at 190°C for 5 minutes in order to attain a nanocomposite sheet. Results show that incorporation of 3 wt% of f-Ti_3C_2 into TPU brought a 31.2% increase in tensile strength, 30.3% decrease in strain to failure, and almost doubled the thermal conductivity.

10.2.4 ROLLER MILLS

The use of roller mills is also among the methods of MXenes incorporation into the polymer hosts. Shah [154] has discovered the benefits of using three-roll mill to integrate MXene nanomaterials into bisphenol A diglycidyl ether, which is a monomer. The three-roll milling was accomplished by using 1-ethyl-3-methylimidazolium dicyanamide (a room-temperature ionic liquid, RTIL). RTIL acts as a dispersant as well as a latent curing agent, as demonstrated by Rahmathullah et al. [157]. Although, on use of RTIL, exfoliation of $Ti_3C_2T_z$ sheets can't attained due to poor shear stress which gets transfer during the milling process and decreases the viscosity. In order to compensate this, pre-curing of epoxy is highly required before the milling process. Still, pre-curing of epoxy could not lead to exfoliation, but it can reduce the size of MXene layers from microns to nanometers [154]. MXene were synthesized using HF as solvent without any fluoride salts. Therefore, none of the cations were found among MXene sheets. Intercalation of cations into HF-only-etched MXene becomes difficult after drying [158]. Roller mills technique can possibly be rectified and optimized in order to exfoliate $Ti_3C_2T_z$ as well as other multilayer MXenes.

In addition, the roller mills technique was also used to develop PP nanocomposites by Shi et al. [56]. The researchers first synthesized MXene nanosheets incorporated with maleic anhydride-grafted polypropylene (MA-g-PP). MXene colloid was prepared by ultrasonication method and was blended with MA-g-PP water-based solution. Thereafter, the mixture solution was dried to attain MA-g-PP/$Ti_3C_2T_z$

in powder form. Afterward, MA-g-PP/Ti$_3$C$_2$T$_z$ powder was melt blended with PP granules in a twin roller mill at 185°C by maintaining roller speed as 50 rpm for 10 minutes. The obtained material was then hot pressed at 190°C under 10 MPa pressure in order to attain the final nanocomposite samples. Results exhibit that mechanical and thermal properties of the obtained nanocomposites were found to be superior than neat polymer. Researchers claimed that the addition of 2.0 wt% of Ti$_3$C$_2$T$_z$ into MA-g-PP through twin roller mill technique has brought up instantaneous improvements in tensile strength, elastic modulus, and ductility by 35.3%, 102.2%, and 676.4%, respectively.

10.2.5 RIR-MAPLE Method

A novel approach known as resonant infrared matrix-assisted pulsed laser evaporation (RIR-MAPLE) was investigated by Ajnsztajn et al. [104] to fabricate a transparent nanocomposite electrode. Through this method, researchers have developed Ti$_3$C$_2$T$_z$ incorporated poly(9,9-di-n-octylflourenyl-2,7-diyl) (PFO). This method is a kind of sequential deposition method in which two different materials having individual suspension or emulsion requirements can be taken. In this method, Ti$_3$C$_2$T$_z$ colloid suspension and a PFO emulsion prepared in trichlorobenzene were taken. Thereafter, material composition of film was maintained by shifting the material either from Ti$_3$C$_2$T$_z$ colloid suspension or from PFO emulsion towards the target. Ti$_3$C$_2$T$_z$ flakes were found to be arbitrarily and uniformly dispersed within PFO nanocomposite film. Such obtained film shows higher areal capacitances that further increase on increasing Ti$_3$C$_2$T$_z$ colloid suspension content, however, the transmittance of visible light decreases [113]. Though the RIR-MAPLE approach is not recommended for large-scale production of bulk MXene-polymer nanocomposites, this method has discrete advantages in construction of novel supercapacitor electrodes and related electronics in which other techniques (like spin coating) are not practically possible due to material restrictions, for instance, solvent stability. The preparation methods and properties of MXene/epoxy thermoset polymer composites are tabulated in Table 10.1.

10.3 BIOMEDICAL APPLICATIONS

In recent times, MXene and its nanocomposites have been found to have a huge potential in the health care sector. Due to unique two-dimensional structure, biocompatibility, good conductivity, sufficient ion intercalation ability, biodegradability, and hydrophilicity nature, MXenes and its nanocomposites with several polymers are widely used in biomedical applications.

The presence of a large number of surface functional groups (i.e., −OH, −F, and −O−) onto MXene makes it suitable to carry drug molecules. Also, being biocompatible, MXene materials are being widely explored for various biomedical applications. However, one of the drawbacks of MXene-based nanomaterials is its deprived water dispersibility, which can deter the effective transport within the blood circulation system and internalization of cells [167, 168]. It has been predicted that during intravenous administration, MXene may accumulate inside the reticuloendothelial

TABLE 10.1

Preparation Methods and Properties of MXene/Epoxy Thermoset Polymer Composites

MXene	Preparation Method	Properties	Ref
Ti_2CT_x	*In situ* intercalative polymerization technology	Thermal + Mechanical + Tribological	90
$Ti_3C_2T_x$	Freeze-drying method	Thermal + Mechanical + Tribological	160
Ti_3CN	Ultrasonic mixing	Mechanical + Thermal	95
$Ti_3C_2T_x$	Ultrasonic mixing	Anticorrosion + Tribological + Mechanical	161
$Ti_3C_2T_x$/graphene	Direct mixing	Anticorrosion + Wear + Mechanical	162
$Ti_3C_2T_x$	Ultrasonic mixing	Thermal	163
$Ti_3C_2T_z$	Direct mixing	Water transport + Mechanical + Thermal stability + Thermo-mechanical	92
$Ti_3C_2T_x$	Ultrasonic mixing	Anticorrosion	164
$Ti_3C_2T_x$	Solution casting method	EMI shielding + Electrical conductivity + Mechanical	165
3D $Ti_3C_2T_x$/C	Vacuum-assisted impregnation	Electrical conductivity + EMI shielding performances + Mechanical	66
rGO-$Ti_3C_2T_x$	Vacuum-assisted impregnation	Electrical conductivity + EMI shielding performance + Dynamic mechanical analyses + Thermal stabilities	166

Source: Reproduced with due permission [159].

systems (RES), which may result in possible toxicity. However, such drawbacks can be addressed up to a certain limit by modifying the surface of nanocomposites, and such nanocomposites may serve various applications, including drug delivery, bone tissue engineering, antimicrobial, photothermal therapy, biosensing, bioimaging, and so on [169]. Major applications of MXene-based polymer nanocomposites in biomedical sector are mentioned in the following sections.

10.3.1 GUIDED BONE REGENERATION (GBR)

The guided bone regeneration (GBR) applications can be fulfiled by MXene-based nanocomposite hybrid materials. Several recent researchers have discovered that MXene-based nanocomposites can be potentially practised for guided bone regeneration purpose. GBR is usually used for oral rehabilitations, for instances, settlement of dental implants and periodontal revival in order to guard the healing bones from interference with nearby soft tissues [170].

In a study, Chen et al. [171] prepared $Ti_3C_2T_z$ incorporated poly(lactic acid) (PLA) membranes by applying interfacial mediations of n-octyl-triethoxysilane. Results exhibit that cell adhesion, proliferation, osteogenic differentiation, and biocompatibility of the prepared membrane were found to be improved after inclusion of $Ti_3C_2T_z$ MXene nanosheets. Prepared nanocomposite membranes were robust in nature along with high biocompatibility, therefore are potential candidates to be used for guided bone regeneration material.

10.3.2   DRUG DELIVERY

Several conventional techniques of cancer treatment, that is, chemotherapy and photodynamic therapy (PDT), may harm the non-malignant cells along with malignant cells, which creates adverse effects to human health. Therefore, stimuli-responsive materials need to be developed which can exactly sense the comparatively lower pH tumour cells [172]. However, a number of investigations have been performed in a couple of years to develop an ideal nanoplatform that can be used as a drug carrier [173, 174]. MXene confirms the collective consequences of both photothermal ablation as well as targeted drug release on their effective photothermal conversion and characteristic pH-sensitivity [170].

In another work, Zhang et al. [67] investigated a nanocomposite hydrogel developed by incorporating Ti_3C_2 into acrylamide through an *in situ* free radical polymerization approach. In this work, Ti_3C_2 was used as a crosslinker to synthesize nanocomposite hydrogels for the very first time. The as-prepared Ti_3C_2 incorporated polyacrylamide (PAM) nanocomposite hydrogels having ultralow Ti_3C_2 concentration show outstanding mechanical behaviour, swelling properties, and excellent sustained drug release kinetics as compare to many conventional organic crosslinked hydrogels. Most prominently, Ti_3C_2 incorporated PAM nanocomposite hydrogels exhibit significant acid resistance at the pH of 1.2 (HCl solution), which shows that such materials are able enough to be practised under acidic conditions.

In a recent report, Xing et al. [175] have mentioned the uses of Ti_3C_2 MXene as a light absorbent incorporated into cellulose hydrogels to be used as biomedicine. In this research, Ti_3C_2 MXenes were crosslinked by epichlorohydrin. It has been found that MXene incorporated cellulose nanocomposite hydrogels show a swift response towards near-infrared-stimulation and such stimulations are found in water as continuous dynamic processes. Therefore, when such nanocomposites were laden with doxorubicin hydrochloride (DOX), an anticancer drug, nanocomposite hydrogels exhibit a meaningful acceleration in release of DOX. Such performance of hybrid hydrogel may be due to pore expansion inside the three-dimensional cellulose architecture and the release kinetics can be faster by illuminating with 808 nm light. Taking advantage of excellent photothermal performance, control, and sustained release of DOX, MXene-cellulose nanocomposite hydrogel material is used as a multipurpose nanoplatform for curing tumours by intra-tumoural injection. Outcomes of this research claimed that PTT and prolonged adjuvant chemotherapy carried out with the help of such multifunctional nanoplatform was extremely effective for immediate tumour obliteration and for oppressing the tumour regeneration, representing the possible use of such nanoplatform in cancer therapy applications.

Thereafter, Liu et al. [176] have prepared Ti_3C_2-based nanoplatforms to be used for chemotherapy, PDT, and PTT. Prepared Ti_3C_2 MXene-incorporated nanosheets showed an outstanding extinction coefficient value as 28.6 $Lg^{-1}cm^{-1}$, excellent efficiency for photothermal conversion as approximately 58.3%, along with effective generation of singlet oxygen on irradiation of 808 nm laser. Moreover, Doxorubicin drug, which is a chemotherapeutic drug, was loaded to the surface of Ti_3C_2 and thereafter coated with hyaluronic acid (HA) to enhance its biocompatibility. This treatment has also improved the specificity in direction of cancerous cells directed by CD44 antigen and therefor allows active-targeting. The *in vitro* as well as *in vivo* results showed that DOX-loaded Ti_3C_2 MXene have advances in the biocompatibility, tumour-specific accumulation property, drug-releasing towards stimuli responses, and makes use of chemotherapy, PDT, and PTT to destroy cancerous and tumour tissues.

10.3.3 ANTIMICROBIAL ACTIVITY

Numerous two-dimensional nanomaterials have already been investigated for their potential antimicrobial properties [177, 178]. It has been found that MXenes exhibit antibacterial property greater than GO, which is regarded as one of the widely studied antimicrobial materials. Research claimed that Ti_3C_2 shows a relatively higher antimicrobial activity than graphene oxide towards both *E. coli* as well as *B. subtilis* [179], though the precise mechanism of antibacterial property of MXenes is yet to be explored. However, most probable conventions include (i) sharp edges of Ti_3C_2 that allow effective absorption of MXenes onto surface of microorganisms [180], (ii) sharp edges of Ti_3C_2 that persuade the punctures to microbial membrane, and (iii) Ti_3C_2 can directly interact with biomolecules present on cell walls and in cytoplasm that leads to the rupture of the microstructure of the cell and subsequent death of the living cell. Rasool et al. [179] have investigated the antibacterial activity of $Ti_3C_2T_x$ MXene incorporated membrane towards both *E. Coli* and *B. subtilis* by growing these bacteria onto the surface of membrane. It has been concluded that the rate of growth of bacteria onto fresh $Ti_3C_2T_x$ membranes was lowered by 67% towards *E. coli* and 73% towards *B. subtilis* as compared to the control specimen. The aged MXene membrane showed greater than 99% downfall in the growth of both bacteria. The results of another research [181] showed that the multilayered $Ti_3C_2T_x$ MXenes can be used as such for antibacterial applications without delaminating it. Researchers claimed that multilayered MXenes showed excellent bactericidal activity. Several characterizations, that is, FE-SEM, EDS, and agar diffusion tests, were carried out in order to analyze chemical structure, responses towards bacteria, and chemical elemental composition. Antibacterial activity was analyzed against *Staphylococcus aureus* and *Escherichia coli*. The images show that procured Ti_3C_2 MXene powder possesses a layered accordion-like structure which supported the fact that MXenes are un-delaminated. Agar diffusion characterizations of Ti_3C_2 powder were carried out against *S. aureus* and *E. coli* bacterial culture. However, multilayered Ti_3C_2 MXenes have shown antibacterial property against only *E. coli* having a zone of inhibition of 4 mm diameter. It has been found that none of the inhibition zones was noticed against *S. aureus*. Also, the zone of inhibition that was formed after 24 hours showed no further change, even after 48 hours.

10.3.4 BIOIMAGING

Fluorescence microscopy (FM) is a sort of light microscope which works on the principle of fluorescence; a substance that absorbs energy of invisible shorter wavelength radiation (i.e., UV light) and emits back longer wavelength radiation (i.e., visible light – green or red light). This fluorescence phenomenon is extensively used in clinical and diagnostic applications in order to carry on quick detection of antibodies, microorganisms, and several other substances. Fluorescence microscopy is user-friendly and uses inexpensive instrumentation, thus it turns out to be one of the vastly privileged bioimaging apparatuses. However, MXenes show low luminescence in aqueous solutions, therefore immobilization of fluorescent molecules onto the surface of MXenes brought up the enhancement in luminescence MXenes [182]. Microscopy image of cells after PTT using Ta_4C_3 soybean phospholipids (SP) as a photothermal agent are shown in Figure 10.1.

Apart from FM, the X-ray computed tomography (CT) scan is another significant bioimaging instrument where MXene is used. Short blood circulation rate, potential kidney damage, and high toxicity shown by traditional CT contrast agents, that is, iodine-containing complexes require the progress of effective and biocompatible materials. Thus, tantalum ($_{73}Ta$)-rich MXenes, that is, Ta_4C_3, is found to be appropriate as CT imaging contrast agents due to their eco-friendly synthesis, appropriate size, and exceptional biocompatibility. Magnetic Resonance Imaging (MRI) is another influential non-invasive method analogous to CT which can be employed on patients for carrying out bioimaging. MRI is applied as a substitute for patients who have any allergic reactions for CT contrast agents. Gadolinium ($_{64}Gd$)-based materials that are typically used as MRI contrast agents show a higher chance of

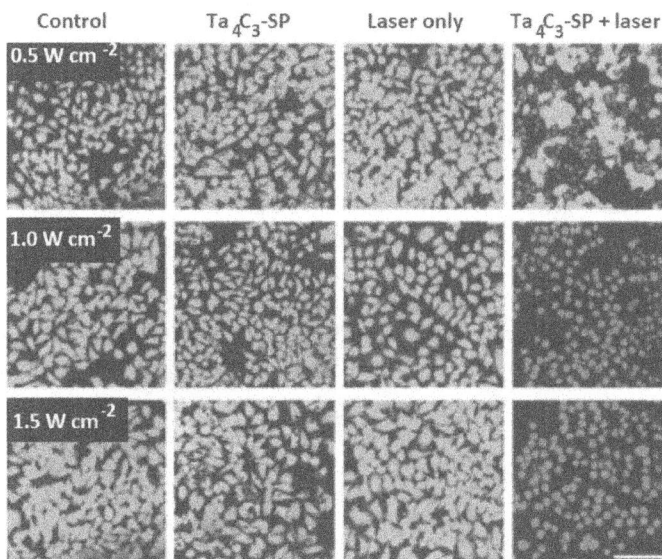

FIGURE 10.1 Microscopy image of cells after PTT using Ta_4C_3 soybean phospholipids (SP) as a photothermal agent. Green objects are living cells (stained with calcein AM); red points are dead cells (stained with propidium iodide). Scale bar, 100 mm.

(Reprinted with permission, from [182]. © Elsevier 2019.)

kidney damage. Therefore, in order to replace hazardous materials, nanocomposites developed by adsorbing nanosized manganese oxide onto MXenes have been found to be an excellent alternative for MRI contrast agent.

10.3.5 Cancer Therapy

Currently, the widely executed treatment methods in cancer therapy include chemotherapy and radiotherapy. Such methods are not precise in target and may end up causing damage to normal tissues together with cancer cells. Such methods may cause severe side effects [185]. By using light-controlled treatments, that is, photothermal therapy (PTT), enhancement in selectivity can be achieved which further diminishes side effects. Light-controlled methods distribute photothermal agents inside cancer tissues, where they change light energy into heat energy. Commonly, cancer cells have low value of heat resistance, so during generation of heat, cancer cells get demolished. Visible light owns low penetration to tissue, so near infra-red radiations are used in PTT. On the basis of wavelength of radiation being used, it can be divided into two segments: first, infra-red radiations bio-window that ranges from 750 to 1000 nm wavelength, and second infra-red radiations bio-window that ranges from 1000 to 1350 nm [186]. Results exhibit that second infra-red radiations bio-window (from 1000 to 1350 nm) possesses several benefits, that is, essential penetration depth to lasers and favourable maximum permissible exposure (MPE), as compare to other method. In spite of that, applications of light-controlled treatments are restricted due to deficiency of materials possessing sufficient near infra-red absorption and excellent photothermal conversion performance. Favourably, it has been found that MXene-based materials exhibit advanced efficiency in both bio-windows. Correspondingly, large surface areas of MXenes afford attaching locations and demonstrate effective accumulation into tumour cells during treatment of cancer [169].

In another research, Lin et al. [187] prepared Nb_2C incorporated PVP hybrid nanocomposites as a photothermal agent to be used in high efficiency *in vivo* photothermal ablation of tumour xenografts in a mouse model at the frequency range matching to both of the bio-windows. Tumour volume was estimated in six groups of mice using calipers on every alternative day. After 16 days of different treatments, it was analyzed that the tumour-infected areas of mice in the control group were persisted large. At same time, tumour-infected areas in mice belonging to (Nb_2C-PVP+near Infra-red-I) and (Nb_2C-PVP+near Infra-red-II) groups were found to be totally eliminated.

10.3.6 Sensors

MXenes show good dispersibility in water, therefore, any polymer which is easily dispersible or soluble in water can easily be blended into MXene through an ordinary aqueous solution mixing method in order to develop an MXene-incorporated polymer nanocomposite-based sensor. Apart from providing excluded volume to MXenes and imparting mechanical strength, polymers also shield MXenes from oxidation. However, particles of elastomeric polymer are typically spherical having a size greater than MXene sheets. Therefore, when such materials are blended together, particles of polymer force MXene sheets into interstitial space available amongst the particles which constructs three-dimensional continuous framework of conductive nanosheets [188]. Existence of such continuous framework of MXenes is important to develop revolutionary sensors.

10.3.7 BIOSENSORS

MXenes have also showed extraordinary applications to develop biosensors as well as gas sensors. In a biosensor, generally, an active protein loses its bioactivity on coming in direct contact to electrode surface. Therefore, some specific materials are designated to immobilize the active proteins in order to maintain their activity. Remarkably, as protein immobilization matrixes, MXenes help to guard active proteins and facilitates direct electron transfer between electrode and enzyme. It yields an opportunity for constructing mediator-free biosensors.

The working of a biosensor is schematically represented in Figure 10.2. Due to the large surface area available for charge transfer, realistic sensitivity, midget size, portability, economic, and quick sensing ability, MXene and MXene-based nanomaterials are integrated into the biosensors.

$Ti_3C_2T_x$ was first to be selected for immobilization of haemoglobin (Hb) in a biosensor (Nafion/Hb/$Ti_3C_2T_x$/GCE) used for detecting the nitrites [189]. Since, $Ti_3C_2T_x$ possesses relatively high conductivity, Nafion/Hb/$Ti_3C_2T_x$/GCE biosensor exhibits an extraordinary biosensor activity in distended detection range with an abridged

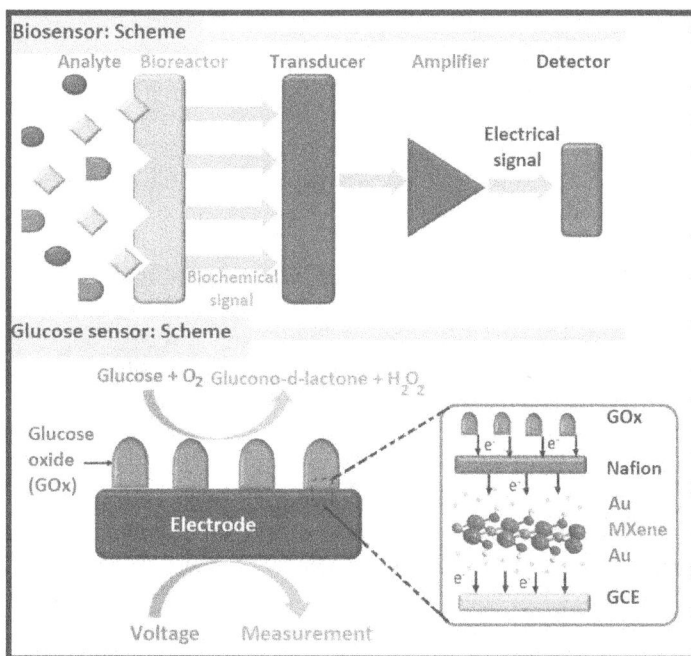

FIGURE 10.2 Schematic representation of the working of a biosensor and glucose biosensor. (Reproduced after due permission [170], © Elsevier 2020.)

detection limit. In a recent investigation [190], it has been found that a TiO_2-modified $Ti_3C_2T_x$ MXene nanocomposite developed through hydrothermal technique can be employed as an immobilization matrix for haemoglobin in a biosensor (Nafion/Hb/ $Ti_3C_2T_x$–TiO_2/GCE) to be used to detect hydrogen peroxide. It is clear from obtained micrographs that $Ti_3C_2T_x$ MXene nanolayers show an organ-like structure having one end closed, whereas another end is opened. Numerous white TiO_2 nanoparticles are also seen being loaded onto $Ti_3C_2T_x$ MXene layers. However, an organ-like structure is advantageous for enzyme immobilization into inner surfaces of $Ti_3C_2T_x$. Furthermore, TiO_2 nanoparticles own good biocompatibility along with chemical stability, therefore, can be used to offer a protective microenvironment for enzymes. Enzyme is immobilized inside the inner surfaces of organ-like structured $Ti_3C_2T_x$ nanolayers and adjacent of TiO_2 that confirms stability and activity of enzyme for prolonged duration. As compared to biosensor Nafion/Hb/$Ti_3C_2T_x$/GCE where $Ti_3C_2T_x$ MXene acts as an immobilization matrix, Nafion/Hb/TiO_2–$Ti_3C_2T_x$/GCE biosensor has a quicker response (a response time of less than 3 seconds) along with a broader linear range from 0.1 to 380 μM. Furthermore, glucose is detected using a biosensor which consists of gold nanostructures in order to maintain the biological activity of an enzyme and decrease the shielding effect of a protein shell for direct electron transfer [191–197]. However, to increase the efficiency of enzymatic glucose biosensor, an MXene-Au nanocomposite system with similar structure to TiO_2–$Ti_3C_2T_x$ is employed as an immobilization matrix in a GOx/Au/MXene/Nafion/GC biosensor as described elsewhere [198]. It is remarkable that developed biosensor exhibits a wide range of detection for glucose having concentration from 0.1 mM to 18 mM, with lowest detection limit of 5.9 μM. Such nanocomposite was found to have good stability and excellent producibility. Gold NPs distributed onto surfaces of MXene nanosheets found to enhance electrical conductivity of MXene nanocomposite that will further highly endorse exchange of electrons between electrodes and enzymes. MXene composites and their desired properties needed for photothermal therapy are mentioned in Table 10.2. In each case, cells were cultured in the form of a monolayer and then irradiated using 808 nm laser.

Two-dimensional materials, that is, graphene and MoS2, are designated as biosensors with great biocompatibility towards analyte recognition. Therefore, MXene materials have also recently been analyzed for analyte recognition applications. MXene-based sensors and their properties are mentioned in Table 10.3.

10.3.8 ARTIFICIAL MUSCLES AND ACTUATORS

Generally, electroactive polymers are employed as artificial muscles and actuators. In such systems, stored electrical energy is changed in mechanical distortion. Basically, to work as an actuator, material must possess adequate bending strength, rapid response time, extended service life in air, and also have small driving voltage. Structure-wise, an actuator contains three basic components: one electrolyte and two metallic electrodes. However, ionic membrane derived from polymers can be applied as electrolyte by sandwiching it between two layers of conductive metal electrodes [213]. As it mentioned earlier, MXene possesses sufficient electrical conductivity, thus they can be employed as electrodes to develop actuators. Nevertheless, low stretchability of such

TABLE 10.2

MXenes Composites and Their Properties for Photothermal Therapy

MXene	Photothermal Conversion Efficiency η (%)	Extinction Coefficient (l.g⁻¹ cm⁻¹)	Laser Power Density for In Vitro PTT (W.cm⁻¹)	Concentration In Vitro (for PTT) (µg.ml⁻¹)	Laser Power Density for In Vivo PTT (W.cm⁻¹)	Administered Dose in Vivo (for PTT) (mg.kg⁻¹)	Blood Circulation Half-Life $T_{1/2}$ (h)	Imaging Technique	Ref
Ti_3C_2-PEG	-	29.1	1.0	50	1.5	-	-	-	199
Ti_3C_2-SP	30.6	25.2	1.0	100	1.5	20	0.76	-	200
MnO_x/Ti_3C_2-SP	22.9	5.0	1.0	160	1.5	2.5	-	MRI	184
Ti_3C_2 QDs	52.2	52.8	-	-	0.5	0.5	-	PA, fluorescence	201
Ta_4C_3-SP	44.7	4.06	1.0	100	1.5	20/4	1.59	PA, CT	202
MnO_x/Ta_4C_3-SP	34.9	8.67	2.0	200	2	20	-	PA, CT, MR	183
Ta_4C_3-IONP-SP	32.5	4.0	1.5	200 ppm	1.5	20	0.5	PA, CT, MR	203
GdW_{10}/Ti_3C_2	21.9	22.5	1.5	200 ppm	1.5	20	0.83	CT, MR	204
Nb_2C-PVP (laser 808 nm vs laser 1064 nm)	36.4	37.6	1.0	100	1.0	20	1.31	PA	187
	45.7	35.4	1.0		1.0				
Ti_3C_2-DOX-HA	58.3	28.6	1.5	100	0.8	2	-	Fluorescence	176
Ti_2C-PEG	87.1	7.39	1.5	62.5	-	-	-	-	205

Source: Reproduced with due permission [182].

TABLE 10.3

Ti$_3$C$_2$ MXene-Based Sensors and Their Properties

Analyte	Detection System	Response Time	Linear Range	Limit of Detection	Sensitivity	Ref.
Nitrite	Amperometric	<3 s	0.5-11800 μM	0.12 μM	-	189
H$_2$O$_2$	Voltammetric	<3 s	0.1-260 μM	20 nM	-	206
H$_2$O$_2$	Chrono amperometric	10 s	-	0.7 nM	596 mA.cm^{-2}. mM^{-1}	207
Glucose	Amperometric	10 s	0.1 – 18 mM	5.9 μM	4.2 μA. M^{-1}. cm^{-2}	198
Volatile organic compounds	Resistive	-	-	9.27 ppm	-	208
Single nucleotide	Electro-chemiluminescent	-	-	1 nM	-	209
Tripropylamine	Electro-chemiluminescent	-	1.0 × 10^{-8} to 1.0 × 10^{-3}	5 nM	-	
Dopamine	Conductometric	-	100 nM to 50 μM	100 nM	-	210
Human bending-release activities	Resistive	<10 ms	-	351 Pa	Gauge factor ~ 180.1	211
Humidity	Resistive, gravimetric	100 s	0-85% relative humidity	0.8% relative humidity	3%	212
Organophosphorus pesticides	Amperometric	-	1 × 10^{-14} to 1 × 10^{-8} M	0.3 × 10^{-14} M	-	

Source: Reproduced with due permission [182].

materials restricts their usability. In order to solve this limit, a few specific polymers, that is, polypropylene, are blended with MXenes, so that such polymer can form ionic bonds with the surface of MXene and initiate the intercalation process. Such nanocomposite systems show quick charge transport and capability of ion intercalation/de-intercalation, along with enhanced stretchability. These properties make ionically crosslinked Ti$_3$C$_2$/PP nanocomposites an extraordinary suitable material to be used to prepare electrodes for actuators. Hydrogen bonds can be created by polymers with −O− and −OH functional groups present onto the surface of Ti$_3$C$_2$ MXene, and works as a pillar to stop Ti$_3$C$_2$ restacking. However, this also eases electrons and ions reversible transportation between electrodes and electrolyte material.

10.4   CHALLENGES

The study of MXenes is still at an early stage as compared to other two-dimensional materials, that is, graphene. As usual, opportunities as well as challenges will coexist in future research of MXenes. At the part of synthesis, (i) several MXenes, for example, Sc$_2$C, Hf$_2$C, W$_2$C, and so on, are predicted to firmly exist. However, precursors for these materials are yet to be explored. Thus, the synthesis of novel MAX phases or other layered carbide as well as nitride precursors are going to be very important for increasing the MXene family. (ii) It is extremely important to design novel approaches for synthesizing MXenes of high quality along with greater lateral dimensions, less defects, and meticulous surface terminations. In specific, uniform terminations can be achieved with only a single type of functional group but it is one of the most challenging tasks for researchers. Eventually, MXenes without any surface terminations should be prepared through chemical vapor deposition (CVD) method or physical methods. At the part of property exploration, a maximum number of known properties are predicted by theoretical calculations and await investigational confirmation. Predominantly, semi-conductivity of MXenes has not been observed despite of theoretical predictions. Furthermore, as such, none of the reports have been received till date that show experimental measurements of band structures of MXenes that may be very significant to comprehend essential properties of MXenes. However, it is also required to realize the structures and characterizations of MXenes blended with different ions and compounds. At the part of end-use applications, even though MXenes exhibit better performance in several domains, various fundamental physical mechanisms are yet to be explored intensively. For instance, in-depth understanding of catalytic and electrocatalytic characteristics of different MXene-based nanocomposites is important to design novel catalysts of high-efficiency and also it can open new paths in the field of clean-energy conversion. Furthermore, working mechanism of high sensitivity of MXenes towards NH_3 remains indefinable, which restricts the applied applications of MXene-based gas sensors.

10.5 FUTURE ASPECTS

This chapter has reviewed the recent development in MXenes and MXene-based · polymer nanocomposites. The advancements in synthesis, properties, and applications of MXene-polymer nanocomposites are discussed. More than 20 various types of MXenes have already been explored in the last couple of years by choosing etching and exfoliation of layered ternary metal carbides, nitrides, or carbonitrides. Synthesis procedure leads to the surface modification of MXenes through different functional groups, that is, −O−, −F, and −OH, where characterization of such functional groups is subjective, as these properties depend upon temperature and etching time. Due to multipurpose surface properties of MXenes, they show several specific mechanical, electronic, magnetic, and electrochemical characteristics. Numerous novel quantum processes such as topological insulation and half-metallicity are also foreseen for a few MXenes and such materials are likable for two-dimensional spintronics. Various properties of MXenes, such as extraordinary flexibility, layered structures, and two-dimensional morphology, are appreciated because all these characteristics of MXenes allow it to form diverse nanocomposites with polymers and other materials. As mentioned earlier, MXene-based composite materials offer an efficient option to tailored the characteristics and performances of prepared MXenes materials by keeping their end-use application in mind.

Furthermore, MXenes and their composites possess excellent biocompatibility and good conductivity, which can immobilize the enzyme and may enable the transfer of electrons between electrode and enzyme in biosensors.

Lastly, associating the experimental and computational experiment results is recommended in order to further investigations of MXene materials. In view of various surface modifications/terminations and chemical compositions, there is a large range of MXene materials. However, previous predictions taken from computational data of characteristics and end-use application potentials of MXenes materials can direct the researchers to develop and explore the utmost promising MXene compounds, which shall significantly decrease the cost of research. Of late, machine learning has been investigated, which can precisely foresee the band gaps of MXene materials [214]. Therefore, it is rational to assume that the machine learning is going to contribute a substantial part in the prediction of properties of MXene materials in the foreseeable future.

REFERENCES

1. Handoko et al. Tuning the basal plane functionalization of two-dimensional metal carbides (MXenes) to control hydrogen evolution activity. *ACS Applied Energy Materials* 1.1 (2017): 173
2. Jun et al. Review of MXenes as new nanomaterials for energy storage/delivery and selected environmental applications. *Nano Research* 12.3 (2019): 471
3. Coleman et al. Two-dimensional nanosheets produced by liquid exfoliation of layered materials. *Science* 331.6017 (2011): 568
4. Anasori et al. 2D metal carbides and nitrides (MXenes) for energy storage. *Nature Reviews Materials* 2.2 (2017): 1
5. Ravikumar and Udayakumar. Preparation and characterisation of green clay-polymer nanocomposite for heavy metals removal. *Chemistry and Ecology* 36.3 (2020): 270
6. Kumar and Singh. Polypropylene clay nanocomposites. *Reviews in Chemical Engineering* 29.6 (2013): 439
7. Xia et al. The effect of temperature and graphene concentration on the electrical conductivity and dielectric permittivity of graphene–polymer nanocomposites. *Acta Mechanica* 231.4 (2020): 1305
8. Chen et al. Significantly improved breakdown strength and energy density of tri-layered polymer nanocomposites with optimized graphene oxide. *Composites Science and Technology* 186 (2020): 107912
9. Singh et al. Polyurethane Nanocomposites for Bone Tissue Engineering. *Engineered Nanomaterials for Innovative Therapies and Biomedicine.* Springer, Cham, 2022
10. Singh et al. Extrusion and evaluation of chitosan assisted AgNPs immobilized film derived from waste polyethylene terephthalate for food packaging applications. *Journal of Packaging Technology and Research* 1.3 (2017): 165
11. Singh et al. Microbial, physicochemical, and sensory analyses-based shelf-life appraisal of white fresh cheese packaged into PET waste-based active packaging film. *Journal of Packaging Technology and Research* 2.2 (2018): 125
12. Kumar et al. Incorporation of nano-Al_2O_3 within the blend of sulfonated-PVdF-*co*-HFP and Nafion for high temperature application in DMFCs. *RSC Advances* 5.78 (2015): 63465
13. Pattanayak et al. Performance evaluation of poly (aniline-co-pyrrole) wrapped titanium dioxide nanocomposite as an air-cathode catalyst material for microbial fuel cell. *Materials Science and Engineering: C* 118 (2021): 111492

14. Singh et al. Nanocellulose Biocomposites for Bone Tissue Engineering. *Handbook of Nanocelluloses: Classification, Properties, Fabrication, and Emerging Applications.* Springer International Publishing, Cham, 2021. 1
15. Singh et al. Surface Functionalizations of Nanocellulose for Wastewater Treatment. *Handbook of Nanocelluloses: Classification, Properties, Fabrication, and Emerging Applications.* Springer International Publishing, Cham, 2021. 1
16. Kuilla et al. Recent advances in graphene-based polymer composites. *Progress in Polymer Science* 35 (2010): 13501375
17. Mukhopadhyay et al. Trends and frontiers in graphene-based polymer nanocomposites. *Plastics engineering* 67.1 (2011): 32
18. Bera and Maji. Graphene-based polymer nanocomposites: materials for future revolution. *MOJ Polymer Science* 1.3 (2017): 00013
19. Silva et al. Graphene-polymer nanocomposites for biomedical applications. *Polymers for Advanced Technologies* 29.2 (2018): 687
20. Barroso-Bujans et al. Permanent adsorption of organic solvents in graphite oxide and its effect on the thermal exfoliation. *Carbon* 48.4 (2010): 1079
21. Guan et al. Three-dimensional graphene-based polymer nanocomposites: preparation, properties and applications. *Nanoscale* 10.31 (2018): 14788
22. Krishnan et al. Graphene-based polymer nanocomposites for sensor applications. *Hybrid Nanocomposites.* Jenny Stanford Publishing, 2019. 1
23. Ling et al. Flexible and conductive MXene films and nanocomposites with high capacitance. *Proceedings of the National Academy of Sciences* 111.47 (2014): 16676
24. Ghidiu et al. Conductive two-dimensional titanium carbide clay with high volumetric capacitance. *Nature* 516.7529 (2014): 78
25. Zhao et al. Highly electrically conductive three-dimensional $Ti_3C_2T_x$ MXene/reduced graphene oxide hybrid aerogels with excellent electromagnetic interference shielding performances. *ACS Nano* 12.11 (2018): 11193
26. Yin et al. Highly efficient catalytic performances of nitro compounds via hierarchical PdNPs-loaded MXene/polymer nanocomposites synthesized through electrospinning strategy for wastewater treatment. *Chinese Chemical Letters* 31.4 (2020): 992
27. Monastyreckis et al. Micromechanical modeling of MXene-polymer composites. *Carbon* 162 (2020): 402
28. Pan et al. Flammability, thermal stability and mechanical properties of polyvinyl alcohol nanocomposites reinforced with delaminated $Ti_3C_2T_x$ (MXene). *Polymer Composites* 41.1 (2020): 210
29. Wan et al. Enhanced dielectric properties of homogeneous $Ti_3C_2T_x$ MXene@ SiO_2/ polyvinyl alcohol composite films. *Ceramics International* 46.9 (2020): 13862
30. Zhang et al. A multidimensional nanostructural design towards electrochemically stable and mechanically strong hydrogel electrodes. *Nanoscale* 12.12 (2020): 6637
31. Xu et al. Lightweight Ti_2CT_x MXene/poly (vinyl alcohol) composite foams for electromagnetic wave shielding with absorption-dominated feature. *ACS Applied Materials & Interfaces* 11.10 (2019): 10198
32. Huang et al. Facile preparation of hierarchical AgNP-loaded MXene/Fe_3O_4/polymer nanocomposites by electrospinning with enhanced catalytic performance for wastewater treatment. *ACS Omega* 4.1 (2019): 1897
33. Mirkhani et al. High dielectric constant and low dielectric loss via poly (vinyl alcohol)/ $Ti_3C_2T_x$ MXene nanocomposites. *ACS Applied Materials & Interfaces* 11.20 (2019): 18599
34. Li et al. New application of MXene in polymer composites toward remarkable anti-dripping performance for flame retardancy. *Composites Part A: Applied Science and Manufacturing* 127 (2019): 105649
35. Habib et al. Heating of $Ti_3C_2T_x$ MXene/polymer composites in response to radio frequency fields. *Scientific Reports* 9.1 (2019): 1

36. Lipton et al. Mechanically strong and electrically conductive multilayer MXene nano-composites. *Nanoscale* 11.42 (2019): 20295

37. Habib et al. Oxidation stability of $Ti_3C_2T_x$ MXene nanosheets in solvents and composite films. *NPJ 2D Materials and Applications* 3.1 (2019): 1

38. Jin et al. Flame-retardant poly (vinyl alcohol)/MXene multilayered films with outstanding electromagnetic interference shielding and thermal conductive performances. *Chemical Engineering Journal* 380 (2020): 122475

39. Carey. On the synthesis & characterization of $Ti_3C_2T_x$ MXene polymer composites. *Dissertation*. Drexel University, 2017. https://core.ac.uk/download/pdf/190330627.pdf

40. Mayerberger et al. Antibacterial properties of electrospun $Ti_3C_2T_z$ (MXene)/chitosan nanofibers. *RSC Advances* 8.62 (2018): 35386

41. Zhang et al. Sulfonated $Ti_3C_2T_x$ to construct proton transfer pathways in polymer electrolyte membrane for enhanced conduction. *Solid State Ionics* 310 (2017): 100

42. Zhou et al. Flexible, robust, and multifunctional electromagnetic interference shielding film with alternating cellulose nanofiber and MXene layers. *ACS Applied Materials & Interfaces* 12.4 (2020): 4895

43. Cai et al. Leaf-inspired multiresponsive MXene-based actuator for programmable smart devices. *Science Advances* 5.7 (2019): eaaw7956

44. Etman et al. $Mo_{1.33}CTz–Ti_3C_2T_z$ mixed MXene freestanding films for zinc-ion hybrid supercapacitors. *Materials Today Energy* 22 (2021): 100878

45. Dong et al. Superlithiated polydopamine derivative for high-capacity and high-rate anode for lithium-ion batteries. *ACS Applied Materials & Interfaces* 10.44 (2018): 38101

46. Hu et al. Flexible and durable cellulose/MXene nanocomposite paper for efficient electromagnetic interference shielding. *Composites Science and Technology* 188 (2020): 107995

47. Shahzad et al. $Ti_3C_2T_x$ MXene core-shell spheres for ultrahigh removal of mercuric ions. *Chemical Engineering Journal* 368 (2019): 400

48. Li et al. Sandwich-structured ordered mesoporous polydopamine/MXene hybrids as high-performance anodes for lithium-ion batteries. *ACS Applied Materials & Interfaces* 12.13 (2020): 14993

49. Wang and Shi. Hydroxide conduction enhancement of chitosan membranes by functionalized MXene. *Materials* 11.11 (2018): 2335

50. Zhang et al. Preparation, mechanical and anti-friction performance of MXene/polymer composites. *Materials & Design* 92 (2016): 682

51. Cao et al. Non isothermal crystallization and thermal degradation kinetics of MXene/ linear low-density polyethylene nanocomposites. *e-Polymers* 17 (2017): 373

52. Carey et al. Dispersion and stabilization of alkylated 2D MXene in nonpolar solvents and their pseudocapacitive behavior. *Cell Reports Physical Science* 1.4 (2020): 100042

53. Ye et al. Six-arm star-shaped polymer with cyclophosphazene core and poly (ε-caprolactone) arms as modifier of epoxy thermosets. *Journal of Applied Polymer Science* 134.2 (2017)

54. Huang et al. Structure and crystallization behavior of poly (ethylene oxide)/$Ti_3C_2T_x$ MXene nanocomposites. *Polymer* 102 (2016): 119

55. Pan et al. 2D MXene-containing polymer electrolytes for all-solid-state lithium metal batteries. *Nanoscale Advances* 1.1 (2019): 395

56. Shi et al. Strengthening, toughing and thermally stable ultra-thin MXene nanosheets/ polypropylene nanocomposites via nanoconfinement. *Chemical Engineering Journal* 378 (2019): 122267

57. Si et al. Functionalization of MXene nanosheets for polystyrene towards high thermal stability and flame retardant properties. *Polymers* 11.6 (2019): 976

58. Carey et al. Nylon-6/Ti$_3$C$_2$T$_z$ MXene nanocomposites synthesized by in situ ring open-ing polymerization of ε-caprolactam and their water transport properties. *ACS Applied Materials & Interfaces* 11.22 (2019): 20425
59. Tanvir et al. Electrically conductive, transparent polymeric nanocomposites modified by 2D Ti$_3$C$_2$T$_x$ (MXene). *Polymers* 11.8 (2019): 1272
60. Hao et al. Novel thin-film nanocomposite membranes filled with multi-functional Ti$_3$C$_2$T$_x$ nanosheets for task-specific solvent transport. *Composites Part A: Applied Science and Manufacturing* 100 (2017): 139
61. Liu et al. Multifunctional, superelastic, and lightweight MXene/polyimide aerogels. *Small* 14.45 (2018): 1802479
62. Li et al. Self-assembled MXene-based nanocomposites via layer-by-layer strategy for elevated adsorption capacities. *Colloids and Surfaces A: Physicochemical and Engineering Aspects* 553 (2018): 105
63. Han et al. Crosslinked P84 copolyimide/MXene mixed matrix membrane with excel-lent solvent resistance and permselectivity. *Chinese Journal of Chemical Engineering* 27.4 (2019): 877
64. Wang et al. 3D Ti$_3$C$_2$T$_x$ MXene/C hybrid foam/epoxy nanocomposites with supe-rior electromagnetic interference shielding performances and robust mechani-cal properties. *Composites Part A: Applied Science and Manufacturing* 123 (2019): 293
65. Wang et al. Robust, lightweight, hydrophobic, and fire-retarded polyimide/MXene aerogels for effective oil/water separation. *ACS Applied Materials & Interfaces* 11.43 (2019): 40512
66. Sobolčiak et al. 2D Ti$_3$C$_2$T$_x$ (MXene)-reinforced polyvinyl alcohol (PVA) nanofi-bers with enhanced mechanical and electrical properties. *PLoS One* 12.8 (2017): e0183705
67. Zhang et al. Fabrication of novel MXene (Ti$_3$C$_2$)/polyacrylamide nanocomposite hydrogels with enhanced mechanical and drug release properties. *Soft Matter* 16.1 (2020): 162
68. Liao et al. Conductive MXene nanocomposite organohydrogel for flexible, healable, low-temperature tolerant strain sensors. *Advanced Functional Materials* 29.39 (2019): 1904507
69. Wu et al. A wearable, self-adhesive, long-lastingly moist and healable epidermal sensor assembled from conductive MXene nanocomposites. *Journal of Materials Chemistry C* 8.5 (2020): 1788
70. Tang et al. Highly conducting MXene–silver nanowire transparent electrodes for flex-ible organic solar cells. *ACS Applied Materials & Interfaces* 11.28 (2019): 25330
71. Shao et al. A novel high permittivity percolative composite with modified MXene. *Polymer* 174 (2019): 86
72. Yue et al. Highly self-healable 3D microsupercapacitor with MXene–graphene com-posite aerogel. *ACS Nano* 12.5 (2018): 4224
73. Fan et al. Plasmonic Ti$_3$C$_2$T$_x$ MXene enables highly efficient photothermal conversion for healable and transparent wearable device. *ACS Nano* 13.7 (2019): 8124
74. Sheng et al. Properties of two-dimensional Ti$_3$C$_2$ MXene/thermoplastic polyurethane nanocomposites with effective reinforcement via melt blending. *Composites Science and Technology* 181 (2019): 107710
75. He et al. Large-scale production of simultaneously exfoliated and functional-ized MXenes as promising flame retardant for polyurethane. *Composites Part B: Engineering* 179 (2019): 107486
76. Seyedin et al. MXene composite and coaxial fibers with high stretchability and con-ductivity for wearable strain sensing textiles. *Advanced Functional Materials* 30.12 (2020): 1910504

77. Singh et al. Fabrication of calcium hydroxyapatite incorporated polyurethane-graphene oxide nanocomposite porous scaffolds from poly (ethylene terephthalate) waste: A green route toward bone tissue engineering. *Polymer* 195 (2020): 122436
78. An et al. Surface-agnostic highly stretchable and bendable conductive MXene multilayers. *Science advances* 4.3 (2018): eaaq0118
79. Ab et al. Highly flexible and stretchable 3D graphene/MXene composite thin film. *Materials Today: Proceedings* 7 (2019): 738
80. Wei et al. Enhanced dielectric properties of a poly (dimethyl siloxane) bimodal network percolative composite with MXene. *ACS Applied Materials & Interfaces* 12.14 (2020): 16805
81. Wu et al. Compressible, durable and conductive polydimethylsiloxane-coated MXene foams for high-performance electromagnetic interference shielding. *Chemical Engineering Journal* 381 (2020): 122622
82. Wang et al. Multifunctional 3D-MXene/PDMS nanocomposites for electrical, thermal and triboelectric applications. *Compos. Part A Appl. Sci. Manuf* 130 (2020): 105754
83. Sliozberg et al. Interface binding and mechanical properties of MXene-epoxy nanocomposites. *Composites Science and Technology* 192 (2020): 108124
84. Zhang et al. Effects of 2-D transition metal carbide Ti_2CT_x on properties of epoxy composites. *RSC Advances* 6.90 (2016): 87341
85. Fang et al. Thermally-induced self-healing behaviors and properties of four epoxy coatings with different network architectures. *Polymers* 9.8 (2017): 333
86. Carey et al. Water transport and thermomechanical properties of $Ti_3C_2T_z$ MXene epoxy nanocomposites. *ACS Applied Materials & Interfaces* 11 (2019): 39143
87. Zhao et al. Preparation and mechanical performances of carbon fiber reinforced epoxy composites by MXene nanosheets coating. *Journal of Materials Science: Materials in Electronics* 30.11 (2019): 10516
88. Kilikevičius et al. Numerical investigation of the mechanical properties of a novel hybrid polymer composite reinforced with graphene and MXene nanosheets. *Computational Materials Science* 174 (2020): 109497
89. Hatter et al. Micromechanical response of two-dimensional transition metal carbonitride (MXene) reinforced epoxy composites. *Composites Part B: Engineering* 182 (2020): 107603
90. Feng et al. Preparation and characterization of epoxy resin filled with $Ti_3C_2T_x$ MXene nanosheets with excellent electric conductivity. *Nanomaterials* 10.1 (2020): 162
91. Wu et al. Highly flexible and low capacitance loss supercapacitor electrode based on hybridizing decentralized conjugated polymer chains with MXene. *Chemical Engineering Journal* 378 (2019): 122246
92. Levitt et al. Electrospun MXene/carbon nanofibers as supercapacitor electrodes. *Journal of Materials Chemistry A* 7.1 (2019): 269
93. Yu et al. MXenes with tunable work functions and their application as electron- and hole-transport materials in non-fullerene organic solar cells. *Journal of Materials Chemistry A* 7.18 (2019): 11160
94. Taloub et al. Improving the mechanical properties, UV and hydrothermal aging resistance of PIPD fiber using MXene ($Ti_3C_2(OH)_2$) nanosheets. *Composites Part B: Engineering* 163 (2019): 260
95. Derradji et al. On the preparation and properties investigations of highly performant MXene ($Ti_3C_2(OH)_2$) nanosheets-reinforced phthalonitrile nanocomposites. *Advanced Composites Letters* 28 (2019): 2633366X19890621
96. Lim et al. Stable colloidal dispersion of octylated Ti_3C_2-MXenes in a nonpolar solvent. *Colloids and Surfaces A: Physicochemical and Engineering Aspects* 579 (2019): 123648

97. Luo et al. Flexible, stretchable and electrically conductive MXene/natural rubber nanocomposite films for efficient electromagnetic interference shielding. *Composites Science and Technology* 182 (2019): 107754

98. Liu et al. Two-dimensional Ti_2CT_x MXene membranes with integrated and ordered nanochannels for efficient solvent dehydration. *Journal of Materials Chemistry A* 7.19 (2019): 12095

99. Raagulan et al. An effective utilization of MXene and its effect on electromagnetic interference shielding: flexible, free-standing and thermally conductive composite from MXene–PAT–poly (p-aminophenol)–polyaniline co-polymer. *RSC Advances* 10.3 (2020): 1613

100. Liu et al. Polyelectrolyte functionalized Ti_2CT_x MXene membranes for pervaporation dehydration of isopropanol/water mixtures. *Industrial & Engineering Chemistry Research* 59.10 (2020): 4732

101. Shi et al. MXene versus graphene oxide: Investigation on the effects of 2D nanosheets in mixed matrix membranes for CO_2 separation. *Journal of Membrane Science* 620 (2021): 118850

102. Mazhar et al. Promising PVC/MXene based flexible thin film nanocomposites with excellent dielectric, thermal and mechanical properties. *Ceramics International* 46.8 (2020): 12593

103. Shamsabadi et al. Pushing rubbery polymer membranes to be economic for CO_2 separation: Embedment with $Ti_3C_2T_x$ MXene Nanosheets. *ACS Applied Materials & Interfaces* 12.3 (2019): 3984

104. Ajnsztajn et al. Transparent MXene-polymer supercapacitive film deposited using RIR-MAPLE. *Crystals* 10.3 (2020): 152

105. Liu et al. $Ti_3C_2T_x$ filler effect on the proton conduction property of polymer electrolyte membrane. *ACS Applied Materials & Interfaces* 8.31 (2016): 20352

106. Boota et al. Interaction of polar and nonpolar polyfluorenes with layers of two-dimensional titanium carbide (MXene): intercalation and pseudocapacitance. *Chemistry of Materials* 29.7 (2017): 2731

107. Xu et al. A MXene based all-solid-state microsupercapacitor with 3D interdigital electrode. *19th International Conference on Solid-State Sensors, Actuators and Microsystems (TRANSDUCERS)*. IEEE, 2017

108. Zhang et al. Quaternary $Ti_3C_2T_x$ enhanced ionic conduction in quaternized polysulfone membrane for alkaline anion exchange membrane fuel cells. *Journal of Membrane Science* 563 (2018): 882

109. Zhou et al. Layer-by-layer assembly of MXene and carbon nanotubes on electrospun polymer films for flexible energy storage. *Nanoscale* 10.13 (2018): 6005

110. Fei et al. Polybenzimidazole/MXene composite membranes for intermediate temperature polymer electrolyte membrane fuel cells. *Nanotechnology* 29.3 (2017): 035403

111. Vahid et al. Thick and freestanding MXene/PANI pseudocapacitive electrodes with ultrahigh specific capacitance. *Journal of Materials Chemistry A* 6.44 (2018): 22123

112. Ren et al. Synthesis of polyaniline nanoparticles deposited on two-dimensional titanium carbide for high-performance supercapacitors. *Materials Letters* 214 (2018): 84

113. Wei et al. $Ti_3C_2T_x$ MXene/polyaniline (PANI) sandwich intercalation structure composites constructed for microwave absorption. *Composites Science and Technology* 169 (2019): 52

114. Boota and Gogotsi. MXene-conducting polymer asymmetric pseudocapacitors. *Advanced Energy Materials* 9.7 (2019): 1802917

115. Cheng and Yang. Preparation of Ti_3C_2-PANI composite as sensor for electrochemical determination of mercury ions in water. *International Journal of Electrochemical Science* 15 (2020): 2295

116. Li et al. An ultrafast conducting polymer@MXene positive electrode with high volumetric capacitance for advanced asymmetric supercapacitors. *Small* 16.4 (2020): 1906851

117. Xu et al. Synthesis of an MXene/polyaniline composite with excellent electrochemical properties. *Journal of Materials Chemistry A* 8.12 (2020): 5853

118. Seroka and Messai. Application of functionalised MXene-carbon nanoparticle-polymer composites in resistive hydrostatic pressure sensors. *SN Applied Sciences* 2.3 (2020): 1

119. Zhang et al. Planar supercapacitor with high areal capacitance based on Ti_3C_2/Polypyrrole composite film. *Electrochimica Acta* 330 (2020): 135277

120. Zhu et al. Highly flexible, freestanding supercapacitor electrode with enhanced performance obtained by hybridizing polypyrrole chains with MXene. *Advanced Energy Materials* 6.21 (2016): 1600969

121. Boota et al. Pseudocapacitive electrodes produced by oxidant-free polymerization of pyrrole between the layers of 2D titanium carbide (MXene). *Advanced Materials* 28.7 (2016): 1517

122. Jian et al. Three-dimensional carambola-like MXene/polypyrrole composite produced by one-step co-electrodeposition method for electrochemical energy storage. *Electrochimica Acta* 318 (2019): 820

123. Wu et al. Enhanced electrochemical performances of organ-like Ti_3C_2 MXenes/polypyrrole composites as supercapacitors electrode materials. *Ceramics International* 45.6 (2019): 7328

124. Le et al. Intertwined titanium carbide MXene within a 3D tangled polypyrrole nanowires matrix for enhanced supercapacitor performances. *Chemistry–A European Journal* 25.4 (2019): 1037

125. Ma et al. Hierarchical porous MXene/amino carbon nanotubes-based molecular imprinting sensor for highly sensitive and selective sensing of fisetin. *Sensors and Actuators B: Chemical* 309 (2020): 127815

126. Chen et al. Charge transfer induced polymerization of EDOT confined between 2D titanium carbide layers. *Journal of Materials Chemistry A* 5.11 (2017): 5260

127. Qin et al. High-performance ultrathin flexible solid-state supercapacitors based on solution processable $Mo_{1.33}C$ MXene and PEDOT: PSS. *Advanced Functional Materials* 28.2 (2018): 1703808

128. Li et al. MXene-conducting polymer electrochromic microsupercapacitors. *Energy Storage Materials* 20 (2019): 455

129. Mannayil et al. Solution processable PEDOT:PSS/multiwalled carbon nanotube composite films for flexible electrode applications. *Phys. Status Solidi* 215 (2018): 201701003

130. Hou et al. Solution-processable $Ti_3C_2T_x$ nanosheets as an efficient hole transport layer for high-performance and stable polymer solar cells. *Journal of Materials Chemistry C* 7.37 (2019): 11549

131. Gund et al. MXene/polymer hybrid materials for flexible AC-filtering electrochemical capacitors. *Joule* 3.1 (2019): 164

132. Wan et al. Lightweight, flexible MXene/polymer film with simultaneously excellent mechanical property and high-performance electromagnetic interference shielding. *Composites Part A: Applied Science and Manufacturing* 130 (2020): 105764

133. Guan et al. Significant enhancement in the seebeck coefficient and power factor of p-type poly (3, 4-ethylenedioxythiophene): poly (styrenesulfonate) through the incorporation of n-type MXene. *ACS Applied Materials & Interfaces* 12.11 (2020): 13013

134. Feng et al. An ultrahigh discharged energy density achieved in an inhomogeneous PVDF dielectric composite filled with 2D MXene nanosheets via interface engineering. *Journal of Materials Chemistry C* 6.48 (2018): 13283

135. Feng et al. High dielectric and breakdown properties achieved in ternary BaTiO$_3$/ MXene/PVDF nanocomposites with low-concentration fillers from enhanced interface polarization. *Ceramics International* 45.6 (2019): 7923

136. Deng et al. Strong interface effect induced high-k property in polymer based ternary composites filled with 2D layered Ti$_3$C$_2$ MXene nanosheets. *Journal of Materials Science: Materials in Electronics* 30.10 (2019): 9106

137. Feng et al. High dielectric and breakdown properties obtained in a PVDF based nano- composite with sandwich structure at high temperature via all-2D design. *Journal of Materials Chemistry C* 7.22 (2019): 6744

138. Li et al. Multilayer-structured transparent MXene/PVDF film with excellent dielectric and energy storage performance. *Journal of Materials Chemistry C* 7.33 (2019): 10371

139. Seyedin et al. Facile solution processing of stable MXene dispersions towards conduc- tive composite fibers. *Global Challenges* 3.10 (2019): 1900037

140. Rajavel et al. 2D Ti$_3$C$_2$T$_x$ MXene/polyvinylidene fluoride (PVDF) nanocomposites for attenuation of electromagnetic radiation with excellent heat dissipation. *Composites Part A: Applied Science and Manufacturing* 129 (2020): 105693

141. Li et al. Hydrophobic and stable MXene-polymer pressure sensors for wearable elec- tronics. *ACS Applied Materials & Interfaces* 12.13 (2020): 15362

142. Tu et al. Large dielectric constant enhancement in MXene percolative polymer compos- ites. *ACS Nano* 12.4 (2018): 3369

143. Tu et al. Solid state MXene based electrostatic fractional capacitors. *Applied Physics Letters* 114.23 (2019): 232903

144. Tu et al. Enhancement of dielectric permittivity of Ti$_3$C$_2$T$_x$ MXene/polymer composites by controlling flake size and surface termination. *ACS Applied Materials & Interfaces* 11.30 (2019): 27358

145. Singh et al. Recent innovations in chemical recycling of polyethylene terephthal- ate waste: A circular economy approach toward sustainability. *Handbook of Solid Waste Management: Sustainability through Circular Economy*. Springer Singapore, Singapore, 2022. 1149

146. Singh et al. Adaptation of 3D printing techniques in bone tissue engineering: an assess- ment of its need, reliability, validity, sustainability, and future scope. *3D Printing Technology*. CRI, New Delhi, 2018.

147. Gong et al. Optimizing the reinforcement of polymer-based nanocomposites by gra- phene. *ACS Nano* 6.3 (2012): 2086

148. Hao et al. High-performance solar-driven interfacial evaporation through molecular design of antibacterial, biomass-derived hydrogels. *Journal of Colloid and Interface Science* 608 (2022): 840

149. Hou and Yu. Modifying the nanostructures of PEDOT: PSS/Ti$_3$C$_2$T$_X$ composite hole transport layers for highly efficient polymer solar cells. *Journal of Materials Chemistry C* 8.12 (2020): 4169

150. Liu et al. Mo$_{1.33}$C MXene-assisted PEDOT: PSS hole transport layer for high-performance bulk-heterojunction polymer solar cells. *ACS Applied Electronic Materials* 2.1 (2019): 163

151. Liu et al. Pebax-based membrane filled with two-dimensional MXene nanosheets for efficient CO$_2$ Capture. *Chemistry-An Asian Journal* 15.15 (2020): 2364

152. Shi et al. Exploration of the synergy between 2D nanosheets and a non-2D filler in mixed matrix membranes for gas separation. *Frontiers in Chemistry* 8 (2020): 58

153. Zou et al. Near-infrared light and solar light activated self-healing epoxy coating hav- ing enhanced properties using MXene flakes as multifunctional fillers. *Polymers* 10.5 (2018): 474

154. Jay Shah. Synthesis of MXene-Epoxy Nanocomposites. Drexel University ProQuest Dissertations Publishing, 2017. 10601635 https://www.proquest.com/openview/5b5d1c 14ccfa0ebc471c8eea8c0439fe/1?pq-origsite=gscholar&cbl=18750

155. Mao et al. MXene quantum dot/polymer hybrid structures with tunable electrical conductance and resistive switching for non-volatile memory devices. *Advanced Electronic Materials* 6.1 (2020): 1900493
156. Kim et al. Graphene/polymer nanocomposites. *Macromolecules* 43.16 (2010): 6515
157. Rahmathullah et al. Room temperature ionic liquids as thermally latent initiators for polymerization of epoxy resins. *Macromolecules* 42.9 (2009): 3219
158. Halim et al. Transparent conductive two-dimensional titanium carbide epitaxial thin films. *Chemistry of Materials* 26.7 (2014): 2374
159. Aghamohammadi et al. Recent advances in developing the MXene/polymer nanocomposites with multiple properties: A review study. *Synthetic Metals* 273 (2021): 116695
160. Meng et al. A review of recent advances in tribology. *Friction* 8.2 (2020): 221–300
161. Yin et al. Spheroidization behaviour of a Fe-enriched eutectic high-entropy alloy. *Journal of Materials Science & Technology* 51 (2020): 173
162. Yan et al. Towards high-performance additive of Ti_3C_2/graphene hybrid with a novel wrapping structure in epoxy coating. *Carbon* 157 (2020): 217
163. Kang et al. Enhanced thermal conductivity of epoxy composites filled with 2D transition metal carbides (MXenes) with ultralow loading. *Scientific Reports* 9 (2019): 9135
164. Yan et al. Ti_3C_2 MXene nanosheets toward high-performance corrosion inhibitor for epoxy coating. *Progress in Organic Coatings* 135 (2019): 156
165. Wang et al. Fabrication on the annealed $Ti_3C_2T_x$ MXene/Epoxy nanocomposites for electromagnetic interference shielding application. *Composites Part B: Engineering* 171 (2019): 111
166. Song et al. Honeycomb structural rGO-MXene/epoxy nanocomposites for superior electromagnetic interference shielding performance. *Sustainable Materials and Technologies* 24 (2020): e00153
167. Qin et al. Two-dimensional transition metal carbides and/or nitrides (MXenes) and their applications in sensors. *Materials Today Physics* 21 (2021): 100527
168. Zhang et al. Natural product interventions for chemotherapy and radiotherapy-induced side effects. *Frontiers in Pharmacology* 9 (2018): 1253
169. Han et al. Applications of nanoparticles in biomedical imaging. *Nanoscale* 11.3 (2019): 799
170. George and Balasubramanian. Advancements in MXene-Polymer composites for various biomedical applications. *Ceramics International* 46.7 (2020): 8522
171. Chen et al. Strong and biocompatible poly (lactic acid) membrane enhanced by $Ti_3C_2T_z$ (MXene) nanosheets for guided bone regeneration. *Materials Letters* 229 (2018): 114
172. Sur et al. Recent developments in functionalized polymer nanoparticles for efficient drug delivery system. *Nano-Structures & Nano-Objects* 20 (2019): 100397
173. Ambekar and Balasubramanian. A polydopamine-based platform for anti-cancer drug delivery. *Biomaterials Science* 7.5 (2019): 1776–1793
174. Ambekar et al. Recent advances in dendrimer-based nanoplatform for cancer treatment: A review. *European Polymer Journal* 126 (2020): 109546
175. Xing et al. Two-Dimensional MXene (Ti_3C_2)-integrated cellulose hydrogels: toward smart three-dimensional network nanoplatforms exhibiting light-induce swelling and bimodal photothermal/chemotherapy. *Anticancer Activity* 10.33 (2018): 27631
176. Liu et al. Surface modified Ti_3C_2 MXene nanosheets for tumor targeting photothermal/photodynamic/chemo synergistic therapy. *ACS Applied Materials & Interfaces* 9.46 (2017): 40077
177. Yadav and Balasubramanian. Polyacrylonitrile/syzygium aromaticum hierarchical hydrophilic nanocomposite as a carrier for antibacterial drug delivery systems. *RSC Advances* 5.5 (2015): 3291
178. Yadav and Balasubramanian. Bioabsorbable engineered nanobiomaterials for antibacterial therapy. *Engineering of Nanobiomaterials*. William Andrew Publishing, 2016. 77

179. Rasool et al. Efficient antibacterial membrane based on two-dimensional $Ti_3C_2T_x$ (MXene) nanosheets. *Scientific Reports* 7 (2017): 1598
180. Rasool et al. Antibacterial Activity of $Ti_3C_2T_x$ MXene. *ACS Nano* 10.3 (2016): 3674
181. Prakash and Balasubramanian. Nanocomposites of MXene for industrial applications. *Journal of Alloys and Compounds* 862 (2021): 158547
182. Szuplewska et al. Future Applications of MXenes in Biotechnology, Nanomedicine, and Sensors. *Trends in Biotechnology* 38 (2020): 264
183. Na et al. Development of a T1 contrast agent for magnetic resonance imaging using MnO nanoparticles. *Angewandte Chemie* 119.28 (2007): 5493
184. Dai et al. Biocompatible 2D titanium carbide (MXenes) composite nanosheets for pH-responsive MRI-guided tumor hyperthermia. *Chemistry of Materials* 29.20 (2017): 8637
185. Piver et al. Sequential therapy for advanced ovarian adenocarcinoma: Operation, chemotherapy, second-look laparotomy, and radiation therapy. *American Journal of Obstetrics and Gynecology* 122.3 (1975): 355
186. Bashkatov et al. Optical properties of human skin, subcutaneous and mucous tissues in the wavelength range from 400 to 2000 nm. *Journal of Physics D: Applied Physics* 38.15 (2005): 2543
187. Lin et al. A two-dimensional biodegradable niobium carbide (MXene) for photothermal tumor eradication in NIR-I and NIR-II biowindows. *Journal of the American Chemical Society* 139.45 (2017): 16235
188. Guo et al. Protein-inspired self-healable Ti_3C_2 MXenes/rubber-based supramolecular elastomer for intelligent sensing. *ACS Nano* 14.3 (2020): 2788
189. Liu et al. A novel nitrite biosensor based on the direct electrochemistry of hemoglobin immobilized on MXene-Ti_3C_2. *Sensors and Actuators B: Chemical* 218 (2015): 60
190. Wang et al. TiO_2 nanoparticle modified organ-like Ti_3C_2 MXene nanocomposite encapsulating hemoglobin for a mediator-free biosensor with excellent performances. *Biosensors and Bioelectronics* 74 (2015): 1022
191. Li et al. Phase and composition controllable synthesis of cobalt manganese spinel nanoparticles towards efficient oxygen electrocatalysis. *Nature Communications* 6.1 (2015): 1
192. Samphao et al. Flow-injection amperometric determination of glucose using a biosensor based on immobilization of glucose oxidase onto Au seeds decorated on core Fe_3O_4 nanoparticles. *Talanta* 142 (2015): 35
193. Devasenathipathy et al. Glucose biosensor based on glucose oxidase immobilized at gold nanoparticles decorated graphene-carbon nanotubes. *Enzyme and Microbial Technology* 78 (2015): 40
194. Xu et al. Graphene/polyaniline/gold nanoparticles nanocomposite for the direct electron transfer of glucose oxidase and glucose biosensing. *Sensors and Actuators B: Chemical* 190 (2014): 562
195. Chowdhury et al. Highly sensitive electrochemical biosensor for glucose, DNA and protein using gold-polyaniline nanocomposites as a common matrix. *Sensors and Actuators B: Chemical* 190 (2014): 348
196. Zhong et al. A nonenzymatic amperometric glucose sensor based on three-dimensional nanostructure gold electrode. *Sensors and Actuators B: Chemical* 212 (2015): 72
197. Wang et al. Preparation of gold nanowires and its application in glucose biosensing. *Materials Letters* 91 (2013): 9
198. Rakhi et al. Novel amperometric glucose biosensor based on MXene nanocomposite. *Scientific Reports* 6.1 (2016): 1
199. Xuan et al. Delamination for the production of functionalized titanium carbide nanosheets with superior photothermal therapeutic performance. *Angewandte Chemie International Edition* 55 (2016): 14569

200. Lin et al. Two-dimensional ultrathin MXene ceramic nanosheets for photothermal conversion. *Nano Letters* 17.1 (2017): 384

201. Yu et al. Fluorine-free preparation of titanium carbide MXene quantum dots with high near-infrared photothermal performances for cancer therapy. *Nanoscale* 9.45 (2017): 17859

202. Lin et al. Theranostic 2D tantalum carbide (MXene). *Advanced Materials* 30.4 (2018): 1703284

203. Liu et al. 2D superparamagnetic tantalum carbide composite MXenes for efficient breast-cancer theranostics. *Theranostics* 8.6 (2018): 1648

204. Zong et al. A polyoxometalate-functionalized two-dimensional titanium carbide composite MXene for effective cancer theranostics. *Nano Research* 11.8 (2018): 4149

205. Szuplewska et al. 2D Ti_2C (MXene) as a novel highly efficient and selective agent for photothermal therapy. *Materials Science & Engineering C* 2 (2019): 874

206. Wang et al. An organ-like titanium carbide material (MXene) with multilayer structure encapsulating hemoglobin for a mediator-free biosensor. *Journal of The Electrochemical Society* 162.1 (2014): B16

207. Lorencova et al. Electrochemical performance of $Ti_3C_2T_x$ MXene in aqueous media: Towards ultrasensitive H_2O_2 sensing. *Electrochimica Acta* 235 (2017): 471

208. Lee et al. Room temperature gas sensing of two-dimensional titanium carbide (MXene). *ACS Applied Materials & Interfaces* 9.42 (2017): 37184

209. Fang et al. Two-dimensional titanium carbide (MXene)-based solid-state electrochemiluminescent sensor for label-free single-nucleotide mismatch discrimination in human urine. *Sensors and Actuators B: Chemical* 263 (2018): 400

210. Xu et al. Field-effect transistors: Ultrathin MXene-micropattern-based field-effect transistor for probing neural activity. *Advanced Materials* 28.17 (2016): 3411

211. Ma et al. A highly flexible and sensitive piezoresistive sensor based on MXene with greatly changed interlayer distances. *Nature Communications* 8.1207 (2017)

212. Zhou et al. Acetylcholinesterase/chitosan-transition metal carbides nanocomposites-based biosensor for the organophosphate pesticides detection. *Biochemical Engineering Journal* 128 (2017): 243

213. Umrao et al. MXene artificial muscles based on ionically cross-linked $Ti_3C_2T_x$ electrode for kinetic soft robotics. *Science Robotics* 4.33 (2019): eaaw7797

214. Zhang et al. Ultrathin nanosheets of MAX phases with enhanced thermal and mechanical properties in polymeric compositions: $Ti_3Si_{0.75}Al_{0.25}C^2$. *Angewandte Chemie* 125.16 (2013): 4457

11 Role of MXene/ Rubber Composites in EMI Shielding

Suman Kumar Ghosh, Krishnendu Nath, and Narayan Chandra Das
Rubber Technology Centre
Indian Institute of Technology
Kharagpur, India

CONTENTS

11.1 INTRODUCTION

In recent years, extensive utilization of communication and information technologically derived electronic and electrical devices causes a serious environmental hazardous effect in terms of electromagnetic interference (EMI) and electromagnetic pollution. These electronic gadgets and electrical equipment emit unwanted harmful electromagnetic radiation which not only affects normal functions of nearby electronic devices adversely but also hampers human life. To overcome this serious issue, many strategies have been adopted to invent highly efficient electromagnetic shielding materials which can attenuate this deleterious EM radiation [1–4]. For a long period, metals have served as electromagnetic shielding material due to their outstanding electrical conductivity. But metals go through some demerits, such as heavy weight, low flexibility, corrosiveness, lower chemical resistance, poor lasting tenure and poor processability. Some ceramic materials such as SiC and Si_3N_4 can be effective alternatives as EMI shields and microwave-absorbing materials. These materials possess multi-scale tunable microstructure and suitable fabrication methods. In this regard, polymer-based shielding material plays a crucial role [5–7]. Conducting polymers are often used for EMI shielding although they possess some

DOI: 10.1201/9781003281511-11

disadvantages like poor processability and mechanical properties; therefore, they are exploited less. In this context, conducting filler containing polymer composites must be utilized to develop commercially viable shielding material. Conducting carbonaceous filler particles are often used in synthesizing conducting polymer composites to achieve attractive electrical properties and good EMI shielding efficiency. Compared to conventional micro and macro filler like carbon fibre (CF), fillers with nanoscale dimensions are more effective in achieving better overall properties and EMI SE due to better polymer-filler interactions. In recent years, various carbonaceous nanofillers such as conducting carbon black (CB), carbon nanofiber (CNF), graphene oxide (GO), carbon nanotube (CNT) and graphene nanoplatelets (GNP) are extensively used as conducting fillers to develop polymer composite-based efficient shielding material. These conductive carbonaceous inclusions-based polymer composites exhibit high flexibility and efficient EMI shielding effectiveness [8–12]. But it is undesirable to achieve superb EMI shielding effectiveness of these polymer composites at ultrathin thickness. Superior EMI SE at the desired thickness of polymer nanocomposites is still a significant challenge.

Very recently, MXene, a new class of inorganic two-dimensional (2D) materials provides sufficient flexibility, superior metallic electrical conductivity and EMI SE when combined with polymer materials. The resultant highly conductive composites are therefore considered as the promising candidates for the next generation EMI shields. These MXene materials generally include transition metal carbides, nitrides and carbonitrides and have a general structural formula of $M_{n+1}X_nT_x$, where M and X denote the transition metal and carbon and/or nitrogen respectively and T_x represents the functionality of surface terminating groups, for example, O, OH and F. These layered structures materials exhibit remarkable metallic electrical conductivity and rich surface chemistry which contribute to efficient EMI SE when they are combined with polymer macromolecules at very low thickness [13, 14]. The most common and widely used MXene to prepare highly flexible, superior conductive polymer composites for efficient EMI shielding application is $Ti_3C_2T_x$. Few studies have been done on the EMI SE of these MXene filled polymer composites. Zhang et al. investigated EMI SE of $Ti_3C_2T_x$/PANI composites and reported EMI SE of 36 dB with a sample thickness of 40 μm [15]. Wang et al. fabricated an excellent EMI shielding material consisting of polypyrrole modified MXene/poly (ethylene terephthalate) textile with exceptional EMI SE of 90 dB at a thickness of 1.3 mm [16]. Types of conductive structure formed by MXene in polymer matrix plays an important role in achieving conductivity and hence EMI SE. Sun et al. reported percolation threshold of 0.26 vol% and EMI SE of 54 dB of $Ti_3C_2T_x$/PS nanocomposites with a low volume of MXene (1.90 vol%) by formation of segregated conductive networking structure. Ultra-thin composite film of $Ti_3C_2T_x$/PEDOT:PSS flexible and lightweight composite with superior EMI shielding effectiveness and having brick-and-mortar structure was prepared by Liu et al. [17]. MXene/polymer composites with multi-interface structure are very promising for highly efficient EMI properties. Some researchers investigated the structure-property relationship in MXene contained polymer composite for improving EMI shielding effectiveness and other performance properties.

Rubbers are extensively used to fabricate light weight, highly flexible and efficient EMI shielding material by combining with various carbonaceous fillers. Currently,

instead of carbonaceous inclusions, MXene is combined with rubber materials for efficient EMI shielding applications and stretchable electronics. Luo et al. investigated
EMI SE of $Ti_3C_2T_x$ MXene/natural rubber (NR) composite film by vacuum-assisted
filtration method and achieved EMI SE of 54 dB at an ultra-low thickness of
240 μm [18]. In another work, Wang et al. assembled MXene-based rubber composite with brick-mortar structure by utilizing NR as soft block. They investigated the
EMI shielding properties of both unvulcanized and DCP vulcanized composite films.
The structural change from brick-mortar to honeycomb after vulcanization formed by
MXene in NR matrix effectively improved EMI SE. Different heterogeneous architectures of MXene/rubber composites affect EMI shielding and mechanical properties.
Attenuation of electromagnetic (EM) waves through these MXene/rubber composites occurs through three different mechanisms: absorption, reflection and multiple
reflections. A very less amount of EM wave transmits through it. These different
ways of EM wave attenuation strongly depend on the conducting network formed by
MXene in the rubber matrix, which is strongly related to the morphology of these
composite materials [19]. Due to some advantageous properties and superior EMI SE,
MXene/rubber composites could be promising alternates for EMI shields and hence
therefore can utilize in stretchable and foldable electronics. In this chapter, the role of
MXene/rubber composites in EMI shielding applications is discussed in detail.

11.2   MXene/RUBBER COMPOSITES FOR EMI
SHIELDING APPLICATIONS

MXenes are 2D inorganic compounds consisting of a few atoms layered structure
of transition metal carbide, nitrides or carbonitrides and having general structural
formula of $M_{n+1}X_nT_x$. They are prepared from selective etching of A layer from
$M_{n+1}AX_n$ materials where M represents the transition metal, A is an element from
group 13 or 14 and X denotes the surface terminating group such as C and/or N.
Presence of a hydrophilic group caused by selective etching process in MXene leads
to the uniform dispersion of these materials in water. The MXene film sheets are
then prepared via spray-coating or extraction filtering of MXene/water suspension.
MXene exhibits metal-like electrical conductivity which could be utilized to fabricate highly efficient EMI shields. But brittleness of these inorganic compounds
limits the high scalability and therefore restricts their utilization in stretchable and
wearable electronic devices. Synthetic polymers and natural macromolecules like
rubber materials are recently combined with MXenes as flexible additives or bind-
ers to improve the mechanical and other performance properties and EMI shielding
effectiveness. MXene ultra-thin films are very prone to oxidation and are unstable
in moisture environments [20]. Therefore electrical conductivity and EMI shielding
properties reduce greatly due to the oxidation process. Combination of long-chain
polymer macromolecules with MXene is a significant way to improve durability
of resulting composites, as these macromolecular chains can encapsulate MXene
and thus greatly reduce permeation of water and air through it. These hydrophobic polymers enrich the hydrophobic property of the composites, which can inhibit
the process of oxidation. The most commonly used MXene to fabricate MXene/
rubber composites is $Ti_3C_2T_x$, which is prepared by selective etching of Ti_3AlC_2 by

concentrated HCl and LiF. Natural rubber (NR) is combined with $Ti_3C_2T_x$ to manufacture flexible, highly efficient EMI shields. Natural rubber in the form of latex, which is a bio-polymer, is mainly composed of cis-1,4-polyisoprene units and has a high molecular weight. Nanocomposites made of NR exhibit high flexibility and improved elasticity and outstanding stress-cracking resistance. NR particles are uniformly distributed in aqueous solution of colloidal NR, and also as NR made up of polyisoprene units, it provides hydrophobicity in MXene/rubber composite films without any oxygen-containing groups. Highly flexible MXene/NR composite films are fabricated via suction filtration of MXene/rubber mixed suspension whereas $Ti_3C_2T_x$ dispersion is prepared by selective etching by acid followed by ultrasonic exfoliation. The natural rubber chains and MXene ($Ti_3C_2T_x$) nanosheets are very well compatible with each other, which contributes to the excellent toughening and strengthening effective in the resulted lightweight, flexible, highly conductive, durable rubber/MXene composite films. In a typical work, pressured-extrusion filtration has been used to prepare NR/$Ti_3C_2T_x$ composite films exhibiting brick-mortar structure where the amount of MXene has been varied from 1.5 to 15 g and 1.5 g (1 wt%) NR latex solution. DCP was used as a curing agent and the NR/$Ti_3C_2T_x$/DCP films were vulcanized under vacuum at 160°C temperature. After the curing process, morphology of resultant composite films changes to a honeycomb structure, which is very beneficial for overall properties improvement and efficient EMI shielding effectiveness of the composite films. In another work, Yang et al. fabricated NR/MXene films where they varied MXene content from 10 to 60 wt% and obtained uniform distribution of exfoliated $Ti_3C_2T_x$ nanosheets in rubber matrix. Other rubber materials such as polyurethane, polydimethylsiloxane (PDMS) are also used to fabricate EMI shielding material with MXene. EMI shielding properties of these composite films (thickness: 11 to 65 μm) were compared with those for pure MXene films. This composite also exhibits excellent thermal stability, and the stability increases with increase in $Ti_3C_2T_x$ loading. These highly electrical conductive composite films show outstanding EMI shielding behavior, and the shielding efficiency combines the absorption and reflection of electromagnetic (EM) waves. Very good hydrophobicity, flexibility, good flame retardancy, durability and thermal stability of these composite films make them very attractive to be applicable in wearable and stretchable electronic devices [21].

11.3   ROLE OF MXene IN EMI SHIELDING BEHAVIOUR OF MXene/RUBBER COMPOSITES

Electrical conductivity plays an important role in achieving efficient EMI shielding effectiveness of the MXene/rubber composite film-based EMI shields. The higher the volume electrical conductivity, the higher will be their EMI SE. Generally, pure MXene films exhibit electrical conductivity of around 250 S/mm, whereas the electrical conductivity of MXene/rubber composite films varies between 15 and 150 S/mm, which is much higher than that for commercial EMI shielding applications. When electromagnetic waves pass through a conductive polymer/filler composite material then the maximum amount of EM waves is reflected or absorbed by the material, and only a very little amount gets transmitted through composite material.

The EMI SE is considered as the ratio of incident to transmitted EM radiation and generally is expressed in terms of decibels, and is mathematically represented by the following equation:

$$SE_T(dB) = 10log_{10}\left(\frac{P_I}{P_T}\right) \tag{11.1}$$

Where P_I and P_T represent the incident power and transmitted power respectively. When EM radiation is projected into the material the relationship between reflection coefficients (R), transmission coefficients (T) and absorption coefficients (A) is given by Eq. 11.2 as follows:

$$R + A + T = 1 \tag{11.2}$$

T, R and A coefficients can be obtained from scattering parameters from VNA (vector network analyzer) using the following expressions:

$$T = |S_{12}|^2 = |S_{21}|^2 \tag{11.3}$$

$$R = |S_{11}|^2 = |S_{22}|^2 \tag{11.4}$$

$$A = 1 - R - T \tag{11.5}$$

The total shielding effectiveness (SE_T) is the sum of shielding effectiveness by reflection (SE_R) and absorption (SE_A) and is evaluated by using the following equations [12, 22]:

$$SE_R(dB) = 10log_{10}\left[\frac{1}{1 - S_{11}{}^2}\right] \tag{11.6}$$

$$SE_A(dB) = 10log_{10}\left[\frac{1 - S_{11}{}^2}{S_{12}{}^2}\right] \tag{11.7}$$

$$SE_T(dB) = 10log\frac{1}{|S_{12}|^2} = 10log\frac{1}{|S_{21}|^2} \tag{11.8}$$

Pure MXene film exhibits excellent EMI SE owing to its high intrinsic electrical conductivity. When it is combined with rubber macromolecules, EMI shielding properties of the resultant composite film are improved significantly compared to neat rubber film. EMI SE values of the composite increase with rise in MXene concentration. When evaluating EM SE of the MXene/rubber composite film, thickness (t) of the films is of great importance. To further evaluate EMI shielding performance of the composite film, specific shielding effectiveness $(SSE = SE / t)$ is calculated. The shielding efficiency is calculated based on Eq. 11.9.

$$Shielding\, efficiency\% = (100 - 1 / 10^{\frac{SE}{10}}) \times 100 \tag{11.9}$$

The reported MXene/rubber composite films show excellent EMI SE greater than -30 dB, which is higher than the commercial requirement (20 dB) in whole X-band frequency region. Also, the shielding efficiencies reach up to 99.99%, indicating that 0.01% radiation can penetrate through the material. Wang et al. investigated the EMI shielding properties of $Ti_3C_2T_x$ MXene/natural rubber composite films in detail with varying concentrations of MXenes. Their composites exhibited outstanding EMI SE at ultra-low thickness. Also, they obtained high SSE (18,989 to 9592 dB.cm^2/g) and shielding efficiency (99.998%) and SE_A was found higher than SE_R, which indicates that absorption is the main shielding mechanism of the brick-mortar structured composite films. Durability or cycling stability is very important when EMI shields of composite films are used in stretchable and wearable electronic devices. Wang et al. performed cycling stability of their composite films and obtained that their materials exhibited excellent mechanical durability and stability and also EMI SE remains almost the same after bending several times (around 78% EMI SE retention). After crosslinking of rubber composites by DCP, they achieved significant property improvements when morphology changes to honeycomb structure constructed by crosslinking networks. The EMI SE increases up to 70% (64 dB) compared to that for unvulcanized samples. Also, shielding by absorption and reflection improves significantly owing to the formation of crosslinking network and honeycomb structure by MXenes. After crosslinking, the composite films produce more conductive interfaces which can shield electromagnetic waves more effectively, and also the formation of a unique honeycomb structure by $Ti_3C_2T_x$ after vulcanization can absorb or reflect more incident microwaves multiple times. Therefore EMI shielding performance of the composite films improves remarkably. Here different heterogeneous architectures formed by MXenes in NR greatly influence the EMI shielding performance. Before the vulcanization process in the case of a brick-mortar structure, the incident EM waves get contacted with the surface of MXenes, few of them are reflected immediately. The remaining waves migrate through the lattice structure of MXene and therefore interact with the electrons in $Ti_3C_2T_x$ MXene. The surviving EM waves then pass through the layers of the MXene and reflect and are absorbed by MXene layers until complete dissipation. When the microstructure changes to the honeycomb structure, EMI SE improves significantly through absorption after a curing process. Unique honeycomb-structured composite consists of several isolated units which repeatedly reflect the EW waves followed by secondary reflection between them when EMWs incident through the composite film. The rest of the EM waves pass through the tightly stacked crosslinking units and then enter into the remaining isolated units, which provides sufficient polarization centres. Therefore electronic and interfacial polarizations induce, which leads to enhancement of electromagnetic wave absorption. After absorption of EM radiations the rest of them, in very little amounts, are dissipated as heat. Also, the more conductive nature of honeycomb structure over brick-mortar structure facilitates more efficient EMI shielding of microwaves, and these structured composite films can be the next generation EMI shields for stretchable electronic applications [23]. Table 11.1 shows a comparison of EMI shielding properties of these highly flexible composite films with that of other polymer composites.

Kim et al. has prepared MXene-based auxetic polyurethane (APU) lightweight composite foam. The APU and MXene composite foams are designated as MX/APU. A

TABLE 11.1

Comparison of EMI SE of Different Polymer Composites with MXene/Rubber Composite Films

Sample	Thickness (mm)	EMI SE (dB)	References
PS/rGO	2.5	45.1	[24]
Epoxy/CNT	2	33	[25]
PE/graphite	2.5	51.6	[26]
PE/MWCNT	3	35.2	[27]
PS/CuNWs	0.2	35	[28]
Paraffin/ $Ti_3C_2T_x$	2	33	[29]
SA/ $Ti_3C_2T_x$	0.008	57	[30]
PS/AgNWs	0.87	33	[7]
CNFs/ $Ti_3C_2T_x$	0.047	24	[31]
SiO_2/rGO	1.5	38	[32]
PS/ $Ti_3C_2T_x$	2	62	[17]
NR/ $Ti_3C_2T_x$ (NRT6)	0.037	45	[23]
NRT6 (after bending)	0.037	44	[23]
NR/ $Ti_3C_2T_x$ (c-NRT10)	0.054	63.5	[20]

simple dip-coating method was employed to fabricate the foam composites. The original polyurethane foam (OPU) has been transformed to APU foams to increase the internal surface area of the foam material so that excessive internal multiple reflections of the electromagnetic wave could be achieved for the foam material. So, OPU is subjected to triaxial volumetric compression (with volumetric compression ratio of 2) at 180°C to get APU. The auxetic structure of the foam pores gave more density with enhanced surface area to the APU compared to OPU. A $Ti_3C_2T_x$ solution is prepared by etching Al from Ti_3AlC_2 precursor by dissolving the latter in a LiF/HCl solution. The APU foam material is dip-coated in $Ti_3C_2T_x$ solution and vacuum dried at 50°C overnight to produce the MX/APU [33]. The schematic representation is given in the Figure 11.1. Non-auxetic PU foams have regular circular foam structures and are designated as NAPU.

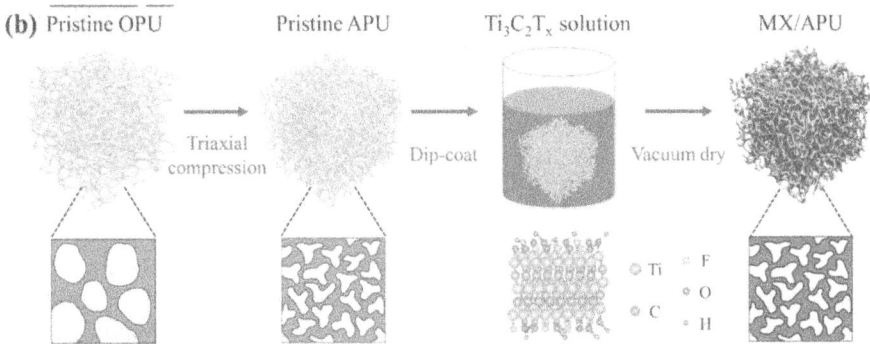

FIGURE 11.1 Schematic representation of fabrication of MX/APU foam composite.

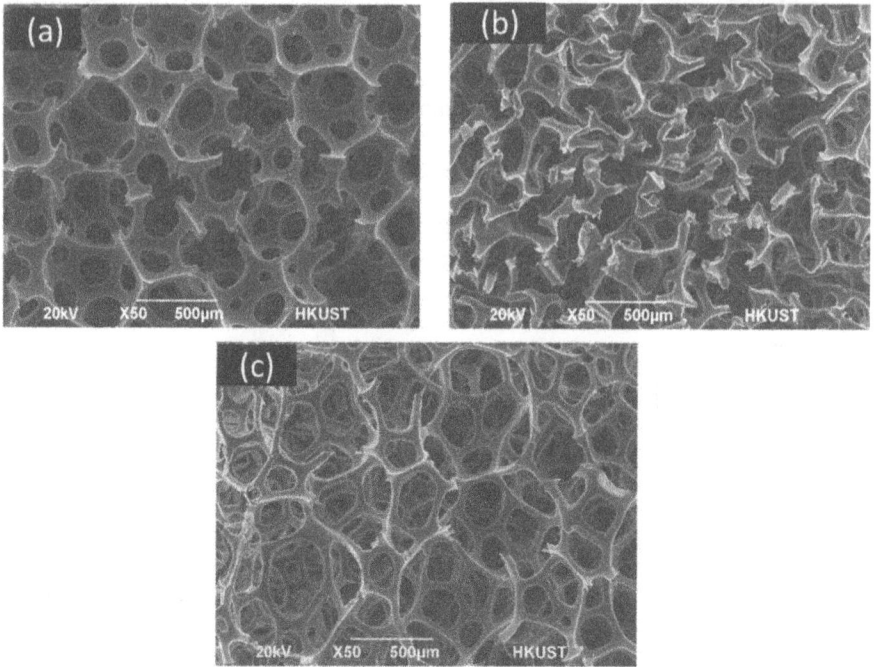

FIGURE 11.2 Scanning electron microscopic (SEM) images of (a) non-auxetic PU (NAPU), (b) APU and (c) OPU.

Scanning electron microscopic images of OPU, APU, and NAPU are shown in Figure 11.2, where (a) represents the NAPU with circular pore PU foams which are structurally similar to OPU with thicker struts and higher density compared that of the OPU (Figure 11.2(c)). Figure 11.2(b) describes the SEM images of APU with clothed and auxetic pore structures and higher PU densities.

The EMI shielding effectiveness of each of MX/NAPU, MX/APU and MX/OPU with different $Ti_3C_2T_x$ content is shown in Figure 11.3.

Figure 11.3(a) shows EMI SE of different foam MXene composites with different MXene content. MX/APU was found to achieve the highest EMI shielding performance at 75 dB with 21 wt% of MXene content, and absorption of EM waves is the most predominant phenomena and contributor to the total EMI effectiveness (Figure 11.3(b)) and absorption is around 33% higher in case of MX/APU compared to MX/NAPU. Figure 11.3(c) describes the effectivity of auxetic pore structure compared to NAPU where the APU pores absorb or internally reflect EM waves compared to the pores of NAPU or OPU.

Duan et al. has prepared PU/MXene-based composites modified with carbon fibre fabric (CFf) to further strengthen the composite samples mechanically [34]. The composite samples are produced by following electrohydrodynamic atomization deposition technique; layer-wise adjustment followed by hot pressing. The

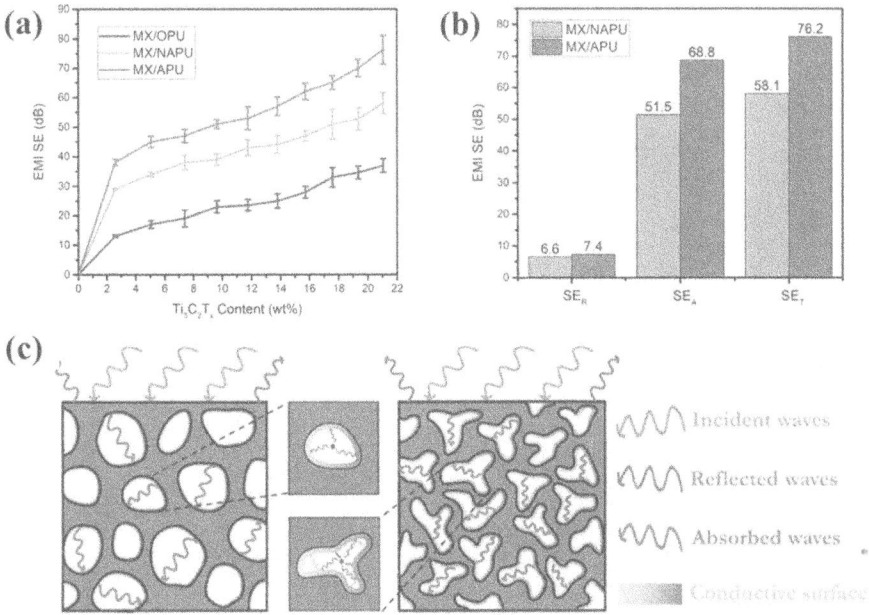

FIGURE 11.3 (a) EMI shielding efficiency study of MX/APU, MX/NAPU and MX/OPU; (b) comparison of different components (SE_A, SE_R) of total EMI shielding efficiency (SE_T) for MX/NAPU and MX/APU; (c) attenuation of EMI shielding performance of foam composites with different pore structures.

CF fabric containing 0.32%, 0.57% and 1.06 wt% of MXenes are designated as MXene/CFf-x (where x equals to 1, 2 and 3 respectively). EMI shielding study of the samples are represented in Figure 11.4. From Figure 11.4(a) it is observed that EMI shielding efficiency increases with increase in MXene wt%, and MXene/CFf-3 is found to be shown EMI SE around 40 dB. The CF/TPU has shown very low EMI effectivness as predicted. Absorption (A), transmission (T) and reflection (R) portions of the total EMI SE of the composites are determined and composites are found to be more reflection-dominated (Figure 11.4(d)). With the increment in $Ti_3C_2T_x$ MXene wt%, the EMI shielding efficiency of the $Ti_3C_2T_x$ MXene/CFf/TPU composites has grown effectively. With an increase of $Ti_3C_2T_x$ MXene wt% of around 0.3%, the SE_T, SE_A and SE_R of $Ti_3C_2T_x$ MXene/CFf/TPU-1 composite have shown values of approximately 24 dB, 20.0 dB and 3.5 dB, respectively. If the $Ti_3C_2T_x$ MXene wt% is gradually changed from 0.32% to 0.57%, the SE_T, SE_A and SE_R of the $Ti_3C_2T_x$ MXene/CFf/TPU-2 composite have shown an increasing trend of approximately around 30 dB, 25 dB and 4 dB, respectively. If the $Ti_3C_2T_x$ MXene wt% is expanded to 1.06%, the SE_T, SE_A and SE_R of the $Ti_3C_2T_x$ MXene/CFf/TPU-3 composite are found to be around 40 dB, 35 dB and 6 dB, respectively. The results have depicted that the expansion of $Ti_3C_2T_x$ MXene wt% in the composites remarkably enlarged the EMI shielding properties of the

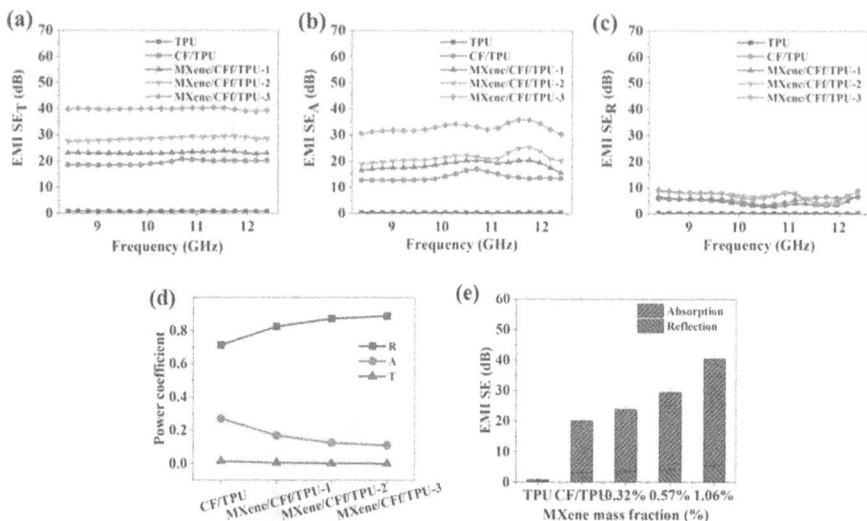

FIGURE 11.4 (a) Total EMI SE (SE$_T$) of the composites within the frequency range of X-band; (b) EMI SE (SE$_A$) values of the composites within the frequency range of X-band; (c) EMI SE (SE$_R$) values of the composites within the frequency range of X-band; (d) power coefficient values of different composites; (e) contribution of absorption/reflection to the total EMI shielding values of the composites with increase in MXene wt%.

Ti$_3$C$_2$T$_x$ MXene/CFf/TPU composite. In addition, the SE$_A$ and SE$_R$ of the composites were also checked, as depicted in Figure 11.4 (b–c). The SE$_A$ value is found to be much enlarged over the SE$_R$ value. For example, the SE$_A$ of the Ti$_3$C$_2$T$_x$ MXene/CFf/TPU-3 composite is found to be approximately around 35 dB, and the SE$_R$ is around 6 dB, which suggests that SE$_A$ deals more than 80% in SE$_T$.

Wu et. al. has developed electrically conductive MXene foam material coated with polydimethylsiloxane (PDMS) to give structural stability to the foam material. The MXene foam material is prepared by carrying out sodium alginate-assisted three-dimensional architecture modeling of MXene aerogels. The mass percentages of Ti$_3$C$_2$T$_x$ MXene in the aerogels are 100, 95.24, 86.95, 74.07 and 62.50 wt%, and the correlated aerogels are designated as MS100, MS95, MS87, MS74 and MS63, respectively. The SEM images of the aerogel MS 74 are given in Figure 11.5. The SEM images show unidirectional pores of the polygonal cellular structure of the MXene-SA aerogels. The unidirectional pores ensure lower electrical resistance and higher conductivity to show good EMI shielding efficiency [35]. EMI shielding efficiency of the composites is represented in Figure 11.6.

EMI shielding efficiency of the aerogels increases with the increment in MXene content (Figure 11.6 (a)). Absorption of electromagnetic waves is the prime mechanism to counter electromagnetic waves (Figure 11.6 (b)). MS74P blending showed EMI SE value around 10 dB, which suggests the conventional melt blending method could not give enough dispersion of MXene throughout the PDMS matrix and MXene is effectively dispersed in foamed PDMS (Figure 11.6 (c)).

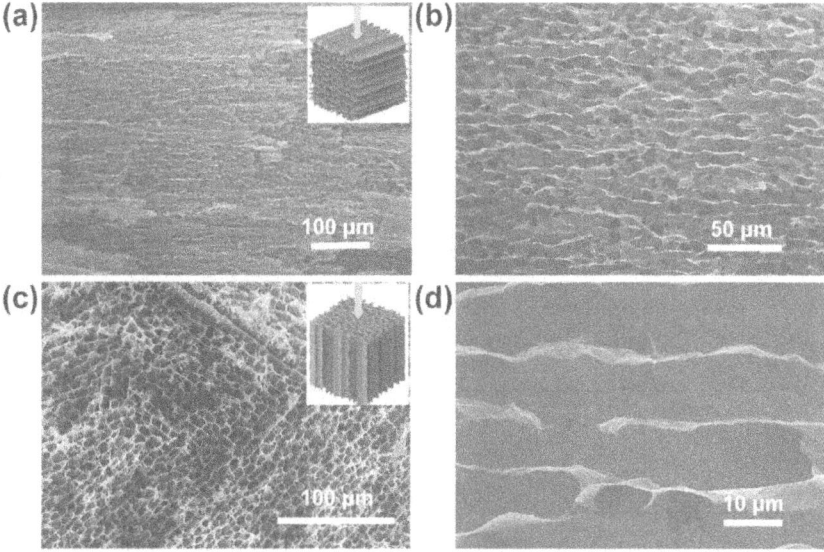

FIGURE 11.5   (a, b, d) Side view of SEM images of the aerogel MS 74, (c) top view of the SEM images of the aerogel MS 74.

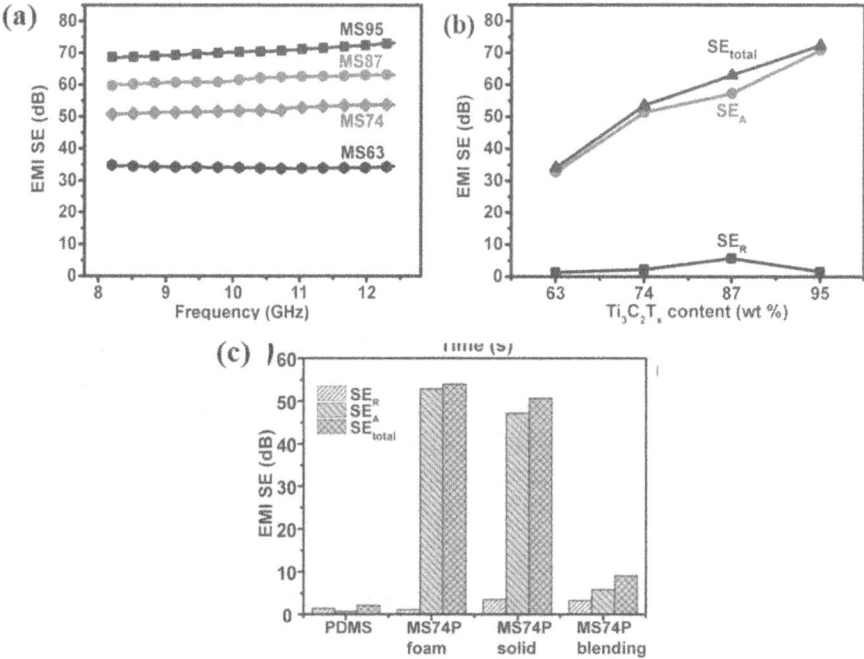

FIGURE 11.6   (a) Total EMI shielding efficiency of the MXene-SA based aerogels; (b) change in the values of SE_{total}, SE_A, SE_R of the MXene-SA based aerogels with the MXene content; (c) total EMI SE of PDMS, MS74P foam, MS74P solid and MS74P blending.

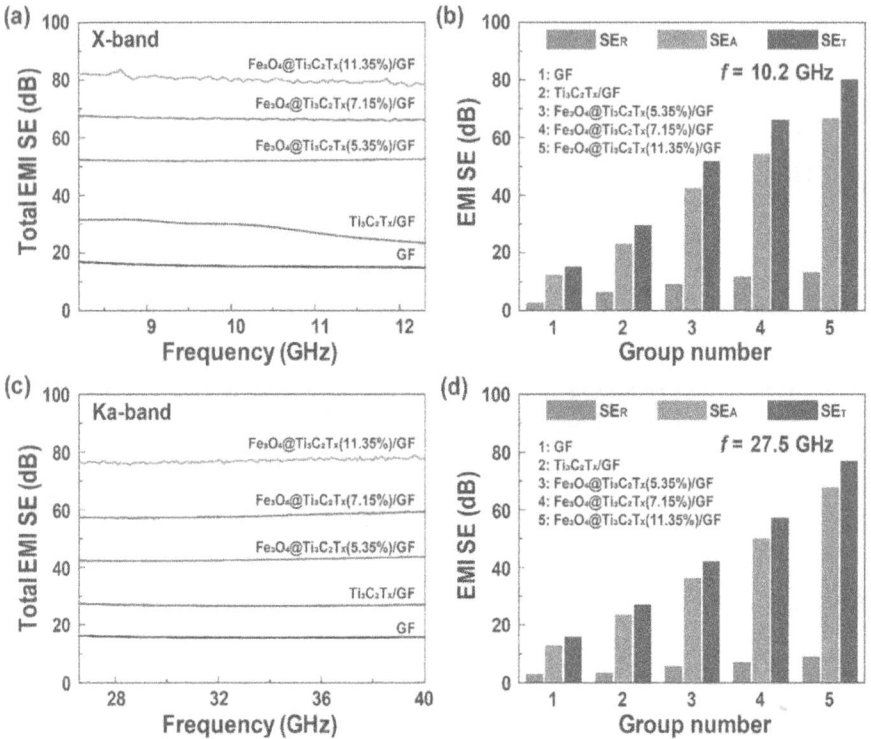

FIGURE 11.7   (a) Total EMI shielding performance of the Fe3O4@Ti$_3$C$_2$T$_X$/GF/PDMS composites with different Ti$_3$C$_2$T$_X$ content (11.35%, 7.15% and 5.35%), Ti$_3$C$_2$T$_X$/GF and GF in the X-band region; (c) total EMI shielding performance of the Fe$_3$O$_4$@Ti$_3$C$_2$T$_X$/GF/PDMS composites with different Ti$_3$C$_2$T$_X$ content (11.35%, 7.15% and 5.35%), Ti$_3$C$_2$T$_X$/GF and GF in the Ka-band region; (b) EMI shielding performance of the composites at a particular frequency of 10.2 GHz in the X-band region; (d) EMI shielding performance of the composites at a particular frequency of 27.5 GHz in the Ka-band region.

Nguyen et al. has prepared EMI shielding skin material comprised of foamed graphene (GF) reinforced PDMS composite material where GF is decorated with Fe$_3$O$_4$ nanoparticles deposited inside intercalated MXene sheets. The composites are designated as Fe3O4@Ti$_3$C$_2$T$_X$ (wt%)/GF. The EMI shielding performance of the composites is represented in Figure 11.7. There are two frequency regions (X-band and Ka-band) within which the EMI shielding experiment is performed [36].

It is observed that in the case of both X-band and Ka-band EMI, shielding performance increases with the increment in MXene content with 11.35 wt% has given around 80 dB. Total EMI shielding efficiency is dominated by absorption mechanism (SE$_A$) in the case of both of X-band and Ka-band for each of the composites.

Yang et al. fabricated multifunctional MXene/NR composite films for efficient EMI shielding applications with excellent flexibility and durability. They varied the weight ratio of NR from 10 to 60% and compared the resultant properties with that of pure MXene film. The thickness of these composite films varied between 0.03 to 0.065 mm with uniform dispersion of exfoliated $Ti_3C_2T_x$ MXene sheets confirmed from the morphological analysis. As NR concentration increases, stacked layers structure changes to interconnected layers for MXene/rubber composite. Natural rubber macromolecules encapsulated the MXene sheets, and also a robust interconnected networking structure is formed due to strong interaction between MXene nanosheets and rubber macromolecular chains. Therefore these composite films are beneficial for efficient stress and heat transfer, indicating more mechanical strength and good thermal stability. Also, these interconnected compact networking pathways impart high electrical conductivity, which results in efficient shielding efficiency of EM waves. The electrical conductivity is in the range of 15 S/mm to 145 S/mm, whereas tensile strength improves remarkably. The $Ti_3C_2T_x$/NR composite films showed excellent flexibility and outstanding folding resistance, which is approximately 60 times higher than that for pure MXene film. The folding durability is also higher than MXene films, even higher than MXene/nanofibre cellulose paper. All composite films show EMI SE > 30 dB in the X-band frequency region and also maximum shielding efficiencies are more than 99.997%, indicating excellent EMI shielding behavior. The specific shielding effectiveness (SSE) which is SE/thickness reported is also much higher than that for CNT, graphene-based polymer composites. The critical thickness of the composite film achieved is 12 μm to impart EMI SE more than the commercial requirement (20 dB). As thickness increases, EMI SE increases as more MXene sheets can interact with EM waves and block more radiations. MXene/natural rubber (60%) film with 130 μm thickness shows EMI SE of 54 dB and can shield 99.996% of incident EM radiation with the negligible transmission, which strongly suggests the effectiveness of these MXene/rubber composites in EMI shielding application. While EMI shielding films of MXene/rubber composite films are utilized in practical applications, environmental durability of these films is an important parameter to evaluate their ability in practical applications. These films showed excellent water resistance and retain EMI SE almost the same, even after soaking in water for 15 days. Also bending test of these composite films under 6000 bending shows good EMI SE, indicating the outstanding durability for practical applications. Yang et al. studied SE_T, SE_A and SE_R to further investigate the EMI shielding mechanism. SE_A was found to be always higher than SE_R which indicates absorption-dominated attenuation of EM waves and also as NR concentration increases shielding by absorption increases. Here the cellular structure of $Ti_3C_2T_x$/NR composite facilitates further multiple internal reflections of EM waves, which finally leads to absorption of more microwaves. Luo et al. also investigated the electrical and EMI shielding properties of MXene/natural rubber nanocomposite films. They varied $Ti_3C_2T_x$ MXene content from 0.5 to 6.7 vol% and DCP was used as a curing agent to prepare MXene/NR composite films. Morphological analysis reveals the selective location of closely connected MXene sheets in the interfaces of rubber particles. As a result of this, electrons can effectively transfer throughout the whole matrix in this continuous network. The strong interaction between rubber chains and $Ti_3C_2T_x$ sheets, and also high quality continuous interconnected conductive network, is very beneficial for improving electrical and other performance properties of these composite films. The electrical percolation threshold obtained is 0.91 vol% of $Ti_3C_2T_x$, and the conductivity reaches up to 1400 S/m, which is one of the reported best results for stretchable electronic applications. Also, their I-V curves result in linear ohmic behavior and increment in slope with the rise in MXene concentration, which indicates the good electrical properties of the composite films. All composite films show very good EMI SE, and with only 2 vol% MXene content composite film exhibits EMI SE > 20 dB, satisfying commercial requirements. EMI shielding effectiveness increases to 54 dB for highest $Ti_3C_2T_x$ content, which is much higher than reported EMI SE of other rubber nanocomposites, for example, 45 dB for NR/CNT (50.6 vol%) with 250 μm thickness, 38.4 dB for CPE/CB (22.0 vol%) with 1 mm thickness and 30 dB for SEBS/rGO (18 vol%) of 223 μm thickness [20]. These reported EMI shielding of other rubber composites strongly suggest the superiority of these MXene/NR composite films as efficient EMI shields. As film thickness increases, EMI SE also increases, and composite film of only 99 μm thickness shows EMI SE of 20.5 dB. They compared EMI shielding performance of MXene/rubber composites with other reported results for stretchable polymer composites in terms of specific shielding effectiveness (EMI SE/t). They found that most of the composite materials show relatively low shielding efficiency and it is very difficult to achieve high EMI shielding performance at lower sample thickness for these composites. The highly flexible MXene filled rubber composite film gives much higher SSE value of 214 dBmm^{-1} at very low MXene content than the other results. This proves the ability of these stretchable $Ti_3C_2T_x$/NR composite films to shield EM radiation effectively. Shielding by absorption increases (SE_A) as $Ti_3C_2T_x$ increases and SE_A is found to be higher than SE_R. The porous network formed by MXene facilitates multiple scattering of EM waves and interfacial polarization, which increases the absorption contribution of EM wave shielding. To study the durability of the composite films, cyclic stretching and folding of films have been performed. Crosslinking of the rubber chains enhances the stability and reproducibility of their conductive networks during these tests. EMI shielding effectiveness retains up to 90% of their original EMI SE, even after 3000 times cyclic folding and stretching. The stable efficient EMI shielding performance and excellent flexibility and foldability ensure the application of these high-performance MXene/rubber-based EMI shields in next-generation stretchable and foldable electronics at very low thickness. Table 11.2 gives a comparative study of EMI shielding results of various MXene/rubber composite films of ultra-low thickness with EMI SE of other polymer-based composites, which clearly indicate the advantages of these MXene-based rubber films as efficient EMI shields.

11.4   MXene/RUBBER COMPOSITES AS NEXT-GENERATION STRETCHABLE, FOLDABLE AND EFFICIENT EMI SHIELDS

Compared to carbonaceous fillers (CNT, CNF, CB, graphene) MXene exhibits high electrical metallic conductivity, and also these carbonaceous inclusions are less prone to achieve efficient EMI shielding properties at ultra-low thickness when combined

TABLE 11.2
EMI Shielding Properties of MXene/Rubber Composite Films and Comparison with Other Reported Polymer Composites and Other Materials

Materials	Filler Concentration	Thickness (mm)	SE (dB)	SSE (dB/cm)	SSE/ρ (dB. cm²/gm)	References
ABS/CNT	10 wt%	1.1	40	364	318	[9]
PVDF/graphene	15 wt%	0.1	22.6	2260	1265	[37]
PMMA/CNT	20 wt%	4.5	30	67	49	[38]
PVA/MG	50 wt%	0.36	20.3	564	329	[39]
PEI/graphene	10 wt%	2.3	40	174	68	[40]
NR/CNT	70 wt%	0.05	21.4	4280	2853	[14]
WPU/CNT	61.5 wt%	0.32	35	1090	779	[13]
MXene/PEDOT:PSS	75 wt%	0.0152	8.99	5914	3585	[15]
CNFs/ $Ti_3C_2T_x$	80 wt%	0.074	26	3510	2154	[16]
CNT sponge	100 wt%	2.38	22	92	4622	[20]
CNF mat	100 wt%	0.29	52.2	1800	1362	[22]
NR/ $Ti_3C_2T_x$ (cMR5)	3.1 vol%	0.021	33	1854	-	[18]
NR/ Ti3C2Tx (cMR7)	6.7 vol%	0.027	54	2250	-	[18]
NR/MXene	60 wt%	0.0337	52.7	15638	9653	[20]
NR/MXene	40 wt%	0.0656	47.8	7287	5693	[20]

with polymeric materials. Among different polymer materials used to fabricate efficient EMI shielding materials, rubbers are widely used now a day. Although rubber materials are insulative in nature and transparent to EM waves, but when combined with conducting fillers show good electrical conductivity as well as EMI SE. The reported results show that a higher amount of filler is required to achieve good shielding properties and need samples of higher thickness. Although these composite materials overcome the issues of employing metals and conducting polymers as efficient EMI shields, but achievement of excellent EMI shielding and other performance properties of these conductive rubber composites at low filler concentration with very low sample thickness is still a major concern. Pure MXene materials exhibit excellent metallic conductivity and MXene films show efficient EMI SE, but the lack of flexibility and brittleness of these films restricts their application in stretchable electronics. Rubber macromolecules are very flexible and they possess some advantageous properties which are important for utilizing these materials to fabricate commercial EMI shielding materials. The combination of MXene and rubber materials gives enough flexibility, durability, very good mechanical properties, excellent thermal stability along with high electrical properties and EMI shielding effectiveness. Also, the resultant composite materials show excellent EMI shielding properties at very low concentrations of MXene and ultra-low thickness. From the reported results which were already discussed, it can be seen that MXene/rubber composite films show very good EMI SE, electrical properties, mechanical properties and excellent flexibility at very low filler concentration. Also, these composite films pass the durability test and have high chemical resistance and thermal stability, which are also required to construct efficient EMI shields in commercial application. Also, these composite films exhibit higher specific shielding effectiveness (SSE) compared to other polymer composites. With an increase in MXene content in composite film conductivity as well as EMI shielding, efficiency increases greatly. The reported MXene/rubber composites show shielding efficiency above 99.99%, which is much higher than that for other carbonaceous filler contained polymer and rubber composites. MXene sheets construct a continuous, strong conducting network with rubber chains, which is beneficial for more shielding of EM waves by absorption than reflection. The reported different heterogeneous structures formed by MXene sheets and uniform distribution of exfoliated MXene in rubber are more favourable for absorption-based shielding than the other polymer composites. So, the various MXene/rubber composite films reported fulfiling the major requirements to be used as efficient EMI shielding materials at ultra-high thickness and with very low MXene concentration compared to those for other filler-contained rubber composites. Therefore it can be concluded that MXene/rubber composites are very promising alternates for high filler contained and thick EMI shielding materials and hence can be used in next-generation stretchable and wearable electronics.

11.5   CONCLUSION

Highly conductive MXene films show excellent EMI shielding effectiveness but possess some disadvantages, and rubber materials are insulative in nature and transparent to electromagnetic radiation. Other carbonaceous filler-contained rubber composites are conductive and show efficient EMI SE but high content of filler is required with higher sample thickness to achieve good EMI shielding properties. Combining the MXene with rubber macromolecules facilitates some advantageous performance properties along with electrical and EMI shielding properties. The resultant MXene/rubber composites show excellent EMI SE at very low MXene loading and ultra-low thickness. Different heterogeneous structures formed by MXene and continuous strong conductive network constructed are very beneficial for good shielding efficiency. These composite films exhibit very good electrical conductivity, and excellent EMI SE and SSE with high flexibility compared to other reported rubber and polymer composites. Also, an MXene network provides sufficient reinforcement to the rubber matrix, which improves mechanical properties greatly. Crosslinking of rubber macro-chains further enhances the stability and reproducibility of the composite films. Absorption-dominated shielding is observed, and also these films retain their electrical and shielding properties even after cyclic bending and folding. So the MXene/rubber composites could be used as efficient EMI shielding materials in next-generation stretchable and foldable electronic devices.

ACKNOWLEDGEMENTS

Narayan Chandra Das thanks to SERB, government of India (CRG/2021/003146), for funding this research work.

REFERENCES

1. Biswas S, Panja SS, Bose S. Tailored distribution of nanoparticles in bi-phasic polymeric blends as emerging materials for suppressing electromagnetic radiation: challenges and prospects. J Mater Chem C 2018;6(13):3120–42.
2. Kumar P, Shahzad F, Yu S, Hong SM, Kim YH, Koo CM. Large-area reduced graphene oxide thin film with excellent thermal conductivity and electromagnetic interference shielding effectiveness. Carbon 2015;94:494–500.
3. Song Q, Ye F, Yin X, Li W, Li H, Liu Y, Li K, Xie K, Li X, Fu Q. Carbon nanotube-multilayered graphene edge plane core-shell hybrid foams for ultrahigh-performance electromagnetic-interference shielding. Adv Mater 2017;29:1701583.
4. Cao M, Song W, Hou Z, Wen B, Yuan J. The effects of temperature and frequency on the dielectric properties, electromagnetic interference shielding and microwave-absorption of short carbon fiber/silica composites. Carbon 2010;48:788–96.
5. Lee TW, Lee SE, Jeong YG. Carbon nanotube/cellulose papers with high performance in electric heating and electromagnetic interference shielding. Compos Sci Technol 2016;131:77–87.
6. Liang C, Song P, Qiu H, Zhang Y, Ma X, Qi F, Gu H, Kong J, Cao D, Gu J. Constructing interconnected spherical hollow conductive networks in silver platelets/reduced graphene oxide foam/epoxy nanocomposites for superior electromagnetic interference shielding effectiveness, Nanoscale 2019;11:22590–8.
7. Arjmand M, Moud AA, Li Y, Sundararaj U. Outstanding electromagnetic interference shielding of silver nanowires: comparison with carbon nanotubes, RSC Adv 2015;5:56590–8.
8. Chaudhary A, Kumari S, Kumar R, Teotia S, Singh BP, Singh AP, et al. Lightweight and easily foldable MCMB-MWCNTs composite paper with exceptional electromagnetic interference shielding. ACS Appl Mater Interfaces 2016;8(16):10600–8.
9. Al-Saleh MH, Saadeh WH, Sundararaj U. EMI shielding effectiveness of carbon based nanostructured polymeric materials: a comparative study. Carbon 2013;60: 146–56.
10. Ghosh SK, Das TK, Ghosh S, Ganguly S, Nath K, Das NC. Physico-mechanical, rheological and gas barrier properties of organoclay and inorganic phyllosilicate reinforced thermoplastic films. J Appl Polym Sci 2020;1–16. https://doi.org/10.1002/app.49735.
11. Ganguly S, Ghosh S, Das P, Das TK, Ghosh SK, Das NC. Poly(N-vinylpyrrolidone)-stabilized colloidal graphene-reinforced poly(ethylene-co-methyl acrylate) to mitigate electromagnetic radiation pollution, Polym Bull 2020;77:2923–43. https://doi.org/10.1007/s00289-019-02892-y.
12. Ghosh SK, Das TK, Ghosh S, Remanan S, Nath K, Das P, et al. Selective distribution of conductive carbonaceous inclusion in thermoplastic elastomer: A wet chemical approach of promoting dual percolation and inhibiting radiation pollution in X-band. Compos Sci Technol 2021;210:108800. https://doi.org/10.1016/j.compscitech.2021.108800.
13. Cao MS, Cai YZ, He P, Shu JC, Cao WQ, Yuan J. 2D MXenes: electromagnetic property for microwave absorption and electromagnetic interference shielding. Chem Eng J 2019;359:1265–302.
14. Liu J, Zhang HB, Sun R, Liu Y, Liu Z, Zhou A, et al. Hydrophobic, flexible, and lightweight MXene foams for high-performance electromagnetic-interference shielding. Adv Mater 2017;29(38):1702367.
15. Zhang Y, Wang L, Zhang J, Song P, Gu J. Fabrication and investigation on the ultra-thin and flexible Ti$_3$C$_2$T$_x$/co-doped polyaniline electromagnetic interference shielding composite films, Compos Sci Technol 2019;183:107833.

16. Wang Q, Zhang H, Liu J, Zhao S, Xie X, Liu L, Yang R, Koratkar N, Yu Z. Multifunctional and water-resistant MXene-decorated polyester textiles with outstanding electromagnetic interference shielding and joule heating performances, Adv Funct Mater 2019;29:1806819.

17. Sun R, Zhang HB, Liu J, Xie X, Yang R, Li Y, et al. Highly conductive transition metal carbide/carbonitride (MXene)@ polystyrene nanocomposites fabricated by electrostatic assembly for highly efficient electromagnetic shielding. Adv Funct Mater 2017;27(45):1702807.

18. Luo JQ, Zhao S, Zhang HB, Deng Z, Li L, Yu ZZ. Flexible, stretchable and electrically conductive MXene/natural rubber nanocomposite films for efficient electromagnetic interference shielding. Compos Sci Technol 2019;182:107754.

19. Hu D, Huang X, Li S, Jiang P. Flexible and durable cellulose/MXene nanocomposite paper for efficient electromagnetic interference shielding. Compos Sci Technol 2020;188:107995.

20. Yang W, Liu JJ, Wang LL, Wang W, Yuen ACY, Peng S, et al. Multifunctional MXene/natural rubber composite films with exceptional flexibility and durability. Compos Part B Eng 2020;188. https://doi.org/10.1016/j.compositesb.2020.107875.

21. Liu R, Miao M, Li Y, Zhang J, Cao S, Feng X. Ultrathin biomimetic polymeric $Ti_3C_2T_x$ MXene composite films for electromagnetic interference shielding. ACS Appl Mater Interfaces 2018;10:44787–95.

22. Ghosh S, Ganguly S, Maruthi A, Jana S, Remanan S, Das P, et al. Micro-computed tomography enhanced cross-linked carboxylated acrylonitrile butadiene rubber with the decoration of new generation conductive carbon black for high strain tolerant electromagnetic wave absorber. Mater Today Commun 2020;24:100989.

23. Wang Y, Liu R, Zhang J, Miao M, Feng X. Vulcanization of $Ti_3C_2T_x$ MXene/natural rubber composite films for enhanced electromagnetic interference shielding. Appl Surf Sci 2021;546:149143.

24. Yan D, Pang H, Li B, Vajtai R, Xu L, Ren PG, Wang JH, Li ZM. Structured reduced graphene oxide/polymer composites for ultra-efficient electromagnetic interference shielding, Adv Funct Mater 2015;25:559–66.

25. Chen Y, Zhang HB, Yang Y, Wang M, Cao A, Yu ZZ. High-performance epoxy nanocomposites reinforced with three-dimensional carbon nanotube sponge for electromagnetic interference shielding. Adv Funct Mater 2016;26:447–55.

26. Jiang X, Yan DX, Bao Y, Pang H, Ji X, Li ZM. Facile, green and affordable strategy for structuring natural graphite/polymer composite with efficient electromagnetic interference shielding, RSC Adv 2015;5:22587–92.

27. Seng LY, Wee FH, Rahim HA, Malek F, You KY, Liyana Z, Jamlos MA, Ezanuddin AAM. Emi shielding based on MWCNTs/polyester composites. Appl Phys A: Mater 2018;124:140.

28. Al-Saleh MH, Gelves GA, Sundararaj U. Copper nanowire/polystyrene Nanocomposites: lower Percolation threshold and higher EMI shielding. Composites, Part A 2011;42:92–7.

29. Liu X, Wu J, He J, Zhang L. Electromagnetic interference shielding effectiveness of titanium carbide sheets, Mater Lett 2017;205: 261–3.

30. Shahzad F, Alhabeb M, Hatter CB, Anasori B, Hong SM, Koo CM, Gogotsi Y. Electromagnetic interference shielding with 2D transition metal carbides (MXenes), Science, 2016;353:1137–40.

31. Cao W, Chen F, Zhu Y, Zhang Y, Jiang Y, Ma M, Chen F. Binary strengthening and toughening of MXene/cellulose nanofiber composite paper with nacre-inspired structure and superior electromagnetic interference shielding properties. ACS Nano 2018;12:4583–93.

32. Wen B, Cao M, Lu M, Cao W, Shi H, Liu J, Wang X, Jin H, Fang X, Wang W, Yuan J. Reduced graphene oxides: light-weight and high-efficiency electromagnetic interference shielding at elevated temperatures. Adv Mater 2014;26:3484–9.

33. Kim E, Zhang H, Lee JH, et al. MXene/polyurethane auxetic composite foam for electromagnetic interference shielding and impact attenuation. Compos Part A Appl Sci Manuf 2021;147. doi:10.1016/j.compositesa.2021.106430.

34. Duan N, Shi Z, Wang Z, et al. Mechanically robust $Ti_3C_2T_x$ MXene/Carbon fiber fabric/thermoplastic polyurethane composite for efficient electromagnetic interference shielding applications. Mater Des 2022;214:110382.

35. Wu X, Han B, Zhang HB, et al. Compressible, durable and conductive polydimethylsiloxane-coated MXene foams for high-performance electromagnetic interference shielding. Chem Eng J 2020;381:122622.

36. Nguyen VT, Min BK, Yi Y, Kim SJ, Choi CG. MXene ($Ti_3C_2T_x$)/graphene/PDMS composites for multifunctional broadband electromagnetic interference shielding skins. Chem Eng J 2020;393:124608.

37. Zhao B, Zhao C, Li R, Hamidinejad SM, Park CB. Flexible, ultrathin, and high-efficiency electromagnetic shielding properties of poly(vinylidene fluoride)/carbon composite films. ACS Appl Mater Interfaces 2017;9(24):20873–84.

38. Das NC, Liu Y, Yang K, Peng W, Maiti S, Wang H. Single-walled carbon nanotube/poly (methyl methacrylate) composites for electromagnetic interference shielding. Polym Eng Sci 2009;49(8):1627–34.

39. Yuan B, Bao C, Qian X, Song L, Tai Q, Liew KM, et al. Design of artificial nacre-like hybrid films as shielding to mitigate electromagnetic pollution. Carbon 2014;75:178–89.

40. Ling J, Zhai W, Feng W, Shen B, Zhang J, Zheng W. Facile preparation of lightweight microcellular polyetherimide/graphene composite foams for electromagnetic interference shielding. ACS Appl Mater Interfaces 2013;5(7):2677–84.

12 Advancement in Nanostructured Carbide/ Nitrides MXenes with Different Architecture for Electromagnetic Interference Shielding Application

Vineeta Shukla
Department of Physics
Indian Institute of Technology Kharagpur
Kharagpur, West Bengal, India

CONTENTS

12.1 INTRODUCTION

Increasing the demand for miniaturized devices led to growth in the electronic and telecommunication sector, aggravating the risk of electromagnetic (EM) pollution. Electromagnetic interference (EMI) is electromagnetic pollution caused by interference of EM that produces electromagnetic noise. The reason for EM pollution may be both anthropogenic and natural. Natural EM pollution results from lighting, rain spray, and solar radiation [1]. On the other hand, man-made miniaturized electronic devices and electrical circuits, working at high-frequency EM wave range not only affect electrical circuits/appliances but also influence living tissues in a living body [2]. The EM noise in electrical circuits is caused by EM coupling and EM induction/conduction. EMI degrades the performance of electrical equipment or circuits and even causes the loss of stored data and limits the lifetime of electronic devices [3]. This pollution becomes a severe problem for human health because it causes headaches, sleeping disorders, eye problems,, and cancer. Moreover, EMI can break the smallest unit of the human body (i.e. DNA), can hinder the development of an infant's brain, and so on [4]. The frequency ranges of 3 kHz to 300 MHz and 300 MHz to 300 GHz in EM spectrum are well-known, as radiofrequency (RF) waves and microwaves (MW) which are very important due to applications in navigation, home appliances, medical equipment, and military assets, as shown in Figure 12.1.

Communication devices and home appliances such as cell phones, computers, Bluetooth devices, laptops, microwave ovens, integrated electrical circuits, military equipment, and medical devices require protection from hazardous microwaves [5]. Thus, current research is mainly focused on the shielding material for these devices and circuits to prevent the EM pollution created by microwaves. The EMI shielding blocks the undesirable EM radiation and protects the electronic gadgets and increases their life [6]. The reflection and absorption of EM waves by materials are believed to be important phenomena in EMI shielding mechanisms that require enough mobile charge carriers and magnetic dipoles in the shielding materials. Traditionally, metals (copper, stainless steel, nickel, aluminum, silver etc.) are the some prevailing materials, owing to good conductivity, for designing the EMI shielding materials [6, 7]. The major drawback with these metal-based shielding materials is their heaviness, corrosiveness, high production cost, and rigidity and processing difficulties. Alternatively, polymers are of low cost and strong corrosion

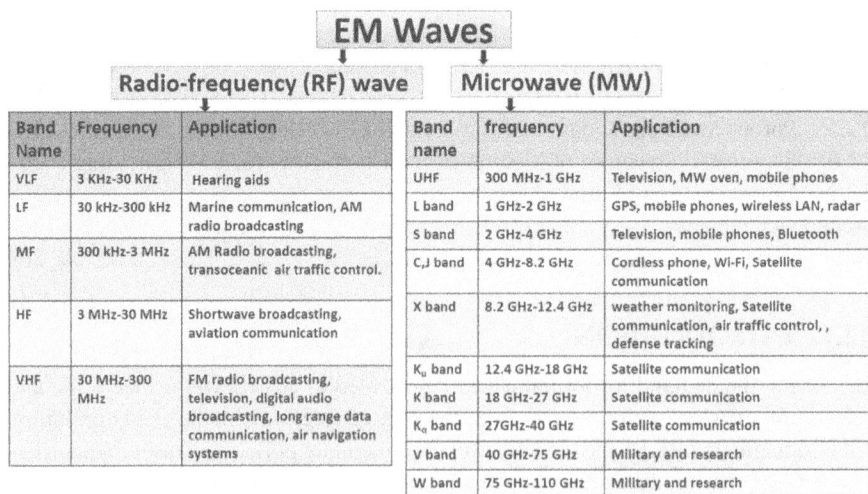

EM Waves

Radio-frequency (RF) wave

Band Name	Frequency	Application
VLF	3 KHz-30 KHz	Hearing aids
LF	30 kHz-300 kHz	Marine communication, AM radio broadcasting
MF	300 kHz-3 MHz	AM Radio broadcasting, transoceanic air traffic control.
HF	3 MHz-30 MHz	Shortwave broadcasting, aviation communication
VHF	30 MHz-300 MHz	FM radio broadcasting, television, digital audio broadcasting, long range data communication, air navigation systems

Microwave (MW)

Band name	frequency	Application
UHF	300 MHz-1 GHz	Television, MW oven, mobile phones
L band	1 GHz-2 GHz	GPS, mobile phones, wireless LAN, radar
S band	2 GHz-4 GHz	Television, mobile phones, Bluetooth
C,J band	4 GHz-8.2 GHz	Cordless phone, Wi-Fi, Satellite communication
X band	8.2 GHz-12.4 GHz	weather monitoring, Satellite communication, air traffic control, , defense tracking
K_u band	12.4 GHz-18 GHz	Satellite communication
K band	18 GHz-27 GHz	Satellite communication
K_a band	27GHz-40 GHz	Satellite communication
V band	40 GHz-75 GHz	Military and research
W band	75 GHz-110 GHz	Military and research

FIGURE 12.1 Application of different RF and MW frequency ranges of EM radiation.

resistance, and are lightweight in comparison with metal-based materials. These features of polymers make them favorable for EMI shielding materials, but their low mechanical strength, low thermal conductivity, and processing-related problems hinder the wide application of polymers [8, 9]. Before choosing the material as EMI shielding material, some characteristics of material should be satisfied, for example material should have high conductivity, high mechanical stability, low density, low cost, excellent thermal conductivity, minimal thickness, and so on [10, 11]. Many low-dimensional materials such as graphene, hexagonal boron nitride [12], transition metal dichalcogenides [13], and transition metal carbides/nitrides [14] have shown excellent performance in the biomedical field [15], as electrodes (in electrical and optical devices) [16–19], in catalysis [20–22], as photodetectors [23], as sensors [24–27], in water purification and environmental remediation [28, 29], in EMI shielding [30, 31], and in many more areas. Among these fascinating two-dimensional (2D) materials, transition metal carbides and nitrides reached the leadership position within a few years due to outstanding electrical conductivity, laminated structure, low density, tunable active surface, thermal stability, large surface area, excellent mechanical strength, and ease of solution processability [32, 33]. Till date, more than 100 possible compositions of MXenes have been studied theoretically [34], but the established MXene family includes Ti, V, Nb, Ta, Cr, Mo, Hf, Zr-based carbide MXenes such as Ti_2C, Ti_3C_2, V_2NT_x, and Mo_2CT_x [35–41]. However, nitride-based MXenes have been less explored. The first 2D titanium carbide ($Ti_3C_2T_x$), in 2011, were exfoliated by a Ti_3AlC_2 MAX precursor in hydrogen fluoride (HF) solution. MXenes are 2D transition metal carbides, nitrides, and carbonitrides with the typical formula represented as $M_{n+1}X_nT_x$ [42]. In the $M_{n+1}X_nT_x$ formula, M refers n+1 layers of early transition metals that are interleaved by n layers of carbon or nitrogen designated as X, and T_x represents surface terminal groups including -F, -O, and -OH [43]. MXenes are synthesized by topochemical selective etching of the A layer from MAX phases (precursor material) in fluorine (F)-containing acid solutions that include HF acid or a combination of fluoride salts lithium fluoride (LiF) with hydrochloric acid (HCl) that continues with the delamination and exfoliation of loosely stacked mono/few-layers 2D MXene flakes [44]. MAX phases are nano-laminated hexagonal crystal structured (P63/mmc) ternary nitrides and carbides where term A refers to main-group sp elements. Interestingly, three types of atom bonding – covalent, ionic, and metallic – in the MAX phase exist simultaneously. The M-X shows the mixed characteristics of valence bonds of ionic, covalent, and metallic bonds. Nevertheless, M-A and A-A bonds depict more metallic characteristics, thus having relatively weaker bond strength than that of the M-X bond. This is the reason that it is easier to peel off the A-layer atoms from the solid structure due to the highest reactivity [45, 46]. According to the difference of the number of layers (n) value, MAX phases are labeled into 211, 312, and 413 series [47]. The electric properties of MXenes occur from metallic to semiconductor, semiconductor to insulator, depending on the number of layers, X (i.e, C or N), and surface terminated groups. $Ti_{n+1}X_n$ is predicted to be metallic in behavior, but increasing n values formation of additional Ti-X bonds weaken the metallic properties [48]. On the other hand, MXene with -F, -OH, -Cl group would be a semiconductor or metal that depends on the orientations and types of these terminated groups. A large number of functional groups on the surface of MXene lowers the magnetic properties of MXene, therefore some MXenes (Ti_2N, Ti_2C and Cr_2C) are found to be paramagnetic [49–51], and some antiferromagnetic (Cr_2N, Mn_2C) [52, 53]. Fortunately, Cr_2C and Cr_2N showed good ferromagnetism, even in the presence of these functional groups. The high conductivity, good ferromagnetism, and lamellar structure of MXenes make them outstanding, in comparison to other 2D materials, for EMI shielding applications. In the present chapter, we highlight the recent advances in the development of titanium carbide/nitrides MXene composites and hybrids against EMI pollution. We discuss the EMI shielding mechanism of EM waves when they interact with shielding materials. The structural form of MXene composites greatly influences the shielding efficiency. This chapter includes the different architecture of MXene-based composites and hybrids along with the future aspects.

12.2   MECHANISMS OF EMI SHIELDING

When the EM waves fall on a compact and laminated structure material, some energy is reflected from the surface of the material, and some part of the energy is absorbed by the material that dissipated in the form of heat energy. The remaining part of EM wave energy is transmitted through the material. Ideally, a shielding material should not transmit EM energy or transmit negligible or zero energy [54]. The three mechanisms – reflection, absorption, and multiple reflections – take place when the shielding material interacts with incoming EM waves. A pictorial illustration of the mechanism of EM shielding from the compact and laminated structure is shown in Figure 12.2. The primary shielding mechanism is the attenuation by reflection. A material that has good conductivity shows the reflection dominant shielding in which mobile charge carriers (i.e. electrons or holes) interacting with

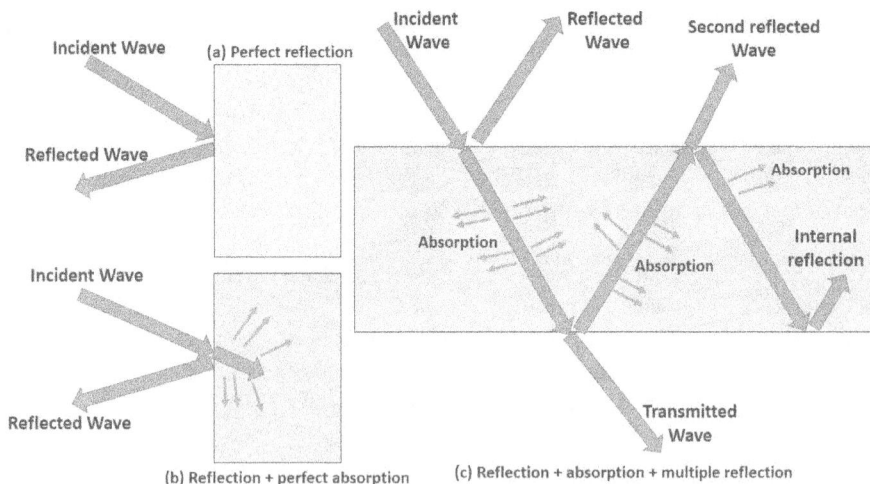

FIGURE 12.2   Mechanism of EMI shielding from compact and/or laminated structures when interacting with EM waves: (a) perfect reflection, (b) reflection + perfect absorption, (c) reflection + absorption + multiple reflection.

the EM wave cause the reflection, owing to impedance mismatching between material and surroundings [55]. Ideally, perfect reflection occurs if there is no loss of energy between the incoming EM wave and the reflected EM wave from the material (Figure 12.2a). Electrically conducting metals are found to have reflection dominant shielding [56]. The second important shielding mechanism is the attenuation by absorption. The absorption occurs due to the dissipation of EM energy in the form of heat energy. The interaction of EM waves with the magnetic dipole of the shielding material causes magnetic loss, including eddy current loss, hysteresis loss, and so on. The material, for good absorption, should have good electrical and magnetic properties. Ferrites are a good example of an absorption-dominant shielding mechanism. Perfect absorption occurs if there is no transmission of EM radiation from the materials (Figure 12.2b). The third mechanism is the multiple reflections that are caused by successive reflections from different surfaces and interfaces of the thin slab. In the case of different structural designs, such as layer-by-layer heterostructure (Figure 12.3a), porous structure (Figure 12.3b), or segregated structure (Figure 12.3c), consideration of internal scattering become important.

The interfaces of alternate layered assemblies of two phases, pores in porous structure and segregated conductive network provided by a segregated structure,

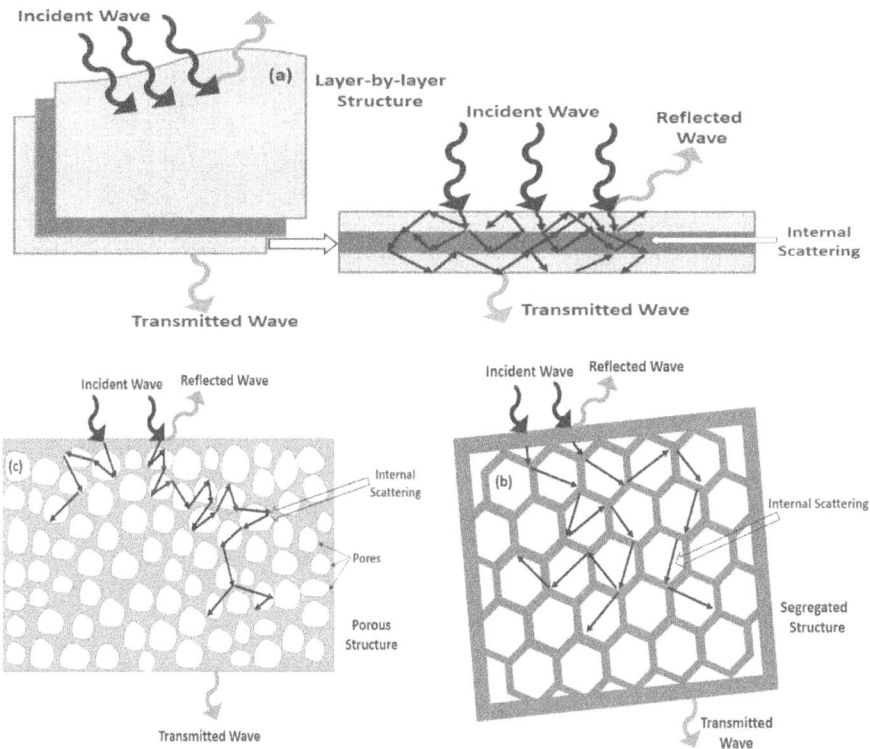

FIGURE 12.3 Architectures for MXene composites and hybrids that promote the internal multiple scattering: (a) layer-by-layer structure, (b) segregated structure, (c) porous structure.

increase the internal multiple reflections and dissipates heat EM energy through absorption, giving the desirable shielding performance [57].

The EM radiation gets trapped inside these types of architecture composites, and successive reflection occurs until the trapped EM wave is absorbed completely within the material. Thus, internal multiple scattering additionally gives rise to the absorption losses within the material that also helps to prevent the primary successive reflection from the material's surface. Multiple reflections are only active in thin materials. It is noteworthy that multiple reflection can be neglected if material's thickness is greater than the skin depth of the material. Skin depth (δ) is that depth of conducting material at which the incident EM wave attenuates to 1/e (i.e. 37%) of its initial value, after interacting with the shielding material. Skin depth is given by following relation

$$\delta = \sqrt{\frac{1}{\pi\sigma\mu f}} \qquad (12.1)$$

The total shielding efficiency (SE_T), or shielding effectiveness (SE), measures the EMI shielding performance in decibel (dB) units. It measures the change in intensity of the EM signal before and after the shielding mechanism. It is given by the ratio of incident field or power strength and transmitted field or power strength in the logarithmic scale. If P_I, E_I, and H_I are the incident intensities of power, electrical field, the magnetic field before attenuation of EM energy respectively. P_T, E_T and H_T are the transmitted intensities of power, electrical field, magnetic field then SE is given by

$$SE = 20 * log_{10} \frac{P_T}{P_I} = SE = 20 * log_{10} \frac{E_T}{E_I} = SE = 20 * log_{10} \frac{H_T}{H_I} \qquad (12.2)$$

In term of shielding effectiveness by reflection, absorption, and multiple reflections, total shielding effectiveness is given by the sum of all three components, that is

$$SE_T = SE_R + SE_A + SE_M \qquad (12.3)$$

Where SE_R, SE_A and SE_M represent shielding effectiveness by reflection, absorption and multiple reflection from the shielding material. Multiple reflection shielding mechanism depends on the material properties and also incident wave properties. In terms of conductivity (σ), frequency (f), and relative permeability (μ) and thickness (d), SE_R, SE_A, and SE_M are [58]:

$$SE_R = 39.5 + 10 * log \frac{\sigma}{2 f \pi \mu} \qquad (12.4)$$

$$SE_A = 8.7 d \sqrt{f \pi \sigma \mu} \qquad (12.5)$$

$$SE_M 20 * log_{10} \left(1 - e^{\frac{-2d}{\delta}} \right) \qquad (12.6)$$

SE$_R$ depends on the fields (electrical or magnetic or both) and varies accordingly, whereas SE$_A$ and SE$_M$ are independent of the kind of electrical or magnetic field. SE$_M$ can be avoided if SE$_A$>10 dB.

12.3   MXene AS EMI MATERIAL

The electrical conductivity of MXene with different compositions occurs in the range of 5 Scm^{-1} to 20,000 Scm^{-1} as reported by Iqbal et al. [59], which is the primary requirement for the reflection as well as for absorption. The multilayered architecture of MXene gives rise to the internal multiple scattering that enhances the attenuation by absorption. A proper combination of permittivity and permeability is very important for effective EMI shielding performance. Theoretically, some MXenes (e.g. Cr_2C and Cr_2N, Fe_2C) are found to possess the ferromagnetic ground states [60, 61], but synthesis procedure, surface terminated groups, number of layers, and defects highly influence the magnetic state of bare MXene [62]. Allen and coworkers employed the different chemical etching conditions to prepare the titanium-based MXenes (i.e. $Ti_3C_2T_x$). Variation in etching time and temperature led to paramagnetic-antiferromagnetic phase transition in this MXene, which manifest the importance of synthesis procedure in magnetic properties of MXene [63]. The poor magnetic properties of MXene can be improved by doping it with other materials such as nickel, flaky carbonyl iron [64, 65]. For example, stable ferromagnetism has been reported in layered 2D Nb-doped Ti_3C_2 MXene by Fatheema et al. [66]. Overall, the tunable electric and magnetic properties along with MXene's unique structure seek future opportunities against EMI pollution. In 2016, Shahzad and coworkers first explored the applicability of $Ti_3C_2T_x$ MXene laminate films (45-micrometer-thick) with exceptional EMI shielding effectiveness of 92 dB [67]. The excellent electrical conductivity of $Ti_3C_2T_x$ films and multiple internal reflection phenomenon from $Ti_3C_2T_x$ flakes in free-standing films are supposed to be responsible for exceptional EMI shielding performance [67]. In comparison with pristine $Ti_3C_2T_x$ MXene, the composites and hybrids of MXene not only improved the electrical properties of MXene, because moderated conductivity is required to prevent forefront reflection, but also improved the magnetic properties of nonmagnetic MXenes that are desirable for magnetic losses [4]. Therefore pristine MXenes can be decorated with different materials, including lightweight carbon materials 1D fillers (e.g., carbon nanofibers [68] and carbon nanotubes [69]); 2D fillers (e.g. graphene [70], reduced graphene oxide [71]); magnetic material such as ion ingredients (Fe_3O_4 [72], carbonyl iron [73]), nickel (Ni) [74], cobalt (Co) [75]; 2D transition metal dichalcogenides such as MoS_2 [76], polymers (polyaniline (PANI) [77], poly(3,4-ethylenedioxythiophene) polystyrene sulfonate (PEDOT:PSS) [78], polypropylene (PP) [79], polystyrene (PS) [80], polypyrrole (PPy) [81], polyvinylidene fluoride (PVDF) [82], polyvinyl alcohol (PVA) [83]), which have been developed to replace the traditional high-density metals in fulfilling the requirements for EMI shielding applications, that are anti-corrosive and light weighted. These composites offer advantageous properties, including good flexibility, better thermal stability, and high mechanical strength, with better environmental stability than the pristine MXene. In the fabrication of $Ti_3C_2T_x$ MXene-based composites and hybrids, structural design

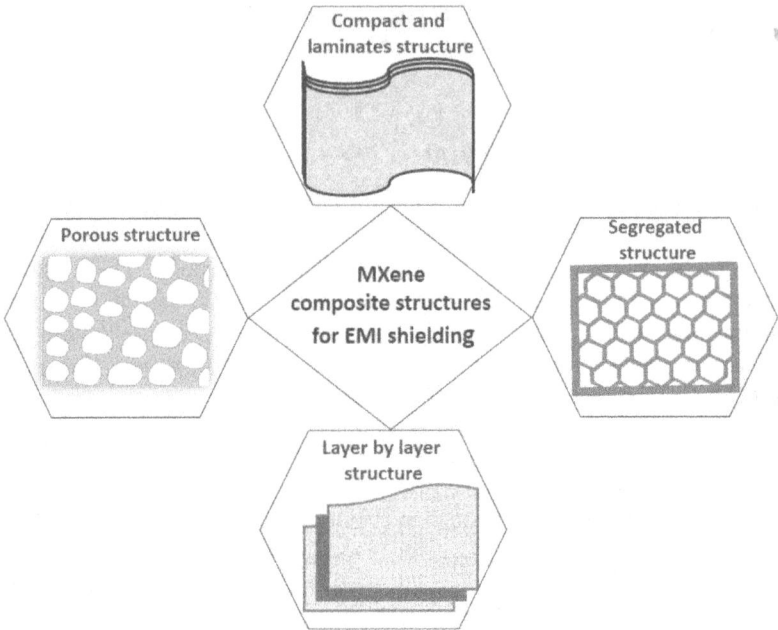

FIGURE 12.4 Schematic diagram of different types of MXene composite structures applicable in EMI shielding.

considerations are very important. MXene is a laminated structure because difficulties arise in peeling off the monolayer. So in the next sections, we have given the brief idea of making the $Ti_3C_2T_x$ MXene-based composites by the introduction of polymer, carbon, metals, and other materials, along with the idea of different architecture for these composites, as shown in Figure 12.4.

12.3.1 TITANIUM CARBIDE MXene-BASED COMPOSITES

Titanium carbide ($Ti_3C_2T_x$) is the most studied MXene by the material scientist due to its low production cost and ease of synthesis, along with a series of advantages over other MXenes [44]. Even though Ti_3C_2 MXene depicts an excellent potential against EMI pollution, there are difficulties of poor impedance matching and incompatibility between permittivity and permeability. Therefore, $Ti_3C_2T_x$ MXenes can be readily integrated with the magnetic nanoparticles, textiles, polymers and 2D graphene, and other materials. Combining the bare Ti_3C_2 MXene with these materials gives the advantage of synergy among them, which led to the proper impedance matching between permittivity and permeability that is crucial for reflection and absorption of EM waves for the shielding material [84]. He and coworkers prepared the $CoFe_2O_4$ nanoparticles decorated Ti_3C_2 MXene composites [85]. The presence of $CoFe_2O_4$ with Ti_3C_2 not only controls the complex permittivity by increasing the percolation threshold but also produces the additional magnetic loss. As a result, a minimum reflection loss value of -30.9 dB is obtained [85]. In comparison to single

2D homogeneous structures, a combination of 2D/2D heterostructures shows the improved properties caused by collective advantages of each 2D material due to a synergistic effect between the two [86]. Quin et al. studied the $Ti_3C_2T_x$/h-BN hybrid films fabricated via coulombic assembly between $Ti_3C_2T_x$ MXene and h-BN nanosheet and ultrasonic blending route. The $Ti_3C_2T_x$/h-BN film showed the improved electrical conductivity of 57.67 S/cm and EMI SE of 37.29 dB [31]. Using $Ti_3C_2T_x$ as a matrix material, Wang et al. fabricated $Ti_3C_2T_x$/Fe_3O_4@PANI composite film with sandwich intercalation structure that was light weight and flexible [77], and maximum EMI SE value of 58.8 dB was achieved even at a low thickness of only 12.1 μm in comparison to pure $Ti_3C_2T_x$ films of the same thickness. It is anticipated that the existence of Fe_3O_4@PANI with $Ti_3C_2T_x$ balances the impedance matching and enhances the magnetic loss mechanism. Surface defects, heterogeneous interfaces, and interlayer structures effectively enhance the dielectric loss in Fe_3O_4@PANI/$Ti_3C_2T_x$. Yoa et al. reported MXene/graphene@Fe_3O_4/PVA films with excellent flexibility and high EMI shielding performance [83]. In the schematic diagram of the synthesis process of MXene/graphene@Fe_3O_4/PVA shown in Figure 12.5a, initially, few-layer (FL) MXene were synthesized by etching-assisted exfoliation in HF solution that is widely used to prepare the monolayer and multilayer MXene for the massive production. The ball milling method is adopted to mix FL MXene and graphene while ternary MXene/graphene@Fe_3O_4 composite was formed by ultrasonication route. Further simple casting method was used to prepare the ternary MXene/graphene@Fe_3O_4/PVA film. The SE_T, SE_R, and SE_A values of PVA, MXene/PVA, MXene/graphene/PVA, and MXene/graphene@Fe_3O_4/PVA are shown in Figure 12.5(b, c, d), respectively. Because of some reflection and absorption of EM wave from pure PVA film, it showed SE value 7–15 dB in the X band (8–12 GHz). Further, with the introduction of MXene, graphene, and Fe_3O_4 particles sequentially, EMI shielding performance of PVA films improved with respect to the former. Maximum EMI SE of 36 dB was achieved in MXene/graphene@Fe_3O_4/PVA film at the thickness of 1 mm (Figure 12.5b), which results from the magnetic-dielectric synergistic effect. In comparison to SE_R (Figure 12.5c), PVA films showed absorption-dominated attenuation, as depicted in Figure 12.5d. MXene and graphene in PVA matrix form a conductive network structure that promotes electronic polarization in MXene/graphene@Fe_3O_4/PVA film. In addition, fillers (MXene, graphene, Fe_3O_4) with multisurface and multi-interface produce the multiple reflections and scattering loss of EM wave within the film (Figure 12.5e).

Also, conductive MXene/graphene and magnetic Fe_3O_4 improved the impedance mismatching that helps to increase the absorption losses of EM waves. SE_A contributes more than SE_R, so absorption mechanism is proposed as the main EMI shielding mechanism. Zheng et al. prepared reduced graphene oxide (RGO)/$Ti_3C_2T_x$ MXenes decorated with cotton fabrics (RMCs) [71]. They used the facile and scalable dip-coating and spray-coating routes to prepare RMCs. It is anticipated that when MXenes functionalize the RGO, strong interfacial interaction takes place through Ti-O-C covalent bonding between RGO and MXene sheets [87]. Sample RMC-4 showed the maximum EMI SE of 29.04 dB due to high electrical conductivity, whereas 4 in RMC-4 represent the number of cycles in the spray-coating method that was used to increase the loading of MXene in RMCs. The SE_A value being

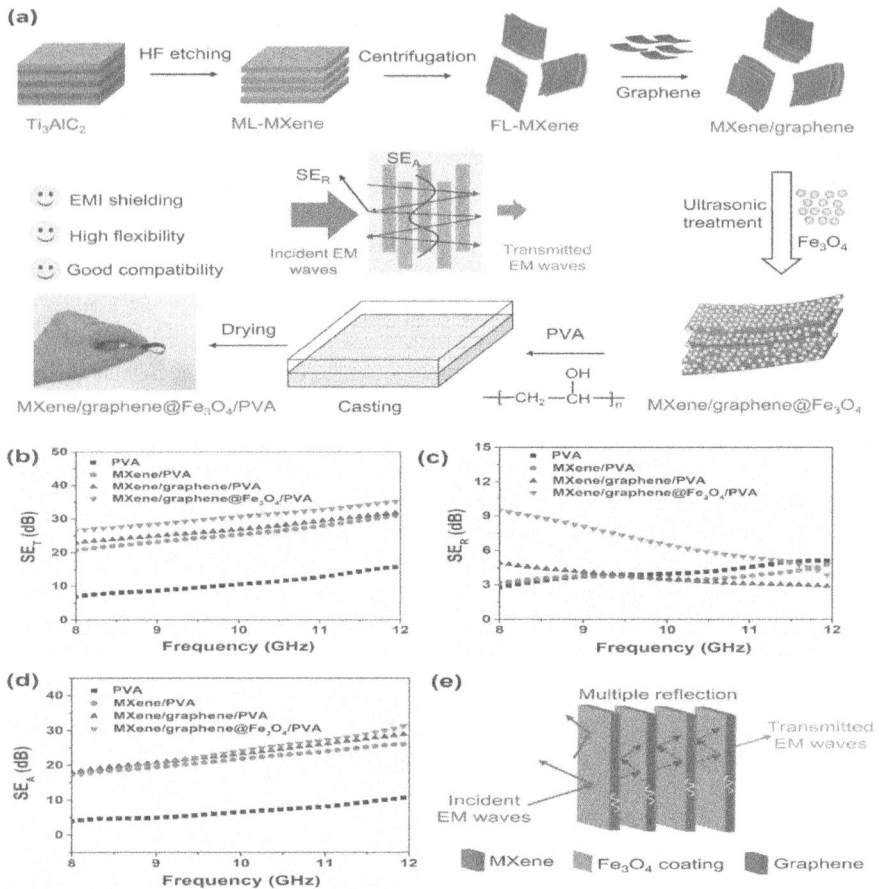

FIGURE 12.5 (a) Schematic of synthesis and mechanism of EMI shielding of MXene/graphene@Fe$_3$O$_4$/PVA film; (b) SE$_T$, (c) SE$_R$, (d) SE$_A$ of PVA and PVA composite films; (e) illustration of shielding mechanism in MXene/graphene@Fe$_3$O$_4$ ternary composites.

([83], reproduced by permission of Elsevier.)

higher than SE$_R$ in RMCs indicates that EM wave absorbed more while reflected less from the surface. The porous structure of RMC forms many scattering centers and interfaces that cause the scattering, internal reflections from the porous RMCs, and then enhance the absorption of EM waves. Therefore, effectively, an absorption-dominant EMI shielding mechanism controls the total shielding efficiency in RMCs. Xia and coworkers fabricated Ti$_3$C$_2$ MXene of loosely packed accordion-like structure and adsorbed by three metal ions (Fe^{3+}/Co^{2+}/Ni^{2+}). They obtained improved EMI SE for Ti$_3$C$_2$:Fe^{3+}/Co^{2+}/Ni^{2+} in comparison with pure Ti$_3$C$_2$ film. Metal ions adsorption on MXene improves the electrical conductivity of Ti$_3$C$_2$:Fe^{3+}/Co^{2+}/Ni^{2+} film. As a result, the EM wave attenuates from the material through the flow of electrons from conducting network. In addition, internal scattering occurs from numerous interfaces that enhance the total shielding performance of Ti$_3$C$_2$:Fe^{3+}/Co^{2+}/Ni^{2+}

films [88]. Li et al. fabricated the uniform yolk-shell hydrogel-core@void@MXene shell poly(N-isopropylacrylamide)@void@polystyrene@MXene designated by PNIPAM@void@PS@$Ti_3C_2T_x$ nanocomposite microspheres via a green and facile plasmolysis-inspired method [89]. The EMI SE value of 72 dB was obtained by PNIPAM@void@PS@$Ti_3C_2T_x$ at a very low 25 μm thickness. It is supposed that the synergistic effect of multilayered microspheres, the 3D network structure of the film, the interaction between conductive microspheres, and numerous void spaces and interior barriers in the 3D network structure of the film are responsible for the superb EMI performance of PNIPAM@void@PS@$Ti_3C_2T_x$ nanocomposite microspheres. Deng et al. fabricated the ultra-light and conductive $Ti_3C_2T_x$ MXene/acidified carbon nanotube (aCNT) anisotropic aerogels (MCAs) by directional freezing and freeze-drying method by varying the mass ratios of aCNTs in the MXene/aCNT mixtures (0, 5, 10, 20, 40, and 100%) and labeled as MA, MCA-1, MCA-2, MCA-3, MCA-4, and CA, respectively [90]. In MCAs, an anisotropic and porous skeleton is constructed by MXene nanosheets while pore walls of these nanosheets are reinforced by aCNTs that make the MCAs compressible along with super elastic. The hydrophilic feature of the aqueous suspension of aCNTs and the aqueous suspension of MXene nanosheets gives the homogeneous suspension when these two are added. The subsequent directional freezing process causes the nucleation of ice crystals that grow vertically from the bottom of the container. In the process of freeze-drying, anisotropic porous MCAs were achieved by subliming these ice pillars, as shown in Figure 12.6a. In comparison to the conductivity of CA (0.5 S m^{-1}), MCAs show

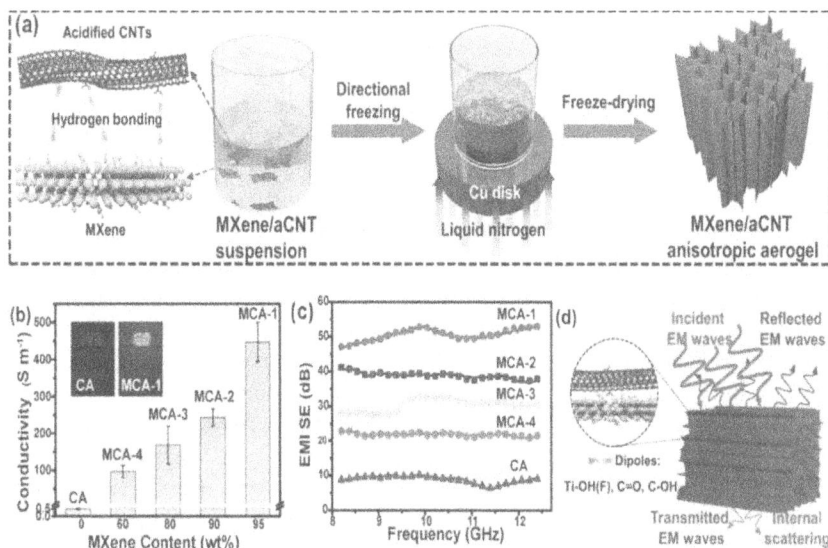

FIGURE 12.6   (a) Schematic representation of the preparation of the MXene/aCNT aerogel; (b) electrical conductivity vs MXene content for MCAs; (c) plots of EMI shielding effectiveness of MCAs; (d) schematic of EMI shielding mechanisms of MCA.

([90], reproduced by permission of the American Chemical Society.)

enhanced conductivities (more than 100 Sm^{-1}), as depicted in Figure 12.6b. Among all, MCA-1, in which 5 wt% aCNT is incorporated with MXene, with low density of 9.1 mg cm^{-3} showed the maximum conductivity of 447.2 Sm^{-1}. The better EMI shielding performances, more than 20 dB, were obtained for all of the MCA samples in comparison to EMI SE 10 dB of CA (Figure 12.6c), and high conductivity MCA-1 showed the maximum SE of 51 dB in the X-band. The high electrical conductivity, porous architecture, and dipole polarization are considered to be responsible for improving the EMI shielding performance in MCAs. When EM waves strike the MCA-1, the functional groups (Ti-OH(F), C=O, C-OH) on the MXene and aCNTs could act as dipoles, giving the dipolar polarization by polarizing in alternating electric field direction and causing the dissipation of EM wave energy. In addition, the scattering of EM waves facilitated from this porous aerogel structure extends the propagation path of EM waves inside MCA and helps in the dissipation of EM energy.

Zhang et al. prepared the MXene nanosheets sandwiched between Ag nanowire (AgNW) polyimide (PI) fiber mats for EMI shielding via electrospinning method [91]. The wet chemical etching method was adopted to synthesize the few layers of thickness Ti$_3$C$_2$T$_x$ MXene films in which solution of HCl and LiF was used as the solvent (Figure 12.7a). The presence of surface-terminated groups such as -OH, -F, and -O on the surface make MXene highly hydrophilic. This is the reason that diluted MXene showed the Tyndall effect, as depicted in Figure 12.7b. XRD pattern Ti$_3$AlC$_2$ shows the intense Al peak at (104) while this Al (104) peak is disappeared for MXene. While the peak at (002) changed to a lower angle side for MXene that indicates the loss of the raw Al layer and increase layer spacing in the MXene, respectively (Figure 12.7c). TEM image shows the clear MXene sheet (Figure 12.7d) of a thickness of 2.25 nm that was confirmed by the AFM image (Figure 12.7e). Combining electrospinning, spraying, and hot pressing processes, AP$_x$M$_y$AP$_x$ films ($x = 1$, 2, 3, 5; $y = 6$, 12, 18, 30) of layer-by-layer architecture were formed (Figure 12.7f). The EMI performance of AP$_x$ films with different AgNW contents in the X-band is depicted in Figure 12.7g. The addition of AgNWs improves the EMI SE of films in contrast to the no shielding effects of polyimide. The 5 wt% AgNW-PI showed EMI SE (5 dB) better than 3 wt% AgNW-PI film SE (3 dB) in the X-band. The EMI SE diagram of the APMyAP films with different amounts of sprayed MXene is shown in Figure 12.7h. The MXene layer in the sandwich film is the main shielding layer. Increasing MXene, the thickness of the MXene layer gradually increases, and the EMI SE showed a gradual increase, accordingly. Maximum EMI SE of 40.73 dB in APMAP film was found at 30 mg cm^{-1} spraying amount MXene. Zhang et al. proposed the mechanism when EM waves strike on the surface of film from the outside; AgNWs in the AP fibers interact with the high-density electron carriers, causing ohmic losses due to induction current, which led to partial attenuation of the striking EM waves. The remaining EM waves enter the interior of the material and interact with the conducting MXene layer that reflected back the EM waves to the AP layer due to the impedance mismatch. Moreover, in the interior of the film, the multiple reflections of EM waves also cause heat dissipation of EM waves and improve the overall performance of the film. In EMI shielding applications, structural design plays a crucial role. As we already explained, multiple internal scattering promotes

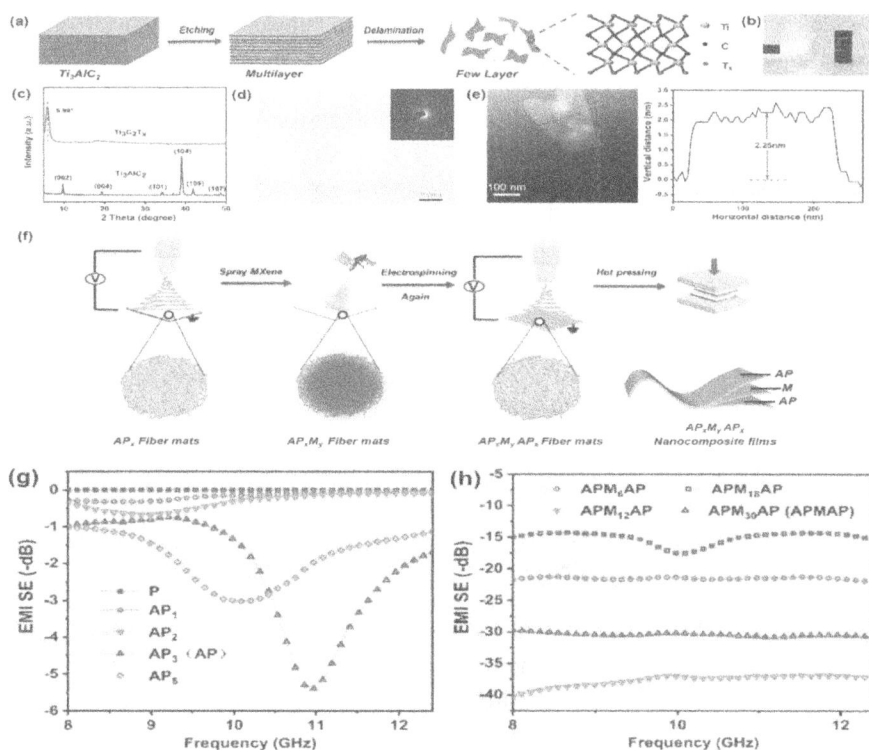

FIGURE 12.7 (a) Preparation of few-layered $Ti_3C_2T_x$ nanosheets by wet chemical etching method; (b) the Tyndall effect was observed in $Ti_3C_2T_x$ MXene solution under laser pointer irradiation; (c) XRD patterns of Ti_3AlC_2 and $Ti_3C_2T_x$; (d) TEM images of $Ti_3C_2T_x$; (e) AFM images and the thickness of $Ti_3C_2T_x$; (g) EMI SE of the AP film with different AgNW contents; (h) EMI SE of APMyAP at different MXene loadings.

([91], reproduced by permission of the American Chemical Society.)

the dissipation of EM wave energy in the form of Joule heating. So, to take advantage of multiple internal reflections of EM waves, different types of MXene-based heterogeneous structures are in trends, such as layer-by-layer structure, segregated structure, and porous structure that includes foam and aerogels.

12.4 LAYER-BY-LAYER ARCHITECTURE (LBL) OF MXene COMPOSITE

Layer-by-layer (LBL) assemblies form through the materials of two different impedances that are enriched with many internal interfaces and promote the internal scattering from these interfaces. The multiple internal scattering, additionally, helps in the attenuation of the EM waves. To make the LBL assemblies of MXene composites, solution casting, spin-coating, dip-coating, and spray-coating methods are widely used [92–95]. Lan and coworkers made the MXene/insulative polymer coating with an alternating structure via a step-wise assembly technique. They found 138.95% enhancement of EMI SE in the multilayered coating in comparison to pure MXene. The improved dielectric properties and multiple internal reflections of EM waves are supposed to be responsible for excellent EMI SE [96]. Zhang et al. observed absorption-dominant performance of $Ti_3C_2T_x$ MXene/non-woven laminated fabrics. They coated three different non-woven fabrics made from polyester, cotton, and calcium alginate on the $Ti_3C_2T_x$ MXene [97]. The presence of $Ti_3C_2T_x$ builds the conductive layer on the fiber. Due to heterogeneous interfaces in $Ti_3C_2T_x$ MXene/non-woven laminated fabrics, multiple reflections among $Ti_3C_2T_x$ MXene sheets are supposed to contribute to attenuation of the EM waves. The maximum EMI SE of 25.26 dB was achieved in calcium alginate/$Ti_3C_2T_x$ MXene at a frequency of 12.4 GHz with a fabric thickness of 3.17mm [97]. Tan and coworkers synthesized chitosan (CS)/$Ti_3C_2T_x$ MXene multilayered film by varying weight percentage of MXene (9.1 wt%, 14.7 wt%, and 25.9 wt%) in the CS/MXene film [98]. They obtained the $Ti_3C_2T_x$ MXene nanosheets by the chemical etching method using LiF/HCl etchant, then synthesized CS/$Ti_3C_2T_x$ composites film by layer-by-layer assembly method, as shown in Figure 12.8a. In contrast to CS conductivity ($3.85*10^{-5}$ S cm^{-1}), the addition of MXene with CS in CS/MXene film gives a sharp rise in conductivity due to the formation of the conductive network. The CS/MXene-25.9 composite film showed the maximum value of electrical conductivity of 969 S m^{-1}, as depicted in Figure 12.8b. The pure CS film shows the SE_T value of 1.6 dB, that is, not sufficient to block EM waves. However, CS/MXene multilayer films showed the SE_T values >20 dB. For CS/MXene-25.9, maximum SE_T of 40.8 dB was found, as shown in Figure 12.8c. The schematic diagram of the EMI shielding mechanism is depicted in Figure 12.8d. EM waves were exposed on CS/MXene alternating film, then, due to the impedance matching, only a few EM waves get reflected from the outermost CS layer while most of the EM waves penetrate within the film and interact with the high electron density of MXene to generate strong eddy current that causes the dissipation of EM waves in the form of heat energy. Meanwhile, the presence of multiple and continuous layer interfaces, in alternating CS/MXene multilayered composites, enhances the interfacial polarization loss, which promotes the high absorption of EM waves. The alternating film architecture provides the continuous layer interface that behaves as a reflected surface to yield successive internal scattering and multiple reflections, so the EM wave does not escape from the material until it is completely absorbed and transformed to heat [98].

Mao et al., initially, prepared tannic acid-modified $Ti_3C_2T_x$ MXene (TA-MXene) by selective etching of the Al layer from bulk Ti_3AlC_2 with LiF/HCl and TA-assisted liquid exfoliation (Figure 12.9a). Subsequently, P, N-co-doped cellulose nanocrystals (CNC) (PA@PANI@CNC) were prepared through the in situ oxidative polymerization method (Figure 12.9b), then alternative deposition of TA-MXene and PA@PANI@CNC onto cotton fabrics were obtained by the LBL self-assembly method (Figure 12.9c), where TA acts as antioxidant and stabilizer and also confers the numerous phenolic hydroxyl groups to $Ti_3C_2T_x$ nanosheets that help to form the hydrogen-bonding interaction with oxygen-containing groups in CNC [99]. Pure cotton shows zero electrical conductivity, indicating that it is transparent to EM waves. After depositing PA@PANI@CNC and TA-$Ti_3C_2T_x$ of 10, 15, and 20 bilayers on

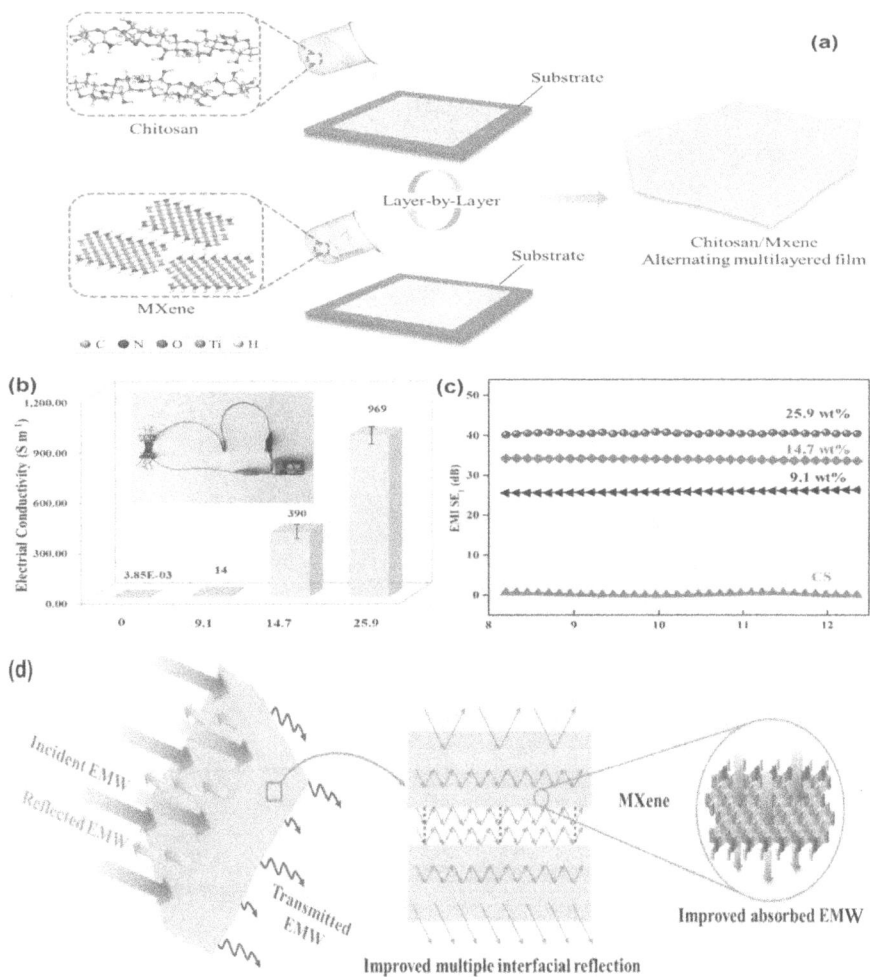

FIGURE 12.8 (a) Schematic representation of the fabrication process of CS/MXene multilayered film using LBL assembly strategy; (b) electrical conductivity of CS/MXene films with the change of MXene fraction; (c) EMI shielding curves of neat CS and CS/MXene films in X band; (d) schematic illustration of the EMI shielding mechanism.

([98], reproduced by permission of Elsevier.)

cotton fabrics, electrical conductivity improved. The cotton-10BL, cotton-15BL, and cotton-20BL showed 30.9, 86.3, and 142.0 S/m electrical conductivity, respectively (Figure 12.9d). Due to the poor conductivity of pure cotton fabric, it exhibits a zero SE value, which is improved subsequently for cotton-10BL, cotton-15BL, and cotton-20BL, and maximum SE ~21 dB was achieved by cotton-20BL over the X-band frequency range (Figure 12.9e). All SE_R, SE_A, and SE_T values are shown in Figure 12.9f, and Figure 12.9g depicts the ratio of SE_A/SE_R with cotton-10BL, cotton-15BL, and cotton-20BL. The smaller SE_R value over SE_A for all treated cotton revealed the

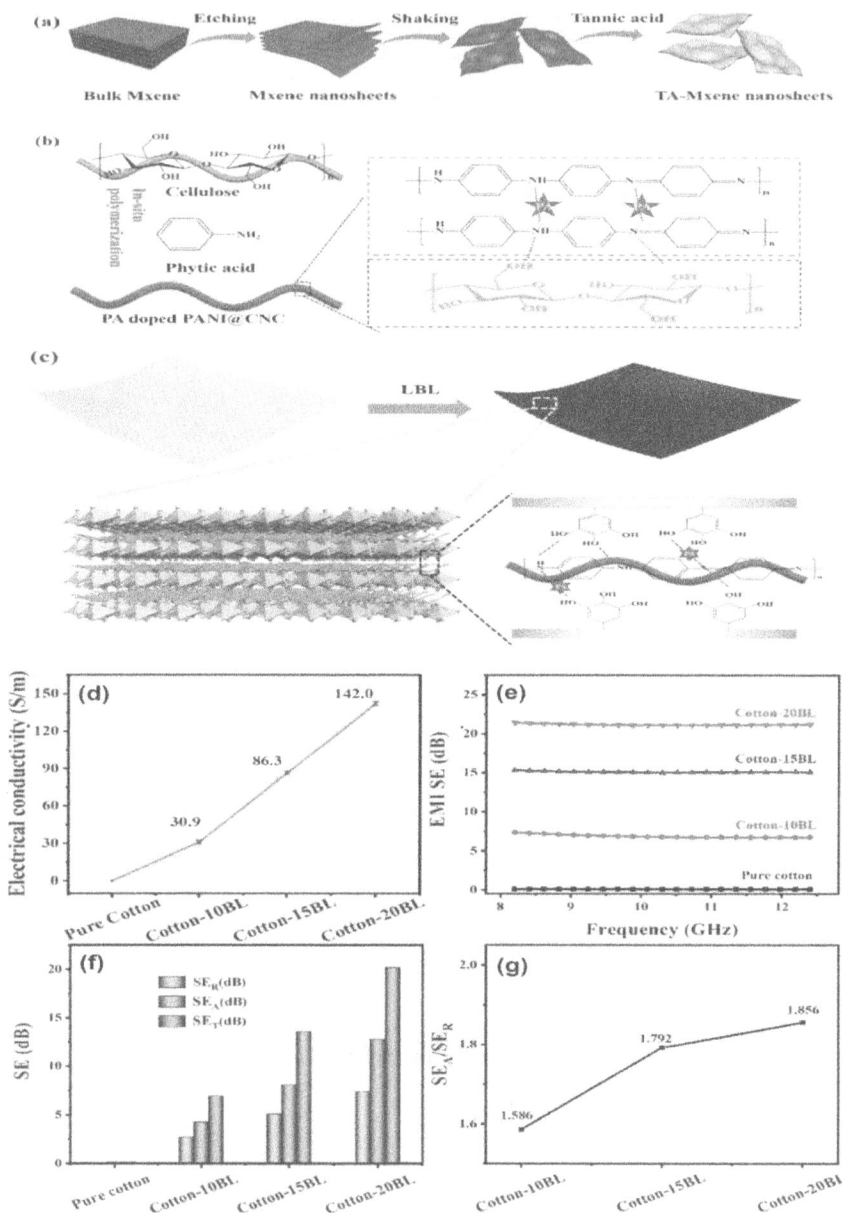

FIGURE 12.9 (a) Schematic illustration of synthesis process of the TA-Ti$_3$C$_2$T$_x$ nanosheets; (b) schematic diagram of procedure for the synthesis of PA@PANI@CNC; (c) schematic diagram of the structure of nanocoatings on cotton fabrics; (d) electrical conductivity versus pure cotton, cotton-10BL, cotton-15BL, and cotton-20BL plot; (e) EMI SE plot as the function of pure cotton, cotton-10BL, cotton-15BL, and cotton-20BL; (f) SE$_R$, SE$_A$, and SE$_T$ for pure and treated cotton fabrics in X-band; (g) the SE$_A$/SE$_R$ ratio for treated cotton fabrics.

([99], reproduced by permission of Elsevier.)

absorption dominant shielding rather than the reflection. Increasing the depositions cycles, the SE_A/SE_R value increases. The interaction of EM wave with PA@PANI@ CNC/$Ti_3C_2T_x$ coated cotton fabric gives rise the some reflected EM from the surface due to the impedance mismatch caused by conductive $Ti_3C_2T_x$. The rest of the microwaves pass through the $Ti_3C_2T_x$ and interact with the charge carriers, such as electrons, holes, and dipoles of $Ti_3C_2T_x$ to generate currents, and attenuates the energy of the waves through the ohmic losses. In PA@PANI@CNC/$Ti_3C_2T_x$ coated cotton fabric, multiple internal reflections between the $Ti_3C_2T_x$ and 1D CNC and dipoles on PANI and $Ti_3C_2T_x$ promote attenuation of EM microwaves and improved overall shielding performance.

12.5 POROUS STRUCTURE (FOAM AND AEROGELS) FOR EMI SHIELDING

In EMI shielding applications, lightweight is the desirable parameter for the material. Two types of architectures of MXenes porous materials – foams and aerogels – have been reported for improved EMI SE in comparison to pristine MXenes. The advantage of pores inside the porous structure not only reduces the density of the material but also increases multiple internal reflections of EM waves, causing higher absorption. Liu and coworkers first synthesized the lightweight, flexible, and hydrophobic MXene foam [100]. It is suggested that the favorable porous structure gives rise to the EMI SE 70 dB in comparison to the unfoamed film counterpart (53 dB) because the porous structure enriched the interfaces for internal scattering that causes the high wave attenuation [100]. Due to their weak interface interaction, difficulties arise in making the freestanding, 3D porous structure of MXene with high flexibility [101]. The addition of a linking agent such as polymer, graphene with MXene forms a well-interconnected and bridged 3D structure. For example, Zhao and coworkers prepared the highly electrically conductive 3D $Ti_3C_2T_x$ MXene/reduced graphene oxide hybrid aerogels with outstanding EMI shielding performances (>50 dB in the X-band). The strong gelation feature of GO helps to form 3D MXene architectures [102]. Fan et al. synthesized MXene ($Ti_3C_2T_x$/graphene (rGO) hybrid (MX-rGO)) foam by processing MXene and GO through drying and reduction heat treatment. The incorporation of $Ti_3C_2T_x$ with rGO gives rise to the electrical conductivity (1000 Sm^{-1}) for MX-rGO in comparison to rGO foam (140 Sm^{-1}). The porous foam structure promotes a higher EM wave attenuation; as a result MX-rGO foam demonstrated an EMI SE value of 50.7 dB [70]. It is proposed that large surface area and interconnected porous conductive network are accountable for the high-quality attenuation of EM waves in MX-rGO foam [70]. On the other hand, Ru and coworkers [101] studied the MXene/reduced graphene oxide (MX/ rGO) aerogels for EMI shielding. They performed unidirectional freezing and subsequent mild chemical reduction treatment in which a mixture of hydroiodic acid and acetic acid glacial was used. The MX/rGO aerogel demonstrated a superior electrical conductivity (467 S/m) and EMI shielding performance (57.67 dB) for the weight ratio of MX/rGO 5:5 in MX/rGO hybrid aerogel. They proposed that the surface of MX/rGO hybrid aerogels consists of abundant free electrons that give a lot of paths themselves. EM waves fall on the surface of the aerogel then some parts get reflected immediately from the surface while the remaining EM waves enter the aerogels and pass through the conductive 3D network, which results in ohmic loss and reduces their energy. Additionally, multiple internal reflections within inner porous structures occur that increase the absorption loss. The existence of heterogeneous interfaces, surface functional groups, and structural defects between the MXene and rGO result in the dipole polarization that helps to attenuate the incident EM radiation. Zhai et al. prepared waste cotton fabric (WCF)/(Ti$_3$C$_2$T$_x$) MXene composite aerogel for EMI shielding through facile freezing-drying and dip-coating method (Figure 12.10a) and studied the EMI performance for WCF, WCF/MXene-3, WCF/MXene-6, WCF/MXene-9, and WCF/MXene-12 where 3,6,9,12 are the number of the dip-coating cycles [103]. The WCF aerogel shows almost zero SE value, which indicates the WCF aerogel is nearly transparent to EM radiation, but the addition of MXene on the surface of WCF aerogel gives rise to the tremendous increment in the EMI shielding performance, as shown in Figure 12.10b. Figure 12.10c depicts

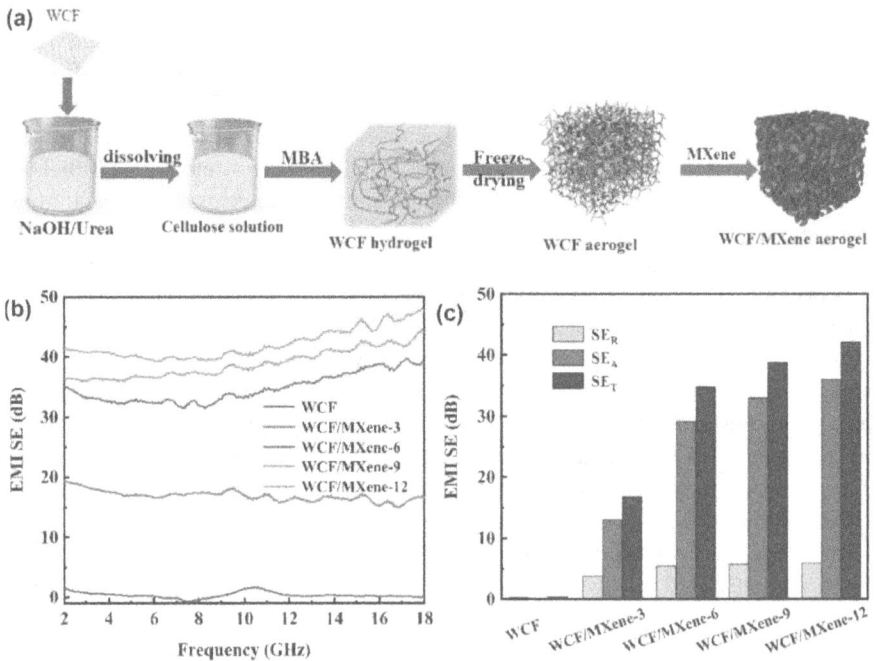

FIGURE 12.10 (a) Schematic illustration of the fabrication of WCF/MXene composite aerogels through facile freezing-drying and dip-coating method; (b) EMI SE of WCF and WCF/MXene aerogels in the frequency limit of 2–18 GHz; (c) average value of SE$_A$, SE$_R$, and SE$_T$ of WCF and WCF/MXene aerogels.

([103], reproduced by permission of Elsevier.)

average SE_A, SE_R and SE_T values of WCF, WCF/MXene-3, WCF/MXene-6, WCF/MXene-9 and WCF/MXene-12. With increasing the number of dip-coating cycles, SE_R, and SE_A shows the enhanced values 0.21dB, 0.17dB to 6.05 dB to 36.06 dB, respectively improving the overall EMI performance of WCF/MXene composites. It is proposed that by increasing the dip-coating cycles, the content of MXene coated on WCF aerogels enhances that make an integrated conductive network to improve the reflection and absorption ability of EM radiation. The absorption dominates over reflection with increasing the content of MXene by increasing dip-coating cycles on the surface of WCF. It is proposed that ohmic loss and energy dissipation of EM waves are caused by the MXene's conductive network due to high electron carrier density. While the 3D skeleton of WCF aerogel coated by MXene nanosheets provides a rich air-solid interface with multiple internal reflections and interface polarization loss. The EM wave, until completely dissipating, continues to reflect inside the 3D conductive network. Moreover, the surface functional group (-F, =O, or -OH) and local defects on the MXene nanoflake also produce dielectric loss owing to the asymmetric distribution of electrons. Including all factors, WCF/MXene aerogels demonstrate the high EMI shielding performance.

Liu et al. adopted the directional-freezing and freeze-drying methods to prepared MXene and MXene/AgNWs aerogels with varying mass ratios of 1:0, 1:0.5, 1:0.75, and 1:1 of MXene/AgNWs, respectively. Then their MXene/AgNWs/ epoxy nanocomposites, along with the aerogel skeleton, were synthesized by the vacuum-assisted infiltration method, as shown in Figure 12.11a. The 3D MXene with AgNWs aerogel provides the interpenetrating double-network structure in epoxy insulating matrix. Few-layer $Ti_3C_2T_x$ nanosheets with a lateral size of about 2–3 µm were seen in TEM image (Figure 12.11b). Figure 12.11c shows SEM image of the AgNWs showing the high aspect ratio that was prepared by the polyol method. With increasing the AgNWs content, electrical conductivity demonstrated the enhanced values owing to formation of various junction points between MXene and AgNWs that construct the numerous conductive network (Figure 12.11d). Electrically insulating epoxy is transparent to EM waves so its EMI SE performance was avoided. EMI SE of MXene/AgNWs/epoxy nanocomposites were found to be in increasing order with increasing the AgNWs content, according to variation of electrical conductivity. The MXAg-3/epoxy demonstrated maximum electrical conductivity (1532 S/m), thus showing a maximum EMI SE of 79.3 dB at the thickness of 2 mm (Figure 12.11e). A possible mechanism of EMI shielding and heat conduction by MXene/AgNWs/epoxy composites is shown Figure 12.11f [104]. The EM waves strike on MXene/AgNWs/epoxy nanocomposites, reflection occurs due to the impedance mismatch that reflects the major part of EM waves from the surface, and the remaining part enter the interior from the outside surface. The porous structure of aerogel provides a large specific surface area and numerous interfaces for multiple reflection or scattering of EM waves inside the material that improved the attenuation by absorption. Meanwhile, ordered structure in the aerogel facilitates electron transfer or migration through giving a conductive path that causes higher ohmic loss in

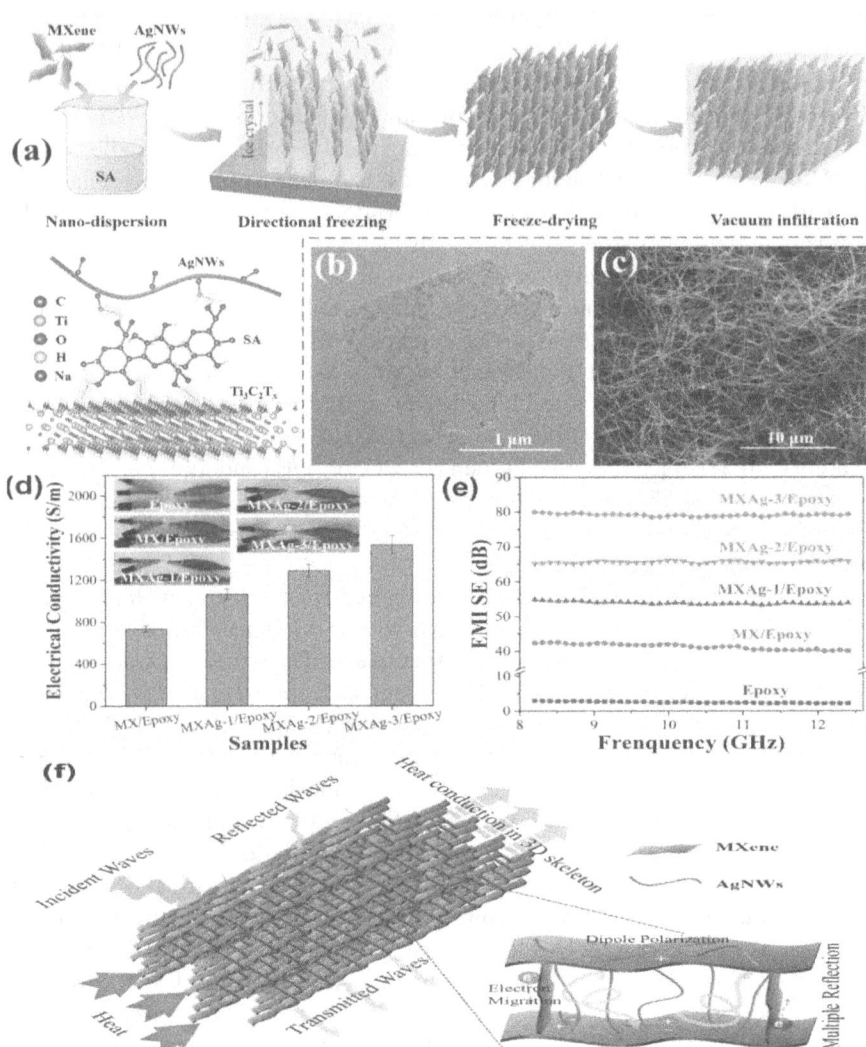

FIGURE 12.11 (a) Schematic illustration of the fabrication of MXene/AgNWs/epoxy composites; (b) TEM image of $Ti_3C_2T_x$ MXene; (c) SEM image of AgNWs; (d) electrical conductivity of MXene/AgNWs/epoxy composites; (e) EMI SE of MXene/AgNWs/epoxy composites with a thickness of 2 mm; (f) schematic illustration of possible mechanism EMI shielding and heat conduction in MXene/AgNWs/epoxy nanocomposites.

([104], reproduced by permission of Elsevier.)

composites with high electrical conductivity. The local defects and the accumulated charges in interface between MXene and AgNWs give rise to polarization losses (dipolar and interfacial) to EM waves. It is proposed that the interconnected path and porous structure of MXene/AgNWs aerogel in epoxy matrix effectively causes EM waves to dissipate heat energy [104].

12.6 SEGREGATED STRUCTURE OF MXene-BASED COMPOSITES FOR EMI SHIELDING

A larger portion of conductive filler volume is incorporated into the polymer matrix for achieving the high electrical conductivity of polymer composites that increases the production cost and leads the complex procedure. The segregated structure is formed at the low electrical percolation threshold value of conductive fillers where all the filler is segregated to form an interconnected framework of the continuous conductive network within an insulating polymer matrix. Lou et al. synthesized the stretchable, flexible MXene/natural rubber (NR) nanocomposite film by the vacuum filtration method and reported the high conductivity 1400 Sm^{-1} at 6.71 vol% of MXene in the NR matrix. As a result, segregated NR/MXene composite demonstrated the high EMI SE of 53.6 dB at 6.71 vol% MXene with NR matrix [105].

Ma and coworkers fabricated the segregated $Ti_3C_2T_x$/PDA-PEI@PP composites and observed their EMI shielding performance. To prepare the segregated structure of polymer-filler composites, two steps – uniform coating on polymer surfaces by conductive fillers and compaction of complex granules – have been performed. Initially, single/few-layer $Ti_3C_2T_x$ sheets were fabricated by etching Ti_3AlC_2 and followed by ultrasound exfoliation in water and ethanol mixed solution. In this etching process, various terminations groups (–OH, -O, and –F) were left on $Ti_3C_2T_x$ sheets that confer negative charge to $Ti_3C_2T_x$ sheets. Meanwhile, a partially cross-linked PDA-PEI shell with functional groups (such as carboxyl and amino) is employed to wrap PP granules by the oxidation polymerization of DA and the crosslinking reaction between DA and PEI [106]. As modified granules become hydrophilic and positively charged, they help to form of hydrogen bonds with polar groups of $Ti_3C_2T_x$ MXene. Further the use of hydrochloric acid (HCl) is supposed to weaken the electronegativity of $Ti_3C_2T_x$ and confer positive charge to PDA-PEI@PP that causes the self-assembling of flocculent $Ti_3C_2T_x$ onto the surface of PDA-PEI@PP granules and giving the $Ti_3C_2T_x$/PDA-PEI@PP granules, as depicted in Figure 12.12a. Subsequently, microwave selective sintering method was used to prepare the segregated $Ti_3C_2T_x$/PDA-PEI@PP composites. The electrical conductivity of the composite versus $Ti_3C_2T_x$ content is shown in Figure 12.12b. The least-square fitting results demonstrate the low percolation threshold of 0.02091 vol% due to formation of conductive path by MXene at the interfaces (Figure 12.12b inset) that is highly desirable of polymer-based composites. Figure 12.12c shows the EMI performance of the composites with different $Ti_3C_2T_x$ content. The EMI SE value for composites shows an increasing trend with increasing $Ti_3C_2T_x$ content. The sintered composite shows the highest EMI SE of 75.12~78.85 dB for 1.138 vol% $Ti_3C_2T_x$ MXene loaded composite. Absorption-dominated mechanism with low SE_R was found in $Ti_3C_2T_x$/PP composites. At low filler content, high EMI SE was achieved,because compact $Ti_3C_2T_x$ fillers generate a high-quality shielding barriers in the segregated composite to prevent the EMI pollution. EM waves experience multiple internal reflections/absorption from the interfacial regions when striking the $Ti_3C_2T_x$ flakes, intensity is reduced substantially, resulting in an overall dissipation of wave in the form of heat [106].

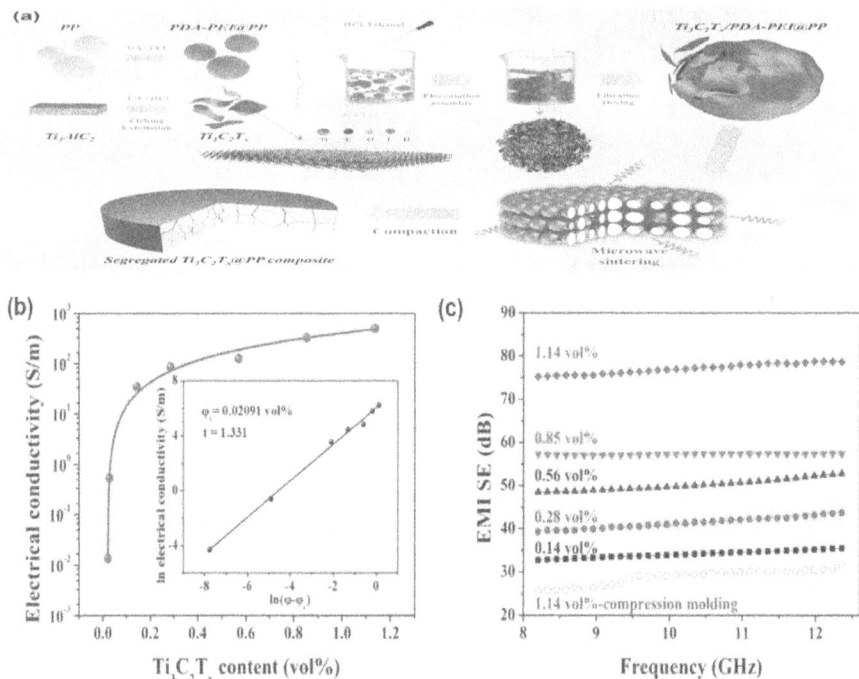

FIGURE 12.12 (a) Schematic diagram of the fabrication of the segregated $Ti_3C_2T_x$@/PP composite through the self-assembly and microwave selective sintering route; (b) electrical conductivity versus $Ti_3C_2T_x$ content for the segregated $Ti_3C_2T_x$/PDA-PEI@PP composites (inset shows ln conductivity versus $ln(\varphi-\varphi_c)$ line fitted by least square fitting); (c) EMI SE of segregated $Ti_3C_2T_x$/PDA-PEI@PP composites as the function of $Ti_3C_2T_x$ content.

([106], reproduced by permission of Elsevier.)

12.7 CONCLUSIONS AND FUTURE ASPECTS

High electrical conductivity, low density, tunable surface chemistry, excellent mechanical strength and easy processing properties, and intrinsic EMI shielding of titanium carbide MXenes have triggered research in the field of EMI shielding application. Different types of MXenes composite structures (such as layer-by-layer assemblies or porous and/or segregated structures) offer the advantage of multiple internal reflections that contribute to the overall shielding performance of MXenes composites and hybrids. Theoretically, the electrical and magnetic properties of more than 100 types of MXene of different compositions have been investigated, but few have been studied for EMI shielding applications. The $Ti_3C_2T_x$ MXene-based composites with 2D materials such as graphene, h-Bn sheet, MoS_2, and polymers have shown their excellency for EMI shielding application. Of course, other pristine MXenes of different compositions, the number of layers, and their composites and hybrids have an opportunity in EMI shielding and need to be explored.

ACKNOWLEDGEMENT

V. Shukla is thankful to the Ministry of Education, government of India, and IIT Kharagpur for providing financial support.

REFERENCES

1. D. Jiang, V. Murugadoss, Y. Wang, J. Lin, T. Ding, Z. Wang, Q. Shao, C. Wang, H. Liu, N. Lu, et al., Electromagnetic interference shielding polymers and nanocomposites – a review, Polymer Reviews 59 (2) (2019) 280–337.
2. T. T. Li, A. P. Chen, P. W. Hwang, Y. J. Pan, W. H. Hsing, C. W. Lou, Y. S. Chen, J. H. Lin, Synergistic effects of micro-/nano-fillers on conductive and electromagnetic shielding properties of polypropylene nanocomposites, Materials and Manufacturing Processes 33 (2) (2018) 149–155.
3. V. Shukla, Review of electromagnetic interference shielding materials fabricated by iron ingredients, Nanoscale Advances 1 (5) (2019) 1640–1671.
4. V. Shukla, The tunable electric and magnetic properties of 2D MXenes and their potential applications, Materials Advances 1 (9) (2020) 3104–3121.
5. S. Sankaran, K. Deshmukh, M. B. Ahamed, S. K. Pasha, Recent advances in electromagnetic interference shielding properties of metal and carbon filler reinforced flexible polymer composites: a review, Composites Part A: Applied Science and Manufacturing 114 (2018) 49–71.
6. D. Wanasinghe, F. Aslani, A review on recent advancement of electromagnetic interference shielding novel metallic materials and processes, Composites Part B: Engineering 176 (2019) 107207.
7. S. Geetha, K. S. Kumar, C. R. Rao, M. Vijayan, D. Trivedi, EMI shielding: Methods and materials – a review, Journal of Applied Polymer Science 112 (4) (2009) 2073–2086.
8. J. Kruželákak, A. Kvasničakóvá, K. Hložeková, I. Hudec, Progress in polymers and polymer composites used as efficient materials for EMI shielding, Nanoscale Advances 3 (1) (2021) 123–172.
9. V. Shukla, Role of spin disorder in magnetic and EMI shielding properties of Fe_3O_4/C/ PPy core/shell composites, Journal of Materials Science 55 (7) (2020) 2826–2835.
10. M. Panahi-Sarmad, M. Noroozi, X. Xiao, C. B. Park, Recent advances in graphene-based polymer nanocomposites and foams for electromagnetic interference shielding applications, Industrial & Engineering Chemistry Research 61 (4) (2022) 1545–1568.
11. L. Wang, Z. Ma, Y. Zhang, H. Qiu, K. Ruan, J. Gu, Mechanically strong and folding-endurance $Ti_3C_2T_x$ MXene/PBO nanofiber films for efficient electromagnetic interference shielding and thermal management, Carbon Energy 4 (2) (2022) 200–210.
12. S. Sankaran, K. Deshmukh, M. B. Ahamed, K. K. Sadasivuni, M. Faisal, S. K. Pasha, Electrical and electromagnetic interference (EMI) shielding properties of hexagonal boron nitride nanoparticles reinforced polyvinylidene fluoride nanocomposite films, Polymer-Plastics Technology and Materials 58 (11) (2019) 1191–1209.
13. D. Zhang, S. Liang, J. Chai, T. Liu, X. Yang, H. Wang, J. Cheng, G. Zheng, M. Cao, Highly effective shielding of electromagnetic waves in MoS_2 nanosheets synthesized by a hydrothermal method, Journal of Physics and Chemistry of Solids 134 (2019) 77–82.
14. L. Wang, L. Chen, P. Song, C. Liang, Y. Lu, H. Qiu, Y. Zhang, J. Kong, J. Gu, Fabrication on the annealed $Ti_3C_2T_x$ MXene/Epoxy nanocomposites for electromagnetic interference shielding application, Composites Part B: Engineering 171 (2019) 111–118.
15. T. Hu, X. Mei, Y. Wang, X. Weng, R. Liang, M. Wei, Two-dimensional nanomaterials: fascinating materials in biomedical field, Science Bulletin 64 (22) (2019) 1707–1727.

16. G. Jo, M. Choe, S. Lee, W. Park, Y. H. Kahng, T. Lee, The application of graphene as electrodes in electrical and optical devices, Nanotechnology 23 (11) (2012) 112001.

17. C. Yang, H. Huang, H. He, L. Yang, Q. Jiang, W. Li, Recent advances in MXene-based nanoarchitectures as electrode materials for future energy generation and conversion applications, Coordination Chemistry Reviews 435 (2021) 213806.

18. N. Joseph, P. M. Shafi, A. C. Bose, Recent advances in 2D-MoS$_2$ and its composite nanostructures for supercapacitor electrode application, Energy & Fuels 34 (6) (2020) 6558–6597.

19. S. Angizi, M. Khalaj, S. A. A. Alem, A. Pakdel, M. Willander, A. Hatamie, A. Simchi, Towards the two-dimensional hexagonal boron nitride (2D h-BN) electrochemical sensing platforms, Journal of The Electrochemical Society 167 (12) (2020) 126513.

20. B. Vellaichamy, P. Prakash, J. Thomas, Synthesis of AuNPs@RGO nanosheets for sustainable catalysis toward nitrophenols reduction, Ultrasonics Sonochemistry 48 (2018) 362–369.

21. M. A. Lukowski, A. S. Daniel, F. Meng, A. Forticaux, L. Li, S. Jin, Enhanced hydrogen evolution catalysis from chemically exfoliated metallic MoS$_2$ nanosheets, Journal of the American Chemical Society 135 (28) (2013) 10274–10277.

22. X. Zhang, Y. Gao, 2D/2D h-BN/N-doped MoS$_2$ heterostructure catalyst with enhanced peroxidase-like performance for visual colorimetric determination of H$_2$O$_2$, Chemistry–An Asian Journal 15 (8) (2020) 1315–1323.

23. N. T. Shelke, B. Karche, Hydrothermal synthesis of WS2/RGO sheet and their application in UV photodetector, Journal of Alloys and Compounds 653 (2015) 298–303.

24. Z. Wu, L. Wei, S. Tang, Y. Xiong, X. Qin, J. Luo, J. Fang, X. Wang, Recent progress in Ti$_3$C$_2$T$_x$ MXene-based flexible pressure sensors, ACS Nano 15 (12) (2021) 18880–18894.

25. V. Semwal, B. D. Gupta, Highly sensitive surface plasmon resonance based fiber optic pH sensor utilizing rGO-Pani nanocomposite prepared by in situ method, Sensors and Actuators B: Chemical 283 (2019) 632–642.

26. J. Zhu, E. Ha, G. Zhao, Y. Zhou, D. Huang, G. Yue, L. Hu, N. Sun, Y. Wang, L. Y. S. Lee, et al., Recent advance in MXenes: a promising 2D material for catalysis, sensor and chemical adsorption, Coordination Chemistry Reviews 352 (2017) 306–327.

27. P. Li, Z. Zhang, Self-powered 2D material-based pH sensor and photodetector driven by monolayer MoSe$_2$ piezoelectric nanogenerator, ACS Applied Materials & Interfaces 12 (52) (2020) 58132–58139.

28. K. Rasool, R. P. Pandey, P. A. Rasheed, S. Buczek, Y. Gogotsi, K. A. Mahmoud, Water treatment and environmental remediation applications of two-dimensional metal carbides (MXenes), Materials Today 30 (2019) 80–102.

29. S. Yu, H. Tang, D. Zhang, S. Wang, M. Qiu, G. Song, D. Fu, B. Hu, X. Wang, MXenes as emerging nanomaterials in water purification and environmental remediation, Science of The Total Environment 811 (2021) 152280.

30. J. Liang, Y. Wang, Y. Huang, Y. Ma, Z. Liu, J. Cai, C. Zhang, H. Gao, Y. Chen, Electromagnetic interference shielding of graphene/epoxy composites, Carbon 47 (3) (2009) 922–925.

31. K. Qian, Q. Zhou, S. Thaiboonrod, J. Fang, M. Miao, H. Wu, S. Cao, X. Feng, Highly thermally conductive Ti$_3$C$_2$T$_x$/h-BN hybrid films via coulombic assembly for electromagnetic interference shielding, Journal of Colloid and Interface Science 613 (2022) 488–498.

32. J. C. Lei, X. Zhang, Z. Zhou, Recent advances in MXene: Preparation, properties, and applications, Frontiers of Physics 10 (3) (2015) 276–286.

33. S. N. Li, Z. R. Yu, B. F. Guo, K. Y. Guo, Y. Li, L -X. Gong, L. Zhao, J. Bae, L. C. Tang, Environmentally stable, mechanically flexible, self-adhesive, and electrically conductive $Ti_3C_2T_x$ MXene hydrogels for wide-temperature strain sensing, Nano Energy 90 (2021) 106502.

34. M. D. Firouzjaei, M. Karimiziarani, H. Moradkhani, M. Elliott, B. Anasori, Mxenes: The two-dimensional influencers, Materials Today Advances 13 (2022) 100202.

35. M. Naguib, O. Mashtalir, J. Carle, V. Presser, J. Lu, L. Hultman, Y. Gogotsi, M. W. Barsoum, Two-dimensional transition metal carbides, ACS Nano 6 (2) (2012) 1322–1331.

36. M. Naguib, J. Halim, J. Lu, K. M. Cook, L. Hultman, Y. Gogotsi, M. W. Barsoum, New two-dimensional niobium and vanadium carbides as promising materials for Li-ion batteries, Journal of the American Chemical Society 135 (43) (2013) 15966–15969.

37. B. Anasori, Y. Xie, M. Beidaghi, J. Lu, B. C. Hosler, L. Hultman, P. R. Kent, Y. Gogotsi, M. W. Barsoum, Two dimensional, ordered, double transition metals carbides (MXenes), ACS Nano 9 (10) (2015) 9507–9516.

38. S. Venkateshalu, J. Cherusseri, M. Karnan, K. S. Kumar, P. Kollu, M. Sathish, J. Thomas, S. K. Jeong, A. N. Grace, New method for the synthesis of 2D vanadium nitride (MXene) and its application as a supercapacitor electrode, ACS Omega 5 (29) (2020) 17983–17992.

39. Y. Guo, S. Jin, L. Wang, P. He, Q. Hu, L. Z. Fan, A. Zhou, Synthesis of two-dimensional carbide Mo_2CT_x MXene by hydrothermal etching with fluorides and its thermal stability, Ceramics International 46 (11) (2020) 19550–19556.

40. J. Zhou, X. Zha, X. Zhou, F. Chen, G. Gao, S. Wang, C. Shen, T. Chen, C. Zhi, P. Eklund, et al., Synthesis and electrochemical properties of two-dimensional hafnium carbide, ACS Nano 11 (4) (2017) 3841–3850.

41. J. Zhou, X. Zha, F. Y. Chen, Q. Ye, P. Eklund, S. Du, Q. Huang, A two-dimensional zirconium carbide by selective etching of Al_3C_3 from nanolaminated $Zr_3Al_3C_5$, Angewandte Chemie International Edition 55 (16) (2016) 5008–5013.

42. B. Anasori, M. R. Lukatskaya, Y. Gogotsi, 2D metal carbides and nitrides (MXenes) for energy storage, Nature Reviews Materials 2 (2) (2017) 1–17.

43. A. VahidMohammadi, J. Rosen, Y. Gogotsi, The world of two-dimensional carbides and nitrides (MXenes), Science 372 (6547) (2021) eabf1581.

44. B. Anasori, U. G. Gogotsi, 2D metal carbides and nitrides (MXenes), Vol. 416, Springer, 2019.

45. F. Qi, L. Wang, Y. Zhang, Z. Ma, H. Qiu, J. Gu, Robust $Ti_3C_2T_x$ MXene/starch derived carbon foam composites for superior EMI shielding and thermal insulation, Materials Today Physics 21 (2021) 100512.

46. S. Bai, M. Yang, J. Jiang, X. He, J. Zou, Z. Xiong, G. Liao, S. Liu, Recent advances of MXenes as electrocatalysts for hydrogen evolution reaction, npj 2D Materials and Applications 5 (1) (2021) 1–15.

47. M. W. Barsoum, M. Radovic, Elastic and mechanical properties of the MAX phases, Annual Review of Materials Research 41 (2011) 195–227.

48. M. Kurtoglu, M. Naguib, Y. Gogotsi, M. W. Barsoum, First principles study of two-dimensional early transition metal carbides, MRS Communications 2 (4) (2012) 133–137.

49. G. Wang, Theoretical prediction of the intrinsic half-metallicity in surface-oxygen-passivated Cr_2N MXene, The Journal of Physical Chemistry C 120 (33) (2016) 18850–18857.

50. C. Si, J. Zhou, Z. Sun, Half-metallic ferromagnetism and surface functionalization-induced metal–insulator transition in graphene-like two-dimensional Cr_2C crystals, ACS Applied Materials & Interfaces 7 (31) (2015) 17510–17515.

51. P. Song, B. Liu, H. Qiu, X. Shi, D. Cao, J. Gu, MXenes for polymer matrix electromagnetic interference shielding composites: a review, Composites Communications 24 (2021) 100653.

52. G. Gao, G. Ding, J. Li, K. Yao, M. Wu, M. Qian, Monolayer MXenes: promising half-metals and spin gapless semiconductors, Nanoscale 8 (16) (2016) 8986–8994.

53. L. Hu, X. Wu, J. Yang, Mn_2C monolayer: a 2D antiferromagnetic metal with high Néel temperature and large spin–orbit coupling, Nanoscale 8 (26) (2016) 12939–12945.

54. K. S. Sista, S. Dwarapudi, D. Kumar, G. R. Sinha, A. P. Moon, Carbonyl iron powders as absorption material for microwave interference shielding: a review, Journal of Alloys and Compounds 853 (2021) 157251.

55. T. Kim, H. W. Do, K. J. Choi, S. Kim, M. Lee, T. Kim, B. K. Yu, J. Cheon, B.W. Min, W. Shim, Layered aluminum for electromagnetic wave absorber with near-zero reflection, Nano Letters 21 (2) (2021) 1132–1140.

56. J. D. Livingston, Electronic properties of engineering materials, Wiley New York, 1999.

57. A. Iqbal, P. Sambyal, C. M. Koo, 2D MXenes for electromagnetic shielding: a review, Advanced Functional Materials 30 (47) (2020) 2000883.

58. V. Shukla, Advances in hybrid conducting polymer technology for EMI shielding materials, in: Advances in Hybrid Conducting Polymer Technology, Springer, 2021, pp. 201–247.

59. A. Iqbal, J. Kwon, M. K. Kim, C. Koo, MXenes for electromagnetic interference shielding: experimental and theoretical perspectives, Materials Today Advances 9 (2021) 100124.

60. Y. Yue, Fe_2C monolayer: an intrinsic ferromagnetic MXene, Journal of Magnetism and Magnetic Materials 434 (2017) 164–168.

61. M. Khazaei, M. Arai, T. Sasaki, C. Y. Chung, N. S. Venkataramanan, M. Estili, Y. Sakka, Y. Kawazoe, Novel electronic and magnetic properties of two-dimensional transition metal carbides and nitrides, Advanced Functional Materials 23 (17) (2013) 2185–2192.

62. B. Scheibe, K. Tadyszak, M. Jarek, N. Michalak, M. Kempiński, M. Lewandowski, B. Peplińska, K. Chybczyńska, Study on the magnetic properties of differently functionalized multilayered $Ti_3C_2T_x$ MXenes and Ti-Al-C carbides, Applied Surface Science 479 (2019) 216–224.

63. K. Allen-Perry, W. Straka, D. Keith, S. Han, L. Reynolds, B. Gautam, D. E. Autrey, Tuning the magnetic properties of two-dimensional MXenes by chemical etching, Materials 14 (3) (2021) 694.

64. Y. Liu, S. Zhang, X. Su, J. Xu, Y. Li, Enhanced microwave absorption properties of Ti_3C_2 MXene powders decorated with Ni particles, Journal of Materials Science 55 (24) (2020) 10339–10350.

65. S. Yan, C. Cao, J. He, L. He, Z. Qu, Investigation on the electromagnetic and broadband microwave absorption properties of Ti_3C_2 MXene/flaky carbonyl iron composites, Journal of Materials Science: Materials in Electronics 30 (7) (2019) 6537–6543.

66. J. Fatheema, M. Fatima, N. B. Monir, S. A. Khan, S. Rizwan, A comprehensive computational and experimental analysis of stable ferromagnetism in layered 2D Nb-doped Ti_3C_2 MXene, Physica E: Low-dimensional Systems and Nanostructures 124 (2020) 114253.

67. F. Shahzad, M. Alhabeb, C. B. Hatter, B. Anasori, S. Man Hong, C. M. Koo, Y. Gogotsi, Electromagnetic interference shielding with 2D transition metal carbides (MXenes), Science 353 (6304) (2016) 1137–1140.

68. W. Xin, G. Q. Xi, W. T. Cao, C. Ma, T. Liu, M. G. Ma, J. Bian, Lightweight and flexible MXene/CNF/silver composite membranes with a brick-like structure and high-performance electromagnetic-interference shielding, RSC Advances 9 (51) (2019) 29636–29644.

69. G. M. Weng, J. Li, M. Alhabeb, C. Karpovich, H. Wang, J. Lipton, K. Maleski, J. Kong, E. Shaulsky, M. Elimelech, et al., Layer-by-layer assembly of cross-functional semi-transparent MXene-carbon nanotubes composite films for next generation electromagnetic interference shielding, Advanced Functional Materials 28 (44) (2018) 1803360.

70. Z. Fan, D. Wang, Y. Yuan, Y. Wang, Z. Cheng, Y. Liu, Z. Xie, A lightweight and conductive MXene/graphene hybrid foam for superior electromagnetic interference shielding, Chemical Engineering Journal 381 (2020) 122696.

71. X. Zheng, W. Nie, Q. Hu, X. Wang, Z. Wang, L. Zou, X. Hong, H. Yang, J. Shen, C. Li, Multifunctional RGO/Ti$_3$C$_2$T$_x$ MXene fabrics for electrochemical energy storage, electromagnetic interference shielding, electrothermal and human motion detection, Materials & Design 200 (2021) 109442.

72. J. Xu, S. Tang, D. Liu, Z. Bai, X. Xie, X. Tian, W. Xu, W. Hou, X. Meng, N. Yang, Rational design of hollow Fe$_3$O$_4$ microspheres on Ti$_3$C$_2$T$_x$ MXene nanosheets as highly-efficient and lightweight electromagnetic absorbers, Ceramics International 48 (2) (2022) 2595–2604.

73. J. Yao, F. Yang, Z. Yao, L. Wan, J. Hou, Z. Jiao, W. Huyan, Y. He, P. Chen, J. Zhou, Ultrathin self-assembly MXene@ flake carbonyl iron composites with efficient microwave absorption at elevated temperatures, Advanced Electronic Materials 7 (12) (2021) 2100587.

74. L. Liang, G. Han, Y. Li, B. Zhao, B. Zhou, Y. Feng, J. Ma, Y. Wang, R. Zhang, C. Liu, Promising Ti$_3$C$_2$T$_x$ MXene/Ni chain hybrid with excellent electromagnetic wave absorption and shielding capacity, ACS Applied Materials & Interfaces 11 (28) (2019) 25399–25409.

75. R. Li, Q. Gao, H. Xing, Y. Su, H. Zhang, D. Zeng, B. Fan, B. Zhao, Lightweight, multifunctional MXene/polymer composites with enhanced electromagnetic wave absorption and high-performance thermal conductivity, Carbon 183 (2021) 301–312.

76. Z. Liu, Y. Cui, Q. Li, Q. Zhang, B. Zhang, Fabrication of folded MXene/MoS$_2$ composite microspheres with optimal composition and their microwave absorbing properties, Journal of Colloid and Interface Science 607 (2022) 633–644.

77. Z. Wang, Z. Cheng, L. Xie, X. Hou, C. Fang, Flexible and lightweight Ti$_3$C$_2$T$_x$ MXene/Fe$_3$O4@PANI composite films for high-performance electromagnetic interference shielding, Ceramics International 47 (4) (2021) 5747–5757.

78. G. Y. Yang, S. Z. Wang, H. T. Sun, X. M. Yao, C. B. Li, Y. J. Li, J. J. Jiang, Ultralight, conductive Ti$_3$C$_2$T$_x$ MXene/PEDOT: PSS hybrid aerogels for electromagnetic interference shielding dominated by the absorption mechanism, ACS Applied Materials & Interfaces 13 (48) (2021) 57521–57531.

79. M. K. Xu, J. Liu, H. B. Zhang, Y. Zhang, X. Wu, Z. Deng, Z. Z. Yu, Electrically conductive Ti$_3$C$_2$T$_x$ MXene/polypropylene nanocomposites with an ultralow percolation threshold for efficient electromagnetic interference shielding, Industrial & Engineering Chemistry Research 60 (11) (2021) 4342–4350.

80. R. Sun, H. B. Zhang, J. Liu, X. Xie, R. Yang, Y. Li, S. Hong, Z. Z. Yu, Highly conductive transition metal carbide/carbonitride (MXene)@ polystyrene nanocomposites fabricated by electrostatic assembly for highly efficient electromagnetic interference shielding, Advanced Functional Materials 27 (45) (2017) 1702807.

81. Q. W. Wang, H. B. Zhang, J. Liu, S. Zhao, X. Xie, L. Liu, R. Yang, N. Koratkar, Z. Z. Yu, Multifunctional and water resistant MXene-decorated polyester textiles with outstanding electromagnetic interference shielding and joule heating performances, Advanced Functional Materials 29 (7) (2019) 1806819.

82. K. Raagulan, R. Braveenth, H. J. Jang, Y. Seon Lee, C. M. Yang, B. Mi Kim, J. J. Moon, K. Y. Chai, Electromagnetic shielding by MXene-graphene-PVDF composite with hydrophobic, lightweight and flexible graphene coated fabric, Materials 11 (10) (2018) 1803.

83. Y. Yao, S. Jin, M. Wang, F. Gao, B. Xu, X. Lv, Q. Shu, MXene hybrid polyvinyl alcohol flexible composite films for electromagnetic interference shielding, Applied Surface Science 578 (2022) 152007.

84. L. Geng, P. Zhu, Y. Wei, R. Guo, C. Xiang, C. Cui, Y. Li, A facile approach for coating $Ti_3C_2T_x$ on cotton fabric for electromagnetic wave shielding, Cellulose 26 (4) (2019) 2833–2847.

85. J. He, S. Liu, L. Deng, D. Shan, C. Cao, H. Luo, S. Yan, Tunable electromagnetic and enhanced microwave absorption properties in $CoFe_2O_4$ decorated Ti_3C_2 MXene composites, Applied Surface Science 504 (2020) 144210.

86. X. Wang, H. Li, H. Li, S. Lin, W. Ding, X. Zhu, Z. Sheng, H. Wang, X. Zhu, Y. Sun, 2D/2D 1T-MoS_2/Ti_3C_2 MXene heterostructure with excellent supercapacitor performance, Advanced Functional Materials 30 (15) (2020) 0190302.

87. T. Zhou, C. Wu, Y. Wang, A. P. Tomsia, M. Li, E. Saiz, S. Fang, R. H. Baughman, L. Jiang, Q. Cheng, Super-tough MXene-functionalized graphene sheets, Nature Communications 11 (1) (2020) 1–11.

88. X. Xia, Q. Xiao, Electromagnetic interference shielding of 2D transition metal carbide (MXene)/metal ion composites, Nanomaterials 11 (11) (2021) 2929.

89. C. Li, X. Ni, Y. Lei, S. Li, L. Jin, B. You, Plasmolysis-inspired yolk–shell hydrogel-core@void@MXene-shell microspheres with strong electromagnetic interference shielding performance, Journal of Materials Chemistry A 9 (47) (2021) 26839–26851.

90. Z. Deng, P. Tang, X. Wu, H. B. Zhang, Z. Z. Yu, Superelastic, ultralight, and conductive $Ti_3C_2T_x$ MXene/acidified carbon nanotube anisotropic aerogels for electromagnetic interference shielding, ACS Applied Materials & Interfaces 13 (17) (2021) 20539–20547.

91. S. Zhang, J. Wu, J. Liu, Z. Yang, G. Wang, $Ti_3C_2T_x$ MXene nanosheets sandwiched between Ag nanowire-polyimide fiber mats for electromagnetic interference shielding, ACS Applied Nano Materials 4 (12) (2021) 13976–13985.

92. H. Cheng, Y. Pan, Q. Chen, R. Che, G. Zheng, C. Liu, C. Shen, X. Liu, Ultrathin flexible poly(vinylidene fluoride)/MXene/silver nanowire film with outstanding specific EMI shielding and high heat dissipation, Advanced Composites and Hybrid Materials 4 (3) (2021) 505–513.

93. H. Liu, R. Fu, X. Su, B. Wu, H. Wang, Y. Xu, X. Liu, Electrical insulating MXene/PDMS/BN composite with enhanced thermal conductivity for electromagnetic shielding application, Composites Communications 23 (2021) 100593.

94. D. Hu, X. Huang, S. Li, P. Jiang, Flexible and durable cellulose/MXene nanocomposite paper for efficient electromagnetic interference shielding, Composites Science and Technology 188 (2020) 107995.

95. B. Zhou, M. Su, D. Yang, G. Han, Y. Feng, B. Wang, J. Ma, J. Ma, C. Liu, C. Shen, Flexible MXene/silver nanowire based transparent conductive film with electromagnetic interference shielding and electro-photo-thermal performance, ACS Applied Materials & Interfaces 12 (36) (2020) 40859–40869.

96. C. Lan, H. Jia, M. Qiu, S. Fu, Ultrathin MXene/polymer coatings with an alternating structure on fabrics for enhanced electromagnetic interference shielding and fire-resistant protective performances, ACS Applied Materials & Interfaces 13 (32) (2021) 38761–38772.

97. H. Zhang, J. Chen, H. Ji, N. Wang, S. Feng, H. Xiao, Electromagnetic interference shielding with absorption-dominant performance of $Ti_3C_2T_x$ MXene/non-woven laminated fabrics, Textile Research Journal 91 (21-22) (2021) 2448–2458.

98. Z. Tan, H. Zhao, F. Sun, L. Ran, L. Yi, L. Zhao, J. Wu, Fabrication of chitosan/MXene multilayered film based on layer-by-layer assembly: toward enhanced electromagnetic interference shielding and thermal management capacity, Composites Part A: Applied Science and Manufacturing 155 (2022) 106809.

99. Y. Mao, D. Wang, S. Fu, Layer-by-layer self-assembled nanocoatings of MXene and P, N-co-doped cellulose nanocrystals onto cotton fabrics for significantly reducing fire hazards and shielding electromagnetic interference, Composites Part A: Applied Science and Manufacturing 153 (2022) 106751.
100. J. Liu, H. B. Zhang, R. Sun, Y. Liu, Z. Liu, A. Zhou, Z. Z. Yu, Hydrophobic, flexible, and lightweight MXene foams for high-performance electromagnetic-interference shielding, Advanced Materials 29 (38) (2017) 1702367.
101. X. Ru, H. Li, Y. Peng, Z. Fan, J. Feng, L. Gong, Z. Liu, Y. Chen, Q. Zhang, A new trial for lightweight MXene hybrid aerogels with high electromagnetic interference shielding performance, Journal of Materials Science: Materials in Electronics 33 (2022) 1–11.
102. S. Zhao, H. B. Zhang, J. Q. Luo, Q. W. Wang, B. Xu, S. Hong, Z. Z. Yu, Highly electrically conductive three-dimensional $Ti_3C_2T_x$ MXene/reduced graphene oxide hybrid aerogels with excellent electromagnetic interference shielding performances, ACS Nano 12 (11) (2018) 11193–11202.
103. J. Zhai, C. Cui, A. Li, R. Guo, R. Cheng, E. Ren, H. Xiao, M. Zhou, J. Zhang, Waste cotton Fabric/MXene composite aerogel with heat generation and insulation for efficient electromagnetic interference shielding, Ceramics International 48 (10) (2022) 13464–13474.
104. H. Liu, Z. Huang, T. Chen, X. Su, Y. Liu, R. Fu, Construction of 3D MXene/silver nanowires aerogels reinforced polymer composites for extraordinary electromagnetic interference shielding and thermal conductivity, Chemical Engineering Journal 427 (2022) 131540.
105. J. Q. Luo, S. Zhao, H. B. Zhang, Z. Deng, L. Li, Z. Z. Yu, Flexible, stretchable and electrically conductive MXene/natural rubber nanocomposite films for efficient electromagnetic interference shielding, Composites Science and Technology 182 (2019) 107754.
106. W. Ma, W. Cai, W. Chen, P. Liu, J. Wang, Z. Liu, Microwave-induced segregated composite network with MXene as interfacial solder for ultra-efficient electromagnetic interference shielding and anti-dripping, Chemical Engineering Journal 425 (2021) 131699.

Index

Note: Locators in *italics* represent figures and **bold** indicate tables in the text.

For Product Safety Concerns and Information please contact our EU
representative GPSR@taylorandfrancis.com
Taylor & Francis Verlag GmbH, Kaufingerstraße 24, 80331 München, Germany

www.ingramcontent.com/pod-product-compliance
Lightning Source LLC
Chambersburg PA
CBHW060343220326
41598CB00023B/2787